THE
RADIUM
GIRLS

KATE MOORE

THE RADIUM GIRLS

THEY PAID WITH THEIR LIVES
THEIR FINAL FIGHT WAS FOR JUSTICE

SIMON &
SCHUSTER

London · New York · Sydney · Toronto · New Delhi

A CBS COMPANY

First published in Great Britain by Simon & Schuster UK Ltd, 2016
A CBS COMPANY

1 3 5 7 9 10 8 6 4 2

Simon & Schuster UK Ltd
1st Floor
222 Gray's Inn Road
London WC1X 8HB

www.simonandschuster.co.uk

Simon & Schuster Australia, Sydney
Simon & Schuster India, New Delhi

The author and publishers have made all reasonable efforts to contact
copyright-holders for permission, and apologise for any omissions or errors
in the form of credits given. Corrections may be made to future printings.

A CIP catalogue record for this book
is available from the British Library

Hardback ISBN: 978-1-4711-5387-7
Trade paperback ISBN: 978-1-4711-4757-9
eBook ISBN: 978-1-4711-5389-1

Typeset in the UK by M Rules
Printed and bound by CPI Group (UK) Ltd, Croydon, CR0 4YY

For all the dial-painters
And those who loved them

I shall never forget you ...
Hearts that know you love you
And lips that have given you laughter
Have gone to their lifetime of grief and of roses
Searching for dreams that they lost
In the world, far away from your walls.

Ottawa High School yearbook, 1925

CONTENTS

LIST OF KEY CHARACTERS

Newark and Orange, New Jersey

The dial-painters
Albina Maggia Larice
Amelia 'Mollie' Maggia, *Albina Maggia Larice's sister*
Edna Bolz Hussman
Eleanor 'Ella' Eckert
Genevieve Smith, *Josephine Smith's sister*
Grace Fryer
Hazel Vincent Kuser
Helen Quinlan
Irene Corby La Porte
Irene Rudolph, *Katherine Schaub's cousin*
Jane 'Jennie' Stocker
Josephine Smith, *Genevieve Smith's sister*
Katherine Schaub, *Irene Rudolph's cousin*
Mae Cubberley Canfield, *instructress*
Marguerite Carlough, *Sarah Carlough Maillefer's sister*
Quinta Maggia McDonald, *Albina and Amelia's sister*
Sarah Carlough Maillefer, *Marguerite Carlough's sister*

For the United States Radium Corporation
Anna Rooney, *forelady*
Arthur Roeder, *treasurer*
Clarence B. Lee, *vice president*

Edwin Leman, *chief chemist*
George Willis, *co-founder with Sabin von Sochocky*
Harold Viedt, *vice president*
Howard Barker, *chemist and vice president*
Sabin von Sochocky, *founder and inventor of the paint*
Mr Savoy, *studio manager*

Doctors
Dr Francis McCaffrey, *New York specialist, treating Grace Fryer*
Dr Frederick Flinn, *company doctor*
Dr Harrison Martland, *Newark doctor*
Dr James Ewing, Dr Lloyd Craver, Dr Edward Krumbhaar, *committee doctors*
Dr Joseph Knef, Dr Walter Barry, Dr James Davidson, *local dentists*
Dr Robert Humphries, *doctor at the Orange Orthopaedic Hospital*
Dr Theodore Blum, *New York dentist*

Investigators
Dr Alice Hamilton, *Harvard School of Public Health, Katherine Wiley's ally and a colleague of Cecil K. Drinker*
Andrew McBride, *commissioner of the Department of Labor*
Dr Cecil K. Drinker, *professor of physiology at the Harvard School of Public Health, husband of Katherine Drinker*
Ethelbert Stewart, *commissioner of the Bureau of Labor Statistics, Washington, DC*
Dr Frederick Hoffman, *investigating statistician, Prudential Insurance Company*
John Roach, *deputy commissioner of the Department of Labor*
Dr Katherine Drinker, *Harvard School of Public Health, wife of Cecil K. Drinker*
Katherine Wiley, *executive secretary of the Consumers League, New Jersey*
Lenore Young, *Orange health officer*
Swen Kjaer, *national investigator from the Bureau of Labor Statistics, Washington, DC*
Dr Szamatolski, *consulting chemist for the Department of Labor*

Ottawa, Illinois

The dial-painters
Catherine Wolfe Donohue
Charlotte Nevins Purcell
Frances Glacinski O'Connell, *Marguerite Glacinski's sister*
Helen Munch
Inez Corcoran Vallat
Margaret 'Peg' Looney
Marguerite Glacinski, *Frances Glacinski O'Connell's sister*
Marie Becker Rossiter
Mary Duffy Robinson
Mary Ellen 'Ella' Cruse
Mary Vicini Tonielli
Olive West Witt
Pearl Payne

For the Radium Dial Company
Joseph Kelly, *president*
Lottie Murray, *superintendent*
Mercedes Reed, *instructress, wife of Rufus Reed*
Rufus Fordyce, *vice president*
Rufus Reed, *assistant superintendent, husband of Mercedes Reed*
William Ganley, *executive*

Doctors
Dr Charles Loffler, *Chicago doctor*
Dr Lawrence Dunn, *physician of Catherine Donohue*
Dr Sidney Weiner, *X-ray specialist*
Dr Walter Dalitsch, *specialist dentist*

PROLOGUE

Paris, France
1901

The scientist had forgotten all about the radium. It was tucked discreetly within the folds of his waistcoat pocket, enclosed in a slim glass tube in such a small quantity that he could not feel its weight. He had a lecture to deliver in London, England, and the vial of radium stayed within that shadowy pocket for the entirety of his journey across the sea.

He was one of the few people in the world to possess it. Discovered by Marie and Pierre Curie late in December 1898, radium was so difficult to extract from its source that there were only a few grams available anywhere in the world. He was fortunate indeed to have been given a tiny quantity by the Curies to use in his lectures, for they barely had enough themselves to continue experiments.

Yet this constraint did not affect the Curies' progress. Every day they discovered something new about their element: 'it made an impression on photographic plates through black paper,' the Curies' daughter later wrote, 'it corroded and, little by little, reduced to powder the paper or the cotton wool in which it was wrapped ... What could it *not* do?' Marie called it 'my beautiful radium' – and it truly was. Deep in the dark pocket of the scientist, the radium broke the gloom with an unending, eerie glow. 'These gleamings,' Marie wrote of its luminous effect, 'seemed suspended in the darkness [and] stirred us with ever-new emotion and enchantment.'

Enchantment . . . it implies a kind of sorcery; almost supernatural power. No wonder the US surgeon general said of radium that 'it reminds one of a mythological super-being'. An English physician would call its enormous radioactivity 'the unknown god'.

Gods can be kind. Loving. Benevolent. Yet as the playwright George Bernard Shaw once wrote, 'The gods of old are constantly demanding human sacrifices.' Enchantment – in the tales of the past, and present – can also mean a curse.

And so, although he had forgotten about the radium, the radium had not forgotten him. As the scientist travelled to that foreign shore, through every second of his journey the radium shot out its powerful rays towards his pale, soft skin. Days later, he would peer in confusion at the red mark blooming mysteriously on his stomach. It looked like a burn, but he had no memory of coming near any flame that could produce such an effect. Hour by hour, it grew more painful. It didn't get bigger, but it seemed, somehow, to get *deeper*, as though his body was still exposed to the source of the wound and the flame was burning him still. It blistered into an agonising flesh burn that grew in intensity until the pain made him suck in his breath sharply and rack his brains for what on earth could have inflicted such damage without him being aware.

And it was then that he remembered the radium.

PART ONE

Knowledge

1

Newark, New Jersey
United States of America
1917

Katherine Schaub had a jaunty spring in her step as she walked the brief four blocks to work. It was 1 February 1917, but the cold didn't bother her one bit; she had always loved the winter snows of her home town. The frosty weather wasn't the reason for her high spirits on that particular icy morning though: today, she was starting a brand-new job at the watch-dial factory of the Radium Luminous Materials Corporation, based on 3rd Street in Newark, New Jersey.

It was one of her close pals who had told her about the vacancy; Katherine was a lively, sociable girl with many friends. As she herself later recalled, 'A friend of mine told me about the "watch studio" where watch-dial numerals and hands were painted with a luminous substance that made them visible in the dark. The work, she explained, was interesting and of far higher type than the usual factory job.' It sounded so glamorous, even in that brief description – after all, it wasn't even a factory, but a 'studio'. For Katherine, a girl who had 'a very imaginative temperament', it sounded like a place where anything could happen. It certainly beat the job she'd had before, wrapping parcels in Bamberger's department store; Katherine had ambitions far beyond that shop floor.

She was an attractive girl of just fourteen; her fifteenth birthday was in five weeks' time. Standing just under five foot four, she was 'a very pretty little blonde' with twinkling blue eyes, fashionably bobbed hair and delicate features. Although she had received only a grammar-school education before she left school – which was 'about all the education that girls of her working-class background received in those days' – she was nevertheless fiercely intelligent. 'All her life,' *Popular Science* later wrote, 'Katherine Schaub had cherished [the] desire to pursue a literary career.' She was certainly go-getting: she later wrote that, after her friend had given her word of the opportunities at the watch studio, 'I went to the man in charge – a Mr Savoy – and asked for a job.'

And that was how she found herself outside the factory on 3rd Street, knocking on the door and gaining admittance to the place where so many young women wanted to work. She felt almost a little star-struck as she was ushered through the studio to meet the forewoman, Anna Rooney, and saw the dial-painters turning diligently to their tasks. The girls sat in rows, dressed in their ordinary clothes and painting dials at top speed, their hands almost a blur to Katherine's uninitiated eyes. Each had a flat wooden tray of dials beside her – the paper dials were pre-printed on a black background, leaving the numerals white, ready for painting – but it wasn't the dials that caught Katherine's eye, it was the material they were using. It was the radium.

Radium. It was a wonder element; everyone knew that. Katherine had read all about it in magazines and newspapers, which were forever extolling its virtues and advertising new radium products for sale – but they were all far too expensive for a girl of Katherine's humble origins. She had never seen it up close before. It was the most valuable substance on earth, selling for $120,000 for a single gram ($2.2 million in today's values). To her delight, it was even more beautiful than she had imagined.

Each dial-painter had her own supply. She mixed her own paint, dabbing a little radium powder into a small white crucible and adding a dash of water and a gum-arabic adhesive: a combination that created a greenish-white luminous paint, which went under the name 'Undark'. The fine yellow powder contained only a minuscule amount of radium; it was mixed with zinc sulphide, with which the radium reacted to give a brilliant glow. The effect was breathtaking.

Katherine could see that the powder got everywhere; there was dust all over the studio. Even as she watched, little puffs of it seemed to hover in the air before settling on the shoulders or hair of a dial-painter at work. To her astonishment, it made the girls themselves gleam.

Katherine, like many before her, was entranced by it. It wasn't just the glow – it was radium's all-powerful reputation. Almost from the start, the new element had been championed as 'the greatest find of history'. When scientists had discovered, at the turn of the century, that radium could destroy human tissue, it was quickly put to use to battle cancerous tumours, with remarkable results. Consequently – as a life-saving and thus, it was assumed, health-giving element – other uses had sprung up around it. All of Katherine's life radium had been a magnificent cure-all, treating not just cancer, but hay fever, gout, constipation ... anything you could think of. Pharmacists sold radioactive dressings and pills; there were also radium clinics and spas for those who could afford them. People hailed its coming as predicted in the Bible: 'The sun of righteousness [shall] arise with healing in his wings, and ye shall go forth and gambol as calves of the stall.'

For another claim of radium was that it could restore vitality to the elderly, making 'old men young'. One aficionado wrote: 'Sometimes I am halfway persuaded that I can feel the sparkles inside my anatomy.' Radium shone 'like a good deed in a naughty world'.

Its appeal was quickly exploited by entrepreneurs. Katherine had seen adverts for one of the most successful products, a radium-lined jar to which water could be added to make it radioactive: wealthy customers drank it as a tonic; the recommended dose was five to seven glasses a day. But as some of the models retailed for $200 ($3,700), it was a product far out of Katherine's reach. Radium water was drunk by the rich and famous, not working-class girls from Newark.

What she did feel part of, though, was radium's all-pervasive entry into American life. It was a craze, no other word for it. The element was dubbed 'liquid sunshine' and it lit up not just the hospitals and drawing rooms of America, but its theatres, music halls, grocery stores and bookshelves. It was breathlessly featured in cartoons and novels, and Katherine – who loved to sing and play piano – was probably familiar with the song 'Radium Dance', which had become a huge hit after featuring in the Broadway musical *Piff! Paff! Pouf!* On sale were radium jockstraps and lingerie, radium butter, radium milk, radium toothpaste (guaranteeing a brighter smile with every brushing) and even a range of Tho-Radia make-up, which offered radium-laced eyeshadows, lipsticks and face creams. Other products were more prosaic: 'The Radium Eclipse Sprayer,' championed one advert, 'quickly kills all flies, mosquitoes, roaches. [It] has no equal as a cleaner of furniture, porcelain, tile. It is harmless to humans and easy to use.'

Not all of these products actually contained radium – it was far too costly and rare for that – but manufacturers from all kinds of industries declared it part of their range, for everyone wanted a slice of the radium pie.

And now, to Katherine's excitement, thanks to her job she would have a prime seat at the table. Her eyes drank in the dazzling scene before her. But then, to her disappointment, Miss Rooney ushered her into a room that was separate to the main studio, away from the radium and the shining girls. Katherine would not be dial-painting that day – nor the day after, as much as she longed to join the glamorous artists in the other room.

Instead, she would be serving an apprenticeship as an inspector, checking the work of those luminous girls who were busy painting dials.

It was an important job, Miss Rooney explained. For although the company specialised in watch faces, they also had a lucrative government contract to supply luminous airplane instruments. Given there was a war raging in Europe, business was booming; the company also used its paint to make gunsights, ships' compasses and more shine brightly in the dark. And when lives were hanging in the balance, the dials had to be perfect. '[I was] to see that the number outlines were even and [thorough] and to correct minor defects,' Katherine recalled.

Miss Rooney introduced her to her trainer, Mae Cubberley, and then left the girls to it, resuming her slow march up and down the rows of painting girls, casting a watchful eye over their shoulders.

Mae smiled at Katherine as she said hello. A twenty-six-year-old dial-painter, Mae had been with the company since the previous fall. Although she was new to the industry when she joined, she already had a reputation as a brilliant painter, regularly turning in eight to ten trays of dials daily (there were either twenty-four or forty-eight dials in each tray, depending on the dial size). She had quickly been promoted to training other girls in the hope that they would match her productivity. Now, in the little side room, she picked up a paintbrush to instruct Katherine in the technique that all the dial-painters and inspectors were taught.

They were using slim camel-hair brushes with narrow wooden handles. One dial-painter recalled: 'I had never seen a brush as fine as that. I would say it possibly had about thirty hairs in it; it was exceptionally fine.' Yet as fine as the brushes were, the bristles had a tendency to spread, hampering the girls' work. The smallest pocket watch they painted measured only three-and-a-half centimetres across its face, meaning the tiniest element for painting was a single millimetre in width. The girls could not go over the edges of these delicate parameters or there would be hell to pay. They had to make the brushes even finer – and there was only one way they knew of to do that.

'We put the brushes in our mouths,' Katherine said, quite simply. It was a technique called lip-pointing, inherited from the first girls who had worked in the industry, who came from china-painting factories.

Unbeknown to the girls, it wasn't the way things were done in Europe, where dial-painting had been in operation for over a decade. Different countries had different techniques, but in none was lip-pointing used. Very likely this was because brushes weren't used either: in Switzerland there were solid glass rods; in France, small sticks with cotton wadding on the ends; other European studios employed a sharpened wooden stylus or metal needles.

However, American girls did not take up the lip-pointing technique with blind faith. Mae said that when she first started, not long after the studio had opened in 1916, she and her colleagues had questioned it, being 'a little bit leery' about swallowing the radium. 'The first thing we asked,' she remembered, '[was] "Does this stuff hurt you?" And they said, "No." Mr Savoy said that it wasn't dangerous, that we didn't need to be afraid.' After all, radium was the wonder drug; the girls, if anything, should benefit from their exposure. They soon grew so used to the brushes in their mouths that they stopped even thinking about it.

But for Katherine it felt peculiar, that first day, as she lip-pointed over and over, correcting defective dials. Yet it was worth persevering: she was constantly reminded why she wanted to work there. Her job involved two types of inspection, daylight and darkroom, and it was in the darkroom that the magic really happened. She would call the girls in to discuss their work and observed, 'Here in the room – daylight barred – one could see evidences of the luminous paint everywhere on the worker. There was a dab here and there on her clothes, on the face and lips, on her hands. As some of them stood there, they fairly shone in the dark.' They looked glorious, like other-worldly angels.

As time went on, she got to know her colleagues better. One was Josephine Smith, a sixteen-year-old girl with a round face, brown bobbed hair and a snub nose. She had used to work at Bamberger's too, as a saleslady, but left to earn the much higher wage of a dial-painter. Although the girls weren't salaried – they were paid piecework, for the number of dials they painted, at an average rate of 1.5 cents a watch – the most talented workers could walk away with an astonishing pay packet. Some earned more than three times the average factory-floor worker; some even earned more than their fathers. They were ranked in the top 5 per cent of female wage-earners and on average took home $20 ($370) a week, though the fastest painters could easily earn more, sometimes as much as double; giving the top earners an annual salary of $2,080 (almost $40,000). The girls lucky enough to gain a position felt blessed.

Josephine, Katherine learned as they talked, was of German heritage, just like Katherine herself. In fact, most dial-painters were the daughters or granddaughters of immigrants. Newark was full of migrants, hailing from Germany, Italy, Ireland and elsewhere; it was one of the reasons the company had opened the studio in the city in the first place, for the large immigrant communities provided a workforce for all sorts of factories. New Jersey was nicknamed the Garden State for its high agricultural production, but in truth it was just as productive industrially. At the turn of the century, the business leadership of Newark had labelled it the City of Opportunity and – as the girls themselves were finding out – it lived up to its name.

It all made for a thriving metropolis. The nightlife after the factories closed was vibrant; Newark was a beer town, with more saloons per capita than any other American city, and the workers made their downtime count. The dial-painters embraced the social bonhomie: they sat together to eat lunch in the workroom at the Newark plant, sharing sandwiches and gossip over the dusty tables.

As the weeks passed, Katherine observed the challenges as well as the attractions of dial-painting: Miss Rooney's constant observation as she walked up and down the studio, and the ever-present dread of being called into the darkroom to be reprimanded for poor work. Above all else, the girls feared being accused of wasting the expensive paint, which could ultimately be a dismissible offence. But although Katherine could see that there were downsides, she still longed to join the women in the main room. She wanted to be one of the shining girls.

A quick learner, Katherine soon excelled at her inspecting, not only perfecting the art of correcting defective dials with her lip-pointed brush, but also becoming adept at brushing off the dust with her bare hand or removing excess paint with her fingernail, as was the technique taught her. She worked as hard as she could, longing for promotion.

Finally, towards the end of March, her perseverance paid off. 'I was asked to paint dials,' she wrote excitedly; 'I said I would like to try it.'

Katherine had achieved her ambition through merit – but there were also wider forces at work in that spring of 1917. Dial-painters were about to be in demand like never before: the company now needed all the women it could get.

For the past two-and-a-half years, the war in Europe had left America mostly untouched, except for the economic boom it brought. The majority of Americans were happy to stay out of the horrific trench warfare happening across the Atlantic; stories of which had reached them undiluted by distance. But in 1917, the neutral position became untenable. On 6 April, just a week or so after Katherine's promotion, Congress voted America into the conflict. It would be, President Wilson declared, 'The war to end all wars.'

In the dial-painting studio on 3rd Street, the impact of the decision was immediate. Demand rocketed. The studio in Newark was far too small to produce the numbers required so Katherine's bosses opened a purpose-built plant just down the road from Newark in Orange, New Jersey, closing the 3rd Street factory. This time there wouldn't only be dial-painters on site; the company had grown so much it was to do its own radium extraction, requiring labs and processing plants. The Radium Luminous Materials Corporation was expanding massively, and the new site comprised several buildings, all located in the middle of a residential neighbourhood.

Katherine was among the first workers through the door of the two-storey brick building that would house the application department. She and the other dial-painters were delighted by what they found. Not only was Orange an attractive, prosperous town, but the second-floor studio was charming, with huge windows on all sides and skylights in the roof. The spring sunshine streamed in, giving excellent light for dial-painting.

An appeal for new workers to help the war effort was made and just four days after war had been declared, Grace Fryer answered the call. She had more reason than most to want to help; two of her brothers would be joining the several million American soldiers heading to France to fight. Lots of dial-painters were motivated by the idea of helping the troops: 'The girls,' wrote Katherine, 'were but a few of the many who through their jobs were "doing their bit".'

Grace was a particularly civic-minded young woman. 'When Grace was just a schoolgirl,' a childhood friend of hers wrote, 'she planned to be a real citizen when she grew up.' Her family was of a political bent; her father Daniel was a delegate to the carpenters' union, and you couldn't grow up in his house without picking up his principles. He was out of work rather a lot, as unionism was not popular at that time, but while the family may not have had much money, they did have a lot of love. Grace was one of ten children – she was number four – and she was especially close to her mother, also called Grace; perhaps because she was the eldest girl. There were six boys and four girls in total and Grace was close to her siblings, especially her sister Adelaide, who was nearest to her in age, and her little brother Art.

Grace was already working when the call-up came, in a position that earned about the same as dial-painting, but she left to join the radium company in Orange, where she lived. She was an exceptionally bright and exceptionally pretty girl, with curly chestnut hair, hazel eyes and clear-cut features. Many called her striking, but her looks weren't of much interest to Grace. Instead, she was career-minded, someone who at the age of eighteen was already fashioning a prosperous life for herself. She was, in short, 'a girl enthused with living'. She soon excelled at dial-painting, becoming one of the company's fastest workers with an average production rate of 250 dials a day.

A young woman called Irene Corby also signed on that spring. The daughter of a local hatter, she was a very cheerful girl aged about seventeen. 'She had a very humorous disposition,' revealed her sister Mary, 'exceptionally so.' Irene instantly got on well with her co-workers – with Grace in particular – and they regarded her as one of the more skilled employees.

It fell to Mae Cubberley and Josephine Smith to train the new girls. The women sat side by side at long tables running the full width of the studio; there was a walkway in between them, so Miss Rooney could continue her over-the-shoulder inspections. The instructresses taught them how to dab a tiny amount of the material (the girls always called the radium 'the material') into their crucibles, 'like a fine smoke in the air', and then mix the paint carefully. Even the softest stirring, however, left most women with splashes on their bare hands.

Then, once the paint was mixed, they instructed them to lip-point. 'She told me to watch her and imitate her,' Katherine remembered of her training. As surely as night follows day, Grace and Katherine and Irene followed the instructions. They put the brush to their lips ... dipped it in the radium ... and painted the dials. It was a 'lip, dip, paint routine': all the girls copied each other; mirror images that lipped and dipped and painted all day long.

They soon found the radium hardened on their brushes. A second crucible was supplied, ostensibly for cleaning the bristles, but the water was changed only once a day and soon became cloudy: it didn't so much clean as spread the bristles, which some workers found a hindrance; they simply used their mouths to dampen the brush instead. Others, however, always used the water: 'I know I done it,' one said, 'because I couldn't stand that gritty taste in my mouth.'

The taste of the paint was a source of debate. 'It didn't taste funny,' Grace observed, 'it didn't have any taste.' Yet some ate the paint specifically because they liked it.

Another new worker tasting the magic element that summer was sixteen-year-old Edna Bolz. 'Here is a person,' *Popular Science* later wrote of her, 'blessed from birth with a sunny disposition.' She was taller than many of her co-workers, though still only five foot five, and had an innate elegance about her. She was nicknamed the 'Dresden Doll' because of her beautiful golden hair and fair colouring; she also had perfect teeth and, perhaps as a result, a beaming smile. Over time she became close to the forelady, Miss Rooney, who described her as 'a very nice type of girl; very clean-living type of very good family'. Edna's passion was music and she was also devoutly religious. She joined in July, at a time when production was rocketing due to wartime demand.

That summer the plant was a ferment of activity: 'The place was a madhouse!' one worker exclaimed. The girls were already doing overtime to keep up with demand, working seven days a week; now, the studio started operating night and day. The dial-painters glowed even brighter from the radium against the dark windows: a workshop of shining spirits labouring through the night.

Though the pace was demanding, the set-up was in many ways fun for the women, who revelled in the drama of the long shifts painting dials for their country. The majority were teenagers – 'merry giggling girls' – and they found time for the odd bit of fun. One favourite game was to scratch their name and address into a watch: a message for the soldier who would wear it; sometimes, he would respond with a note. New girls were joining all the time, which made the job even more sociable. In Newark, perhaps seventy women had worked in the studio; during the war that number more than tripled. The girls now sat crammed in on both sides of the desks, only a few feet apart.

Hazel Vincent was among them. Like Katherine Schaub, she came from Newark; she had an oval face with a button nose and fair hair that she set in the latest styles. Another new worker was twenty-one-year-old Albina Maggia, the daughter of an Italian immigrant, who came from a family of seven girls; she was the third. She was a somewhat round and diminutive woman of only four foot eight, with classic Italian dark hair and eyes. She was pleased to get back into the world of work – as the eldest unmarried daughter, she'd quit her hat-trimming job to nurse her mother, who'd died the year before – but she discovered she was not the fastest dial-painter. She found the brushes 'very clumsy' and painted only a tray and a half a day. Nonetheless, she tried as hard as she could, later remarking, 'I always did my best for that company.'

Joining Albina at the long wooden desks was her little sister Amelia, whom everyone called Mollie. She seemed to have found her calling at the studio, being unusually productive. A foot taller than Albina, she was a sociable nineteen-year-old with a broad face and bouffant brown hair, often seen laughing with her colleagues. She got on particularly well with another newcomer, Eleanor Eckert (nicknamed Ella): the two were as thick as thieves. Ella was popular and good-looking, with blonde, slightly frizzy hair and a wide smile; a sense of fun was never far from her, whether she was at work or play. The girls would socialise and eat lunch together, barely stopping work as they shared food across the crowded desks.

The company also organised social events; picnics were a favourite. The dial-painters, dressed in white summer dresses and wide-brimmed hats, would eat ice-cream cones while sitting on the narrow makeshift bridge that lay across the brook by the studio, swinging their legs or holding on to one another as they tried not to fall in the water. The picnics were for all employees – so at these events the girls got to mix with their co-workers, whom they rarely saw: the men who worked in the laboratories and refining rooms. It wasn't long before the odd 'office romance' began; Mae Cubberley started walking out with Ray Canfield, a lab worker: one of many blossoming relationships among the girls, though most were not with colleagues. Hazel Vincent, for one, was in love with her childhood sweetheart, a mechanic called Theodore Kuser, who had baby-blue eyes and fair hair.

The company's founder, Sabin von Sochocky, an Austrian-born, thirty-four-year-old doctor, could often be seen holding court at these picnics, sat among his workers on a rug, his jacket off and a beaker of cold drink in one hand. The girls seldom saw him in their studio – he was usually too busy working in his lab to grace them with his presence – so it was a rare opportunity for their paths to cross. It was he who had invented the luminous paint they used, back in 1913, and it had certainly been a success for him. In his first year, he had sold 2,000 luminous watches; now, the company's output ranked in the millions. In many ways he was an unlikely entrepreneur, for his training had been in medicine; initially, he'd intended the paint to be a 'potboiler' to fund medical research, but the growing demand had necessitated a more businesslike approach. He had met a 'kindred soul' in Dr George Willis and the two physicians had founded the company.

Von Sochocky was, according to his colleagues, a 'remarkable man'. Everyone called him simply 'the doctor'. He was indefatigable: 'someone that liked to start late, but is then willing to go on and on until all hours'. *American* magazine called him 'one of the greatest authorities in the world on the subject of radium' – and he had studied under the best: the Curies themselves.

From them, and from the specialist medical literature he had studied, von Sochocky understood that radium carried great dangers. Around the time he studied with the Curies, Pierre was heard to remark that 'he would not care to trust himself in a room with a kilo of pure radium, as it would burn all the skin off his body, destroy his eyesight and probably kill him'. The Curies, by that time, were intimately familiar with radium's hazards, having suffered many burns themselves. Radium could cure tumours, it was true, by destroying unhealthy tissue – but it was indiscriminating in its powers, and could devastate healthy tissue too. Von Sochocky himself had suffered its silent and sinister wrath: radium had got into his left index finger and, when he realised, he hacked the tip of it off. It now looked as though 'an animal had gnawed it'.

Of course, to the layman, all this was unknown. The mainstream position as understood by most people was that the effects of radium were all positive; and that was what was written about in newspapers and magazines, championed across product packaging and performed on Broadway.

Nonetheless, the lab workers in von Sochocky's plant in Orange were provided with protective equipment. Lead-lined aprons were issued, along with ivory forceps for handling tubes of radium. In January 1921, von Sochocky would write that one could handle radium 'only by taking the greatest precautions'.

Yet despite this knowledge, and the injury to his own finger, von Sochocky was apparently so transfixed by radium that all reports say he took little care. He was known to play with it, holding the tubes with his bare hands while watching the luminosity in the dark or immersing his arm up to the elbow in radium solutions. Company co-founder George Willis was also lax, picking up tubes of radium with his forefinger and thumb, not bothering with forceps. Perhaps understandably, their colleagues learned from them and copied what they did. No one heeded the warnings of Thomas Edison, working just a few miles away in sight of the Orange plant, who once remarked, 'There may be a condition into which radium has not yet entered that would produce dire results; everybody handling it should have a care.'

Yet in the sunny second-floor studio, the girls working there had not a care in the world. Here there were no lead aprons, no ivory-tipped forceps, no medical experts. The amount of radium in the paint was considered so small that such measures were not deemed necessary.

The girls themselves, of course, had no clue that they might even be needed. This was radium, the wonder drug, they were using. They were lucky, they thought, as they laughed among themselves and bent their heads to their intricate work. Grace and Irene. Mollie and Ella. Albina and Edna. Hazel and Katherine and Mae.

They picked up their brushes and they twirled them over and over, just as they had been taught.

Lip ... Dip ... Paint.

Wars are hungry machines – and the more you feed them, the more they consume. As the fall of 1917 wore on, demand at the factory showed no signs of slowing; at the height of operations, as many as 375 girls were recruited to paint dials. And when the firm announced it needed more women, the existing workers eagerly promoted the job to their friends, sisters and cousins. It wasn't long before whole sets of siblings were sat alongside each other, merrily painting away. Albina and Mollie Maggia were soon joined by another sister, sixteen-year-old Quinta.

She was an extremely attractive woman with large grey eyes and long dark hair; she considered her pretty teeth her best feature. Down-to-earth and kind, her favourite pastimes included card games, checkers and dominoes. She also confessed cheekily, 'I don't go to church half as often as I should.' She hit it off brilliantly with Grace Fryer and the two became 'inseparable'.

Grace was another who brought her little sister to work: Adelaide Fryer adored the social side of it, being a very gregarious girl who loved to be around people, but she wasn't quite as sensible as her big sister; in the end, she was fired for talking too much. The girls may have been sociable, but they still had a job to do, and if they didn't knuckle down and do it, they were out.

It could be tough. As Katherine Schaub had observed in Newark, the girls were under a lot of pressure. If a worker failed to keep up, she was criticised; if she fell short repeatedly, she eventually lost her job. The only time the girls really saw Mr Savoy, whose office was downstairs, was when he came to scold them.

The biggest issue was the wasting of the paint. Each day, Miss Rooney issued a set amount of powder to the girls for completing a particular number of dials – and they had to make it last. They could not ask for more, but neither could they eke it out; if the numerals were not sufficiently covered by the material, it would show up during inspection. The girls took to helping each other out, sharing material if one found she had a little extra left over. And there were also their water dishes, filled with the radium sediment. Those, too, could be a source of extra material.

But the cloudy water hadn't gone unnoticed by the company bosses. Before too long, the crucibles for cleaning the brushes were removed with the explanation that too much valuable material was wasted in the water. Now the girls had no choice but to lip-point, as there was no other way to clean off the radium that hardened on the brush. As Edna Bolz observed, 'Without so doing it would have been impossible to have done much work.'

The girls themselves were also targeted in the drive to limit waste. When a shift was over and they were about to leave for home, they were summoned to the darkroom to be brushed off: the 'sparkling particles' were then swept from the floor into a dustpan for use the next day.

But no amount of brushing could get rid of all the dust. The girls were covered in it: their 'hands, arms, necks, the dresses, the underclothes, even the corsets of the dial-painters were luminous'. Edna Bolz remembered that even after the brushing down, 'When I would go home at night, my clothing would shine in the dark ... you could see where I was – my hair, my face.' The girls shone 'like the watches did in the darkroom', as though they themselves were timepieces, counting down the seconds as they passed. They glowed like ghosts as they walked home through the streets of Orange.

They were unmissable. Unassailable. The residents of the town noticed not just the wraith-like shine but also the expensive, glamorous clothes, for the girls dressed in silks and furs, 'more like matinee idlers than factory workers', a perk of their high wages.

Despite the attractions of the job, however, it wasn't for everyone. Some found the paint made them sick; one woman got sores on her mouth after just a month of working there. Though the girls all lip-pointed, they did so at different intervals, which perhaps accounted for the varying reactions. Grace Fryer found that 'I could do about two numbers before the brush dried', whereas Edna Bolz lip-pointed on every number, sometimes even two or three times per number. Quinta Maggia did the same, even though she hated the taste: 'I remember chewing [the paint] – gritty – it got between my teeth. I remember it distinctly.'

Katherine Schaub was one of the more infrequent pointers; only four or five times per watch would she slip the brush between her lips. Nonetheless, when she suddenly broke out in pimples – which could have been due to her hormones, for she was still only fifteen – she was perhaps mindful of some of her colleagues' adverse reactions, as she decided to consult a doctor.

To her concern, he asked her if she worked with phosphorus. This was a well-known industrial poison in Newark and it was a logical suspicion – but it made Katherine feel anything but logical and calm. For it wasn't only her acne that caused the doctor concern: there were changes, he noted, in Katherine's blood. Was she *sure* she didn't work with phosphorus?

The girls weren't entirely clear what was in the paint. Flummoxed by her doctor's questions, Katherine turned to her colleagues. When she told them what her physician had said, they became frightened. En masse, they confronted Mr Savoy, who tried to allay their fears, but this time his words about the paint being harmless fell on deaf ears.

And so, as any middle manager would do, he went to *his* managers. Soon after, George Willis came over from New York to lecture the girls on radium and convince them it was not dangerous; von Sochocky also participated. There was nothing hazardous in the paint, the doctors promised: the radium was used in such a minuscule amount that it could not cause them harm.

And so the girls turned back to their work, their shoulders a little lighter, Katherine probably feeling a bit sheepish that her teenage spots had caused such bother. Her skin cleared up, and so too did the minds of the dial-painters. When one of the greatest radium authorities in the world tells you that you have no need to worry, quite simply, you don't. Instead, they laughed about the effect the dust had on them. 'Nasal discharges on my handkerchief,' Grace Fryer remembered, 'used to be luminous in the dark.' One dial-painter, known as a 'lively Italian girl', painted the material all over her teeth one night before a date, wanting a smile that would knock him dead.

Those budding romances of the girls were now coming into full flower. Hazel and Theo were as close as ever, and Quinta started courting a young man called James McDonald – but it was Mae Cubberley who became a winter bride on 23 December 1917. As was traditional, she wanted to leave work right away, but Mr Savoy asked her to stay a little longer, so she was still in the studio when Sarah Maillefer signed on that same month.

Sarah was a little different to the other girls. For a start, she was older at twenty-eight: a shy, matronly woman who often seemed a little separate from the teenagers, though they were inclusive of her. Sarah was broad-shouldered with short dark hair – and those shoulders needed to be broad, for she was also a single mother. She had a six-year-old daughter, Marguerite, named after Sarah's little sister.

Sarah had married, back in 1909. Her husband, Henry Maillefer, was a tall, French-Irish sexton with black hair and black eyes. But Henry had disappeared; where he was now, nobody knew. And so Sarah and Marguerite still lived with her mom and pop, Sarah and Stephen Carlough, as well as her sister Marguerite, who was sixteen. Stephen was a painter and decorator, and the family were 'hard-working, reasonable' people. Sarah, too, was hard-working, and would become one of the most loyal employees the radium company had.

For Mae Cubberley Canfield, however, her loyalty had come to an end. Soon after she married, she fell pregnant and therefore handed in her notice in the early months of 1918. That chapter of her life was over.

Her place was quickly filled. That year, an estimated 95 per cent of all the radium produced in America was given over to the manufacture of radium paint for use on military dials; the plant was running at full capacity. By the end of the year, one in six American soldiers would own a luminous watch – and it was the Orange girls who painted many of them. Jane Stocker (nicknamed Jennie) was a new recruit, and in July a slim, elfin-featured girl called Helen Quinlan joined. She was an energetic woman whom the company rather sniffily described as 'the type that did altogether too much running around for her own good'. She had a boyfriend she often brought to the girls' picnics, a smart, blond young man who wore a shirt and tie to the gatherings. He and Helen posed for a picture at one of them: Helen had her skirts flapping around her knees, always on the move, while he stared at her rather than the camera, looking utterly besotted with this playful creature he had somehow been lucky enough to meet.

The women were still encouraging their families to join them in their work. In September 1918, Katherine wrote proudly, 'I obtained a position for Irene at the factory.' Irene Rudolph was her orphaned cousin, the same age as Katherine; she lived with the Schaubs. Perhaps understandably given her early life, Irene was a cautious, thoughtful young lady. Rather than spending her wages on silks and furs as some of the other girls did, she squirrelled it away in a savings account. She had a narrow face and nose with dark eyes and hair; the only picture of her that survives shows her somewhat downcast.

A month after Irene started, another new employee began work. But this was no dial-painter striding into a new job: this was Arthur Roeder, a highly successful businessman who was the company's new treasurer. He'd already demonstrated a skill for seizing career opportunities: though he had left university without a degree, he'd ascended rapidly through the ranks of his chosen career. A round-faced, smart-looking man with a Roman nose and thin lips, he favoured bow-ties and pomade, which he slicked through his dark hair to press it close to his skull. He was based at the head office in New York and now took on responsibility for the dial-painters. Though he said he was in the studio on numerous occasions, his presence there was an exception, as most of the executives rarely went inside. In fact, of the firm's top men, Grace Fryer remembered von Sochocky passing through her place of work just once. She didn't pay it much attention at the time, but it would come to take on a great significance.

She was at her desk as usual that day, lipping and dipping her brush; as were all the other girls. Von Sochocky, as per *his* usual, had his head full of ideas and complex science as he walked briskly about his work. On this occasion, as he passed swiftly through the studio, he stopped and looked straight at her – and at what she was doing, as though seeing her actions for the first time.

Grace glanced up at him. He was a memorable-looking man, with a dominant nose and close-cut dark hair above his slightly protruding ears. Conscious of the pace of work around her, she bent again to her task and slipped the brush between her lips.

'Do not do that,' he said to her suddenly.

Grace paused and looked up, perplexed. This was how you did the job; how all the girls did it.

'Do not do that,' he said to her again. 'You will get sick.'

And then he was on his way.

Grace was utterly confused. Never one to back down from something she thought needed further investigation, she went straight over to Miss Rooney. But Miss Rooney merely repeated what the girls had already been told. 'She told me there was nothing to it,' Grace later recalled. 'She told me it was not harmful.'

So Grace went back to her work: *Lip ... Dip ... Paint.* There was a war on, after all.

But not for much longer. On 11 November 1918, the guns fell silent. Peace reigned. More than 116,000 American soldiers had lost their lives in the war; the total death toll for all sides was around 17 million. And in that moment of the Armistice, the radium girls, the company executives and the world gave thanks that the brutal, bloody conflict was over.

Enough people had died. Now, they thought, it was time for living.

Exactly one month after the Armistice, Quinta Maggia put those seize-the-day principles into action, marrying James McDonald. He was a cheery man of Irish heritage who worked as a chain-store manager. The newlyweds set up home in a two-storey cottage; to begin with, Quinta was still dial-painting, but that didn't last long. She left the firm in February 1919 and soon fell pregnant; her daughter, Helen, would be born two days after Thanksgiving.

Nor was she the only dial-painter to depart. The war was over; the girls were growing up. Irene Corby also resigned, having landed a job as an office girl in New York City. Later, she would marry the rather dashing Vincent La Porte, a man with piercing blue eyes who worked in advertising.

Those who left were quickly replaced. Sarah Maillefer managed to get a position for her little sister, Marguerite Carlough, in August 1919. She was a dynamic young woman who wore rouge and lipstick and liked dramatic clothes: smart tailored coats with outsized lapels and wide-brimmed hats with feathered edges. Marguerite became best friends with the little sister of Josephine Smith, Genevieve, who had also started working there; another close friend was Albina Maggia, who was still slaving away over her trays of dials, having seen her younger sister marry ahead of her. Albina didn't resent Quinta's happiness, but she couldn't help but wonder when her time might come; that summer, she too decided to leave, returning to hat-trimming.

It was a time of change all round. That same summer, Congress passed the Nineteenth Amendment, giving women the right to vote. Grace Fryer, for one, couldn't wait to make it count. At the plant, too, change was afoot: soon a new chemist – and future vice president – Howard Barker, together with von Sochocky, started playing around with the recipe of the luminous paint, substituting mesothorium for radium. A memo revealed: 'Barker would just mix whatever he had around the place and sell it, 50–50 or 10 per cent [mesothorium] and 90 per cent [radium], or whatever.' Mesothorium was an isotope of radium (dubbed radium-228 to denote its difference to the 'normal' radium-226): also radioactive but with a half-life of 6.7 years, in contrast to radium-226's 1,600 years. It was more abrasive than radium and – crucially for the firm – much cheaper.

In the studio, meanwhile, the girls, for some unknown reason, were asked to try a new technique. Edna Bolz remembered, 'They passed little cloths around: we were supposed to wipe the brush on this little cloth instead of putting it in our mouth.' But within a month, Edna said, 'They were taken away from us. We were not allowed to use the cloths; it wasted too much of the radium.' She concluded, 'The lips were resorted to as the better way.'

It was important to the company that the production process was as efficient as possible because the demand for luminous products showed no signs of slowing, even now the war was over. In 1919, much to new treasurer Arthur Roeder's delight, there was a production high: 2.2 million luminous watches. No wonder Katherine Schaub was feeling tired; that fall she noticed 'cracking and stiffness of her legs'. She was feeling in poor spirits generally, as her mother had passed away that year; Katherine became closer to her father, William, as they grieved.

Yet life, as Katherine's orphaned cousin Irene Rudolph knew too well, goes on, even when those we love die. She and Katherine could do nothing but knuckle down to their jobs, alongside their fellow workers still toiling in the dust-filled studio: Marguerite Carlough and her sister Sarah Maillefer, Edna Bolz and Grace Fryer, Hazel Vincent and Helen Quinlan, Jennie Stocker and – still making everyone laugh – Ella Eckert and Mollie Maggia, who were in fact two of the fastest workers, despite their high jinks at the company socials. They played hard, but they were tough workers too. You had to be to keep your job.

Still the endless orders came. The company began to consider its post-war strategy. It resolved to expand its presence in the field of radium medicine; Arthur Roeder also oversaw the trademarking of 'Undark'. Peacetime frivolity meant there were numerous products customers wanted to make glow in the dark: the company now sold its paint directly to consumers and manufacturers, who performed their own application. All this gave the radium firm another idea – they planned to set up in-house studios for watch manufacturers. This would dramatically reduce the dial-painting workforce in Orange, but the company would still profit by supplying the paint.

In fact, the firm had a compelling reason to want to leave Orange, or at least condense operations. The site's position in the middle of a residential neighbourhood was proving problematic now that the fervour of wartime patriotism had gone. The local residents started to complain that factory fumes discoloured their laundry and affected their health. Company officials took the unusual step of appeasing the residents themselves: one executive gave a neighbour $5 ($68.50) compensation for her damaged washing.

Well, that was a mistake. It opened the floodgates. Next

thing, *all* the residents wanted money. People in this poor community were 'anxious to take advantage of the company'. The firm learned its lesson: it immediately drew the company purse strings tight and not a single further dollar was paid out.

The executives turned their attention back to the watch-company studios. The demand for them was obvious; in 1920, luminous-watch production would surpass 4 million units. Arrangements were soon put in place and everyone was happy; everyone, it seemed, but the original dial-painters.

For while the *company* was doing well through the new agreement, it left the women out in the cold. There was simply not enough work coming in to keep them all employed. Demand dwindled until the Orange studio was running only on a part-time basis.

For the dial-painters, who were paid by the number of dials they did, it was an unsustainable situation. Their numbers decreased until there were fewer than a hundred women remaining. Helen Quinlan was one of those who got out, as did Katherine Schaub, in search of better-paid employment. Helen became a typist, while Katherine found a job in the office of a roller-bearing factory – and found that she loved it. 'The girls at the office,' she wrote, 'were a sociable crowd; they had a club which they invited me to join. Most of the girls embroidered or crocheted, making things for their hope-chests.'

Hope-chests were also called dowry chests, and contained items collected by young single women in anticipation of marriage. In the spring of 1920 Katherine was eighteen, but she didn't seem in any hurry to settle down; she liked the nightlife too much for that. 'I wasn't making anything for my hope-chest,' she wrote, 'so while the girls worked, I played the piano and sang songs that were popular in those days.'

Grace Fryer was also canny enough to see the writing on the wall. Dial-painting, for her, could only ever have been a stopgap job: it was something important to help the war effort, but it wasn't a long-term proposition for someone of her skills. She set her sights high and was thrilled to secure a position at the Fidelity, a high-end bank in Newark. She loved travelling to her office, her dark hair neatly set and an elegant string of pearls around her neck, ready to get to grips with work that challenged her.

As with Katherine's new colleagues, the girls at the bank were sociable; Grace was 'the kind of girl who loved dancing and laughing' and she and her new work friends often hosted dry parties, for Prohibition had begun in January 1920. Grace also liked to swim in her spare time, propelling her lithe body through the local swimming baths as she kept fit. She thought the future looked bright – and she wasn't the only one. In Orange, Albina Maggia had met her man at last.

It felt so wonderful to be courting after waiting all this time. Just when she'd started to feel really old – she was twenty-five, years older than most girls wed at that time; and with a suddenly creaky left knee to boot – he had finally come along: James Larice, a bricklayer and Italian immigrant who'd moved to the United States aged seventeen. He was a war hero, awarded a Purple Heart and one oak leaf cluster. Albina started to allow herself to dream of marriage and children; of finally moving out of the family home.

Her sister Mollie, meanwhile, wasn't waiting for any knight in shining armour to come for her. Independent-minded and confident, although unmarried she left her family to board in an all-female house on Highland Avenue, a tree-lined street in Orange with beautiful detached homes. Mollie was still working at the radium company; one of the few girls left, but she was brilliant at her job and didn't want to leave. Every morning she went to work full of energy and enthusiasm, which was more than she could say for some of her colleagues. Marguerite Carlough, who could normally be relied on for a laugh, kept saying she felt tired all the time; Hazel Vincent, meanwhile, felt so run-down that she decided to leave. She and Theo weren't yet married, so she got herself a job with the General Electric Company.

But her new surroundings didn't improve her condition. Hazel had no idea what was wrong with her: the weight was dropping off her, she felt weak and her jaw ached something rotten. She was so concerned that in the end she asked the company doctor at her new firm to examine her, but he was unable to diagnose her illness.

The one thing she could be assured of, at least, was that it wasn't her work with radium that was the cause. In October 1920, her former employer was featured in the local news. The residue from radium extraction looked like seaside sand and the company had offloaded this industrial waste by selling it to schools and playgrounds to use in their children's sandboxes; kids' shoes were reported to have turned white because of it, while one little boy complained to his mother of a burning sensation in his hands. Yet, in comments that made reassuring reading, von Sochocky pronounced the sand 'most hygienic' for children to play in, 'more beneficial than the mud of world-renowned curative baths'.

Katherine Schaub certainly had no qualms about returning to work for the radium firm when she was head-hunted at the end of November 1920 to train the new workers in the watch-company studios. These were mostly based in Connecticut, including at the Waterbury Clock Company. Katherine taught scores of girls the method she herself had learned: 'I instructed them,' she said, 'to put the brush in their mouth.'

The new girls were excited to be working with radium, for the unstoppable craze continued; brought to fever pitch by a visit of Marie Curie to the United States in 1921. In January of that same year, as part of the constant press coverage of the element, von Sochocky penned an article for *American* magazine. In it, he opined gravely: 'Locked up in radium is the greatest force the world knows. Through a microscope, you can see whirling, powerful, invisible forces, the uses of which' – he admitted – 'we do not yet understand.' He added, as a cliffhanger on which to leave his readers: 'What radium means to us today is a great romance in itself. But what it may mean to us tomorrow, no man can foretell . . .'

In fact, no man can foretell much, von Sochocky included. And there was one event in particular that the doctor didn't see coming: in the summer of 1921, he was frozen out of his own company. His co-founder, George Willis, had sold a large share of his stock to the company treasurer, Arthur Roeder; not long after that, both Willis and von Sochocky were unceremoniously ousted in a corporate takeover. The newly named United States Radium Corporation (USRC) seemed destined for great things in the post-war world, but von Sochocky wouldn't be at the helm to guide it through whatever lay ahead.

Instead, it was Arthur Roeder who slipped graciously into the vacant president's chair.

5

Mollie Maggia poked her tongue gingerly into the gap where her tooth had once been. *Ouch.* The dentist had removed it for her a few weeks ago, after she'd gotten a toothache, but it was still incredibly sore. She gave herself a little shake and turned back to her dials.

The studio was very quiet, she reflected; so many girls had gone. Jennie Stocker and Irene Rudolph had been laid off, and Irene's cousin Katherine had left for the second time. She and Edna Bolz both went to dial-paint for the Luminite Corporation, another radium firm based in Newark. Of the original girls, it was really only the Smith and Carlough sisters left now – and Mollie herself. Saddest of all, in her opinion, was that Ella Eckert had quit to go to Bamberger's. It sure wasn't the same place anymore, not since Roeder took over.

Mollie completed her tray of dials and stood to take it up to Miss Rooney. Despite herself, she found her tongue flicking back to that hole. The pain was just so nagging. If it didn't get better soon, she thought, she was going to go to the dentist again – and a different one this time, someone who really knew what he was doing.

It didn't get better soon.

And so, in October 1921, she made an appointment with a Dr Joseph Knef, a dentist who'd been recommended to her as an expert on unusual mouth diseases. For Mollie, the appointment couldn't come soon enough. For several weeks, the pain in her lower gum and jaw had become so intense that it was almost unbearable. As Knef ushered her into his office, she found herself hoping he would be able to help her. The other dentist had only seemed to make things worse.

Knef was a tall, middle-aged man with tortoiseshell glasses and an olive complexion. He gently probed Mollie's gums and teeth, shaking his head as he examined the place where her tooth had been removed by the previous dentist. Although it had been over a month since it had been taken out, the socket had failed to heal. Knef observed her inflamed gums and softly touched her teeth, several of which seemed a little loose. He nodded briskly, certain he had found the cause of the trouble. 'I treated her,' he later said, 'for pyorrhea.' This was a very common inflammatory disease, affecting the tissues around the teeth; Mollie appeared to have all the symptoms. Knef was sure that, with his expert care, her condition would soon improve.

Yet it didn't improve. 'Instead of responding to the treatment,' Knef recalled, 'the girl became steadily worse.'

It was so terribly, terribly sore. Mollie had more teeth out, as Knef tried to stop the infection in its tracks by removing the source of her pain – but none of the extractions ever healed. Instead, ever-more agonising ulcers sprouted in the holes left behind, hurting her even more than the teeth had.

Mollie struggled on, continuing to work at the studio, even though using her mouth on the brush was extremely uncomfortable. Marguerite Carlough, who was feeling completely well again, tried to engage her in chatter, but Mollie barely responded. It wasn't just the pain of her gums, which seemed to take up all her concentration, but the bad breath that came with it. There was a disagreeable odour whenever she opened her mouth, and she was embarrassed by it.

At the end of November 1921, her sister Albina married James Larice. The wedding was held the day before Quinta's daughter's second birthday, and the bride found herself drinking in her niece's funny antics with a newly maternal air. Soon, she thought, she and James would have their own little ones running about.

There was one cloud on the horizon though, darkening the newlyweds' bliss: Mollie. Even though Albina now rarely saw her as the two lived distantly, all of Mollie's sisters couldn't help but be concerned by her deteriorating condition. For as the weeks passed, it wasn't just her mouth that became sore; she started to have aches and pains in completely unconnected places. 'My sister,' Quinta remembered, 'began having trouble with her teeth and her jawbone and her hips and feet. We thought it was rheumatism.' The doctor administered aspirin and sent her home to Highland Avenue.

At least she lived with an expert. One of the women she boarded with, fifty-year-old Edith Mead, was a trained nurse, and she cared for Mollie as best she could. But nothing in her training could make sense of this disease; she had never seen anything like it. Neither Knef, nor Mollie's family doctor nor Edith seemed able to make her better. Every appointment brought forth another expensive doctor's bill, but no matter how much Mollie spent there was no cure.

In fact, the more Knef tried to help – and he employed some 'extreme methods of treatment' – the worse Mollie became: the worse her teeth were, and the ulcers, and her gums. Sometimes, Knef didn't even have to pull her teeth anymore; they fell out on their own. Nothing he did arrested the disintegration in the slightest degree.

And disintegration was the word for it. Mollie's mouth was literally falling apart. She was in constant agony and only superficial palliatives brought her any relief. For Mollie, a girl who had always loved to joke around, it was unbearable. Her smile, which had once been a toothy grin that beamed across her face, was unrecognisable as more and more of her teeth came out. Well, no matter; she was in too much pain to smile anyway.

As Christmas passed and the New Year began, the doctors finally thought they had diagnosed her mysterious condition. Sores in the mouth … joint pain … extreme tiredness … a young single woman living outside the family home … Well, it was obvious, really. On 24 January 1922, her physicians tested her for syphilis, or Cupid's disease – a sexually transmitted infection.

But the test came back negative. The doctors would have to think again.

By now Dr Knef had noticed certain things about her case that made him doubt his initial diagnosis. It was, it appeared, an 'extraordinary affliction'; it was almost like something was attacking her from the inside, though he knew not what it could be. As well as the seemingly unstoppable disintegration of her mouth, to his trained nostrils the noticeable smell coming from her seemed 'peculiar': 'it differed decidedly from the odour commonly associated with the usual forms of necrosis of the jaw'. Necrosis meant bone decay. Mollie's teeth – those that were left – were literally rotting in her mouth.

After conducting further research, Knef reached a conclusion. She was, he determined, suffering from a condition not unlike phosphorus poisoning. It was the same suggestion Katherine Schaub's doctor had made, when she'd had her outbreak of teenage spots a few years before.

'Phossy jaw' – as the victims of phosphorus poisoning had grimly nicknamed the condition – had very similar symptoms to those that Mollie was enduring: tooth loss, gum inflammation, necrosis and pain. And so, at her next appointment, Knef asked Mollie how she was employed.

'Painting numbers on watches so that they will shine at night,' she responded, wincing as her tongue formed the words and touched the ulcers in her mouth.

With that, his suspicions increased. Knef decided to take matters into his own hands. He visited the radium plant – but received little cooperation. 'I asked the radium people for the formula of their compound,' he remembered, 'but this was refused.' Undark was, after all, a highly lucrative commercial property; the company couldn't share the top-secret formula with just anyone. Knef was, however, told that no phosphorus was used and assured that work in the factory could not have caused the disease.

His own tests seemed to back up the firm's assertion. 'I thought phosphorus might have been in the paint and caused her trouble,' he later said, 'but all the tests I made failed to show it.' They were, it seemed, still in the dark.

None of this helped Mollie. By now, the pain was excruciating. Her mouth had become a mass of sores; she could barely speak at all, let alone eat. It was horrifying for her sisters to watch. She suffered such agony, said Quinta, that it 'has unnerved me every time I recall [it]'.

Anyone who has ever endured an abscessed tooth may be able to imagine some small degree of her suffering. By now, Mollie's entire lower jaw, the roof of her mouth and even the bones of her ears might be said to be one huge abscess. There was no way in the world that she could work in such a condition. She quit her job at the Orange studio, where she had spent so many happy hours painting dials, and was confined to her home. Surely, one day soon, the doctors would determine what was wrong, and cure her, and she could get on with her life again.

But no cure came. In May, Knef suggested that she come in again to his office, so that he could examine her and see what progress had been made. Mollie limped into his office; the rheumatism in her hips and feet had grown worse, and she was almost lame. But it was her mouth that took up all her thoughts, all her time, and consumed *her*. There was no escape from the agony.

She hobbled over to Dr Knef's dental chair and then leaned back. Gingerly she opened her mouth for him. He bent over her and prepared to probe inside.

There were barely any teeth left now, he saw; red-raw ulcers peppered the inside of her mouth instead. Mollie tried to indicate that her jaw was hurting especially, and Knef prodded delicately at the bone in her mouth.

To his horror and shock, even though his touch had been gentle, her jawbone broke against his fingers. He then removed it, 'not by an operation, but merely by putting his fingers in her mouth and lifting it out'.

A week or so later, her entire lower jaw was removed in the same way.

Mollie couldn't bear it – but there was no relief. All the doctors could offer were analgesic drugs that barely helped. Her whole face beneath her bouffant brown hair was just pain, pain, *pain*. She became anaemic, weakening further. Knef, even though he wasn't a physician (nor proficient at the procedure), tested her for syphilis again on 20 June – and this time the results came back positive.

Mollie would probably have been devastated if she'd been told, but many doctors at that time kept diagnoses from their patients and it is very likely Knef didn't tell her, wanting her to concentrate on getting well. She would have known, had she heard the news, that it couldn't *possibly* be that. But she had no idea what the real cause could be. If anything, she should be full of health – not only was she young, in her twenties, but she had worked with radium for years, for goodness' sake. Only that February, the local paper had declared: 'Radium may be eaten ... it seems that in years to come we shall be able to buy radium tablets – and add years to our lives!'

But, for Mollie, time it seemed was running out. After her jaw had gone, an important discovery was made. Knef had always hoped that by removing a tooth, or a piece of infected bone, the progress of the mysterious disease would be halted. But now it became evident that 'whenever a portion of the affected bone was removed, instead of arresting the course of the necrosis, it speeded it up'. Over the summer, Mollie's condition deteriorated even further. She was getting painfully sore throats now, though she knew not why. Her jaw, at times, would spontaneously bleed, and Edith would press white cotton bandages to her face, trying to stem the flow.

September 1922. In Newark Mollie's former colleague Edna Bolz was preparing to wed. Her husband-to-be was Louis Hussman, a plumber of German heritage with blue eyes and dark hair. He was 'devoted' to her. She laid out her accoutrements with touching anticipation: the bridal gown, her stockings, her wedding shoes. Not long now.

Not long now. They are words of excitement. Expectation. And reassurance – to those in pain.

Not long now.

In September 1922, the peculiar infection that had plagued Mollie Maggia for less than a year spread to the tissues of her throat. The disease 'slowly ate its way through her jugular vein'. On 12 September, at 5 p.m., her mouth was flooded with blood as she haemorrhaged so fast that Edith could not staunch it. Her mouth, empty of teeth, empty of jawbone, empty of words, filled with blood, instead, until it spilled over her lips and down her stricken, shaken face. It was too much. She died, her sister Quinta said, a 'painful and terrible death'.

She was just twenty-four years old.

Her family knew not what to do with themselves; knew not what could have happened to take her from them so suddenly. 'She died and the doctors said they didn't just know from what,' remembered Albina.

The family tried to find out. Albina added, 'My elder sister went down to Dr Knef's office; we were told after her death that she died of syphilis.'

Syphilis. What a shameful, sad little secret.

The final medical bills came in, addressed to the girls' father,

Valerio, and labelled 'for Amelia'. Her family doctor reduced, on request, the amount he charged them. But though it was a welcome gesture, it couldn't bring Mollie back.

They buried her on Thursday 14 September 1922, in Rosedale Cemetery, Orange, in a wooden coffin with a silver nameplate. It was inscribed simply 'Amelia Maggia'.

Before they said their goodbyes, her loved ones laid out her clothes: her white dress, her stockings, her black leather pumps. Gently, they clothed her body in the garments, and then Mollie was laid to rest.

Her family hoped that now, at last, she would find peace.

6

Ottawa, Illinois
United States of America
September 1922

Two days after Mollie's funeral, and 800 miles away from Orange, a small advertisement appeared in the local paper of a little town called Ottawa, in Illinois. GIRLS WANTED it declared. And then continued: 'Several girls, 18 years or over, for fine brushwork. This is a studio proposition, the work is clean and healthful, surroundings pleasant. Apply to Miss Murray, old high school building, 1022 Columbus Street.'

It sounded *wonderful*.

Ottawa was a tiny town – population 10,816 – located 85 miles south-west of Chicago. It billed itself as a 'genuine American community' in its town directory, and those words were on the money. It was the kind of place where its banks proclaimed themselves to be 'where friendliness reigns' and local businesses advertised their location as being 'one block north of Post Office'. Ottawa was in the heart of rural Illinois, surrounded by farmland and the impossibly wide skies of the Midwest. It was a place where folks were happy simply to get on with life: raise their families, do good work, live decent lives. The community was close-knit and emphatically religious; Ottawa was

'a small [town] of many churches' with the majority of residents
Catholic. 'The citizens of Ottawa,' chirped the town directory,
'are liberal-minded, prosperous and progressive.' The perfect
populace, then, for this new dial-painting opportunity.

It wasn't the United States Radium Corporation hiring,
although they knew their competitor well. The employer was
the Radium Dial Company; its president was Joseph A. Kelly.
He was based at the head office in Chicago, however, so it was
to Miss Murray, the studio's superintendent, the Ottawa girls
applied.

Lottie Murray was an immensely loyal employee, a slim
single woman forty-four years old who had been with the company five years, as it moved its studio around various locations
before settling now in Ottawa. One of her very first successful
applicants was nineteen-year-old Catherine Wolfe. She was
Ottawa born and bred, a devoted parishioner of St Columba
Church, which was located diagonally opposite the studio.
Despite her young age, Catherine had already had some hard
knocks in life. When she was only six, her mother Bridget had
passed away; just four years later, in 1913, her father Maurice
died from 'lung trouble'. As a result, ten-year-old Catherine
was sent to live with her elderly aunt and uncle, Mary and
Winchester Moody Biggart, sharing their home at 520 East
Superior Street.

Catherine was a shy, quiet person who was very unassuming.
She had thick, jet-black hair and very pale skin; a rather neat
woman with tidy limbs who didn't go in for showy gestures.
The job at the studio would be her first, painting the dials of
timepieces and aeronautical instruments. 'It was fascinating
work, and the pay was good,' she enthused, 'but every line had
to be just so.'

And there was only one known way to get the necessary
point on the 'Japanese brushes the size of a pencil' that the
Ottawa girls used. 'Miss Lottie Murray,' Catherine remembered,
'taught us how to point camel-hair brushes with our tongues.

We would first dip the brush into water, then into the powder, and then point the ends of the bristles between our teeth.'

It was the 'lip, dip, paint routine' all over again – but with an entirely new cast.

Joining Catherine at the studio was sixteen-year-old Charlotte Nevins. The advert had said '18 years or over', but she wasn't going to let a little thing like that stop her: all her friends were there and she wanted to join them. Charlotte was the youngest of six siblings and perhaps she just wanted to grow up fast. She was a cheerful, caring girl who, like Catherine, was a devout Catholic. Though she was generally quiet, she could be pretty outspoken when she needed to be.

Charlotte wasn't the only one to tweak the truth about her age; another employee who did the same – though the company must have known – was Mary Vicini, a sweet Italian girl who had come to America as a baby. She was only thirteen in 1922 but nonetheless made it into the coveted workforce. In truth, the nimble fingers of prepubescent girls suited the delicate work of dial-painting; records show that some were as young as eleven.

Assisting Miss Murray with the applicants were Mr and Mrs Reed. Rufus Reed was the assistant superintendent, a thirty-nine-year-old New Yorker who was a company man to his bones. Tall and bald-headed, he was medium-built and wore dark-framed glasses. He was in fact deaf, but it didn't hinder him in his work; perhaps his disability made him all the more grateful to the firm that had treated him well. Like Miss Murray, Reed and his wife, Mercedes, who worked there as an instructress, had been with the company for years.

Mercy Reed was famed for her instructing: 'she ate the luminous material with a spatula to show the girls that it was "harmless"', licking it right in front of them. Charlotte Nevins remembered: 'When I was working in the plant painting dials they always told me the radium would never hurt me. They even encouraged us to paint rings on our fingers and paint our dress buttons and buckles.'

And the girls did exactly as instructed. They were 'a happy, jolly lot' and they frequently practised their painting, especially in the fields of fashion and art. Lots of them took paint home; one woman even painted her walls with it for interior decoration with a difference. Radium Dial seems not to have been as concerned as USRC had been about wasting material: former employees report that the radium was handled carelessly and, in contrast to the brushing-down in Orange, 'washing was a voluntary procedure and not many of the workers made use of the washing facilities'.

Why would they – when they could go home glowing like angels? 'The girls were the envy of others in the little Illinois town when they stepped out with their boyfriends at night, their dresses and hats and sometimes even their hands and faces aglow with the phosphorescence of the luminous paint,' a newspaper reported. A young local girl recalled, 'I used to *wish* I could work there – it was the elite job for the poor working girls.' When the dial-painters visited the drugstore for home-made candies or carbonated ice cream, they left a trail of glowing dust behind them. Catherine remembered, 'When I went home and washed my hands in a dark bathroom, they would appear luminous and ghostly. My clothes, hanging in a dark closet, gave off a phosphorescent glare. When I walked along the street, *I* was aglow from the radium powder.' The women were 'humorously termed the ghost girls'.

They worked six days a week, using a similar greenish-white paint to that used in Orange, with identical ingredients, and the girls 'were expected to work, work, work'. They did get a lunch break, but Mrs Reed ate her food at her painting desk and although a number of the girls popped home or went to nearby coffee shops, the majority chose to stay in the studio, following their instructress's example. Catherine recalled, 'We used to eat our lunches right beside the work benches near the luminous paint and brushes which we used; we hurried as fast as we could.' After all: 'We made more money that way.'

The girls declared, 'We were extremely happy in our work,' and Radium Dial was equally content. It followed the attitude of its main client, Westclox. Their Manual for Employees read: 'We expect you to work hard, and the pay is accordingly large … If you do not expect to work hard and carefully, you are in the wrong place.'

But for Catherine and Charlotte and Mary, this place felt very, very right indeed.

7

Newark, New Jersey
November 1922

'Miss Irene Rudolph?'

Irene got tentatively to her feet as she heard her name being called by Dr Barry and shuffled into his office. Her trouble had first started in her feet, though they were currently the least of her worries; she could just about get by if she took things slow. Her family, including her cousin Katherine Schaub, helped out a lot. It was now her mouth that was the real problem.

She had been attending this dental practice since August, though she'd been having tooth trouble since the spring of 1922. Despite the attentions of various dentists, her condition had worsened; so much so that, in May, she'd had to give up her job in a corset factory. Without a job, yet with increasing medical bills, Irene soon found her financial position precarious. She'd been sensible when she'd worked as a dial-painter, squirrelling away her high wages, but her mysterious condition had exhausted her hard-earned savings.

With every expensive appointment, she hoped for improvement. As she levered herself into Dr Barry's chair, she opened her mouth wide and prayed that, this time, he would have good news to offer.

Walter Barry, an experienced dentist of forty-two, examined Irene's mouth with deepening confusion. He and his partner, Dr James Davidson, had been operating on Irene since the summer. Yet every course of treatment they tried, such as cutting out the diseased bone in her mouth and removing teeth, seemed only to increase her suffering. Their surgery was located at 516 Broad Street, practically opposite the Newark Public Library, yet it appeared no textbook or medical journal on the shelves of the library or their own practice contained the solution. As Barry examined Irene's butchered mouth at this latest appointment, on 8 November 1922, he could see that, still, there was only more infection, inflaming her empty gums with an unhealthy yellow sheen.

James Davidson had experience of treating phossy jaw and he and Barry now became convinced this was Irene's trouble. 'I immediately started to question [Irene] as to what [her] occupation was,' Barry recalled. 'I made an effort to ascertain whether there was any phosphorus in this material she was using.'

Unknowingly, he was following in the footsteps of Dr Knef, who had treated Mollie Maggia – but the two investigations didn't cross, nor did Knef have opportunity to share his own discovery: how Mollie's jaw was destroyed faster and faster, the more of her he removed. The same accelerated decline was now affecting Irene.

Barry told his patient it was his opinion she was suffering from 'some occupational trouble'. But as Katherine Schaub later noted, 'The word radium was never brought into it.' Radium was such an established medical boon that it was almost beyond reproach; people didn't question it. And so, although there was some suspicion that the luminous paint was to blame for Irene's condition, culprit number one was phosphorus.

In December, Irene took a turn for the worse and was admitted to hospital. She was shockingly pale and found to be anaemic. And it was while in hospital that she decided she was not going to lie down and suffer quietly.

For although Irene's dentists may not have crossed paths with Knef, the dial-painters' friendships were a stronger network. By now, Irene had heard about Mollie Maggia's death. The gossip-mongers were saying syphilis had killed her, but the girls who knew her found that hard to believe. And so, while in hospital, Irene told her doctor that there had been another girl, who'd had symptoms *just like hers*, who'd died only a few months before. The Maggia family were trying to move on with their lives without their sister – that winter, Quinta was pregnant again and Albina was hoping that any month now she would have the same good news to announce – but for Irene, sitting weakly in her hospital room, Mollie's death was definitely not something in the past, but somehow horribly present.

And then she told the doctor something else. Another girl, she said, was sick.

She could have meant Helen Quinlan, who had been taken ill with a severe sore throat and swollen face, which had inflamed her pixie features. She, too, had had trouble with a tooth and was beginning to show signs of anaemia. But Helen appears not to have moved in the same social circles as Irene; it was Hazel Vincent to whom she referred.

Since Hazel had left USRC she had become more and more ill. She had been told that she was suffering from anaemia and pyorrhea; her doctor, too, suspected phossy jaw from the black discharge with its 'garlic odour' exuding from her nose and mouth. Hazel's childhood sweetheart, Theo, was worried sick about her.

In Irene's opinion, Hazel's case and hers were just too similar to be discounted as mere coincidence. Carefully, she set out the parallels during a consultation with Dr Allen at the hospital, trying to make him see that there was something more going on. And as the doctor listened to his patient talk, he'd heard enough. All the evidence suggested that it was an occupational problem. On 26 December 1922, Allen reported Irene

Rudolph as a case of phosphorus poisoning to the Industrial Hygiene Division – and asked them to investigate. The authorities launched straight into action and within days an inspector was at the Orange plant looking into the claims of industrial poisoning.

The inspector was escorted up to the dial-painting studio by Harold Viedt, a vice president of USRC who had responsibility for operations. Together, they quietly observed the girls at work. There were not many of them there – dial-painting in Orange had become almost a seasonal occupation so the girls did not work continuously anymore – yet the inspector noted with some incredulity the universal practice of lip-pointing. This was called to the attention of Mr Viedt, who was quick to address the concerns. Viedt told him, the inspector reported, that 'he has warned [the girls] time and time again of this dangerous practice, but he could not get them to stop it'.

Had the dial-painters overheard this conversation, they would probably have been stunned. Other than Sabin von Sochocky's one-off warning to Grace Fryer that lip-pointing would 'make her sick', not a single other dial-painter, including the instructresses and forewomen, ever reported a warning being issued; and certainly not one that included reference to lip-pointing being a 'dangerous practice'. On the contrary, they had received countless assurances of the exact opposite; when the company deigned to concern itself with their work processes, that was. On the whole, the firm left them to get on with their work without interference. It didn't, in truth, really seem to care just *how* the women dial-painted, as long as the material wasn't wasted and the work got done.

The inspector continued to observe the girls. One, he noticed, a rather matronly woman who was older than the rest, appeared to be limping as she carried her dials up to the new forelady, Josephine Smith, who had recently been promoted; Miss Rooney would be leaving to join the Luminite Corporation.

Sarah Maillefer *was* limping. She was getting old, she

supposed; she was now thirty-three, and you should expect a few more aches and pains as you got older. Plus, it was exhausting being a working mother. She didn't have the energy to keep up with her sister Marguerite, let alone her eleven-year-old daughter. She felt blessed that the company was so understanding about her limp; 'a foreman of the company [took] her to and from work each day because of this trouble'.

The inspection concluded with the official taking a paint sample for testing; he sent it to John Roach, the deputy commissioner of the New Jersey Department of Labor, with a recommendation that Roach's team 'make a survey of this plant as it is outside our jurisdiction'. Consequently, an additional inspection took place in the next few weeks, with the inspector, Lillian Erskine, delivering her findings to Roach on 25 January.

Erskine took a rather different approach from the first inspector. As part of her investigation, she spoke with a radium authority and informed Roach that 'no reports of necrosed bones as a result of radium treatment exist'. She therefore concluded: 'This case [Irene Rudolph] and the reported second case [Hazel Vincent] are probably an accidental coincidence, resulting from abscessed teeth and incompetent dental surgery.'

Roach arranged for the paint to be tested by Dr Szamatolski, a chemist. Szamatolski was an educated man, and thought it extremely unlikely any phosphorous would be in the paint, as this had never been hinted at as an ingredient. Without having run a single test, he wrote sagely to Roach on 30 January 1923: 'It is my belief that the serious condition of the jaw has been caused by the influence of radium.'

This was a radical idea – yet Szamatolski's off-the-wall suggestion did have some science to back it up. In a bibliography of radium studies that USRC itself had published just four months before, there was an article headed 'Radium Dangers – Injurious Effects'. In fact, the bibliography contained articles as far back as 1906 on the damage radium could cause. The company conceded in an internal memo that there were a 'considerable' number of articles dealing with the hazards from the early twentieth century. A woman had even died in Germany in 1912 after being treated with radium; her doctor had said 'one cannot doubt for a moment' that radium poisoning was the cause.

Yet the flip side of the coin was all the *positive* literature about radium. As early as 1914, specialists knew that radium could deposit in the bones of radium users and that it caused changes in their blood. These blood changes, however, were interpreted as a good thing – the radium appeared to stimulate the bone marrow to produce extra red blood cells. Deposited inside the body, radium was the gift that kept on giving.

But if you looked a little closer at all those positive publications, there was a common denominator: the researchers, on the whole, worked for radium firms. As radium was such a rare and mysterious element, its commercial exploiters in fact controlled, to an almost monopolising extent, its image and most of the knowledge about it. Many firms had their own radium-themed journals, which were distributed free to doctors, all full of optimistic research. The firms that profited from radium medicine were the primary producers and publishers of the positive literature.

Szamatolski's opinion, therefore, was a lone, unheard and hypothetical voice, set against the flamboyant roar of a well-funded campaign of pro-radium literature. Szamatolski himself, however, was a conscientious as well as smart man. Given his tests would take a few months, and mindful of the fact that work was continuing in the dial-painting studio, he took care to add a special note to his letter of 30 January. Though his radical theory had not yet been proven, he wrote plainly, 'I would suggest that every operator be warned through a printed leaflet of the dangers of getting this material on the skin or into the system, especially the mouth, and that they be forced to use the utmost cleanliness.'

Yet, for some reason, this did not happen. Perhaps the message was never passed on.

Perhaps the company chose to ignore it.

As 1923 drew on and Szamatolski ran his tests, Irene Rudolph, who had been sent home from hospital, continued to endure the horrific ulcers and sores that had tortured Mollie Maggia. Irene's anaemia grew more serious; as did Helen Quinlan's. They were pale, weak creatures, with no energy to them; no life. Doctors treated them first for one thing and then another – but not a single treatment helped.

And they weren't the only ones who were sick. Since George Willis, the co-founder of the Orange radium firm, had been ousted from his company, things had deteriorated for him. It seemed a long time ago that he had thoughtlessly carried tubes of radium with his bare hands every day at work – but all time is relative. With a half-life of 1,600 years, radium could take its time to make itself known.

As the months had passed since his departure from the company, Willis had sickened and in September 1922, the same month Mollie Maggia died, he'd had his right thumb amputated; tests revealed it was riddled with cancer. Willis didn't keep his sickness to himself; instead, he published his findings. In February 1923, in the *Journal of the American Medical Association* (*JAMA*), he wrote: 'The reputation for harmlessness enjoyed by radium may, after all, depend on the fact that, so far, not very many persons have been exposed to large amounts of radium by daily handling over long periods . . . There is good reason to fear that neglect of precautions may result in serious injury to the radium workers themselves.'

What his former company thought of his article is not recorded. They probably dismissed it: he didn't work for them anymore; thus, he did not matter. And they weren't the only ones to ignore it. No one, it seemed, took much notice of the small article in the specialist publication.

By the April of 1923, Szamatolski had completed his tests. As he had suspected, there was not a single trace of phosphorus in the luminous paint.

'I feel quite sure,' he therefore wrote, on 6 April 1923, 'that the opinion expressed in my former letter is correct. Such trouble as may have been caused is due to the radium.'

8

Ottawa, Illinois
1923

The radium, the girls in Ottawa thought, was one of the best things about their new job. Most women who worked in the town at that time were shop girls, secretaries or factory workers – this was something a little different. No wonder it was the most popular gig in town.

Girls from all walks of life tried their hand at it, drawn by radium's allure. Some dial-painters 'were what I call "slumming"', said one of their colleagues with more than a hint of disapproval. '[One worker] was the darling daughter of a prominent physician – one of the better people. She and her friend were just there for a very few days.' The well-off women simply wanted to see what it was like to be one of the ghost girls: a kind of voyeuristic life tourism. Perhaps as a result of their interest, 'Mrs Reed had [her training] room all dolled up like a kindergarten': it boasted curtains at the windows and flowers in a porcelain vase.

Radium Dial had initially advertised for fifty girls, but would eventually employ as many as two hundred. More workers were needed to keep up with demand: in 1923, Westclox, Radium Dial's main client, had a 60-per-cent share of the US alarm-clock market, worth $5.97 million ($83 million). So many girls wanted to become dial-painters that the company could afford to be choosy. 'The practice,' a former employee recalled, 'was to hire about ten girls at a time and try them out. Out of the ten, they would usually keep about five.'

One who made the grade was Margaret Looney, whose family called her Peg. She was good friends with Catherine Wolfe: they had gone to the same parish school and Peg also attended St Columba across the way; as did the vast majority of Radium Dial workers.

Everyone knew the Looney family. At the time Peg started at Radium Dial in 1923, there were eight kids in the clan; that number would eventually grow to ten. The whole family lived in a cramped house right on the railroad tracks, where the roar of the trains was so frequent that they didn't even think about it anymore. 'It was a very tiny house: one-storey, wood frame [and] four rooms, basically,' said Peg's niece Darlene. 'It had two bedrooms [and] the big bedroom, where the kids slept, had blankets hanging from the ceiling, separating the girls' side from the boys'; there'd be three to four kids in a bed. They were dirt poor, just as poor as you could be.'

But they were close-knit; you had to be, at such close quarters. And they had fun. Peg, a slender freckled redhead who was tiny in stature, was known for her giggling fits and as the eldest girl – she was seventeen – her siblings looked up to her and followed her lead. In the summer, the Looney children would run around barefoot because they couldn't afford shoes, but it didn't stop their games with their neighbourhood friends.

Given her family background, Peg was thrilled to land a job as a well-paid dial-painter. She earned $17.50 ($242) a week – 'good money for a poor Irish girl from a large family' – and gave her mom most of it. The job meant she had to park her ambition to become a schoolteacher, but she was still young; there was plenty of time to teach later on in life. She was a very intelligent girl, so scholarly that at high school her favourite hobby had been 'trying to hide behind the dictionary', which she used to read in a 'delightful sunny nook'. She had the smarts to make it as a schoolteacher, but she would dial-paint for a time to help out at home.

Anyway, she had a good time at work, painting with her friends. Peg started out, as did all the new girls, by painting the Big Ben alarm clocks that Westclox produced. He was 'a rugged handsome fellow' of a clock, with a dial that measured about 10 centimetres across, giving him nice big numbers for the less experienced girls to paint. As they gained in skill, they were moved on to the Baby Bens, smaller clocks about half the size, and eventually to the pocket watches: the Pocket Ben and the Scotty, which were just over three centimetres wide.

Peg held the dials in her hand as she carefully traced the numbers with the greenish-white paint, lipping and dipping her camel-hair brush as she had been taught. The paper dial was mounted on a slim metal disc, cool to the touch. It had little ridges on it at the back, by which it would later be attached to the rest of the clock.

Sitting alongside Peg in the studio was another new joiner: Marie Becker. 'I was working for the bakery downtown,' Marie later recalled, 'and I thought, "Gee, I believe I'll go over there [to the radium studio], see how much you get." Well, I was gonna get about twice as much. And in them days you liked money 'cause you got very little of it.' She was sold.

Like Peg, Marie came from an underprivileged background. After her father had died of dropsy, her mother remarried and when Marie was only thirteen her stepfather set her to work. She'd done all sorts of jobs since then: the bakery, factory work and being a sales girl at the dimestore. Her stepfather's instruction was something Marie had taken in her stride – as she did all things in life. 'Her attitude was wonderful,' a close relative said. 'I don't ever remember her being in a bad mood. You know how other people get grudges or they get a chip on their shoulder – never. She was a ball of wax. She laughed a lot. She laughed loud. Her laugh would make you laugh.'

She was an instant hit in the dial-painting studio. Marie was a real character, full of opinions and wisecracks. She was a 'skinny minny' with dimples who, despite her German heritage, had almost a Spanish look to her, with striking dark eyes and long brown hair, which she wore in a bun, sometimes with a spit curl on her forehead. She became good pals with Charlotte Nevins and declared Peg Looney her 'closest friend'.

At first, however, Marie wasn't sure about staying there. 'My first day at work,' she recalled, 'I said I didn't like it ... they learn you to put the thing in your mouth, that's the first thing they taught you. I said when I went home at noon that I'd never come back [but] I went back the next morning. I didn't like the work but I thought, "There's money involved, see." I just stayed.' Those high wages were so hard to say no to.

Not that Marie saw them. 'Well, you know where my first pay cheque went,' she commented drily. 'It went to my stepfather ... you turned your cheque over to him and he spent the money. I never liked it.' It was especially hard for her because most of the other girls at work could keep their money – and they spent it on the latest fashions at T. Lucey & Bros, where girls could buy 'corsets, gloves, laces, ribbons, fancy goods and notions'.

Marie dreamed of spending her wage packet on high heels, which she loved. One day, she'd had enough: *she'd* earned the money, not her stepfather, and she thought to herself, 'Gee, this week when I get paid I'm gonna go in [that shoe shop] and cash my cheque.' And she did. 'I bought my first high-heeled slippers and I told the guy to leave them right on because I was gonna wear them home!' She had a 'ruckus' with her stepfather about not handing over her cheque and eventually moved out when she was seventeen. Thanks to her wages and her spirit, she was well able to.

Marie's new feistiness was a sign of the times. It was the Roaring Twenties, after all, and even in a tiny town like Ottawa the breeze of female independence and fun times was stirring the sidewalks, blowing the winds of change. Peg's sister said of the dial-painters, 'They were all pretty young girls just raring to go out and lick the world.'

And what a time to be doing it … 'Prohibition was huge [in Ottawa],' remarked a resident. 'There was a lot of drinking and gambling joints.' And not just that: big bands and good times. The dial-painters were among those dancing to the Twentieth Century Jazz Boys and, later, Benny Goodman. The year 1923 was when the Charleston dance craze took America by storm and the Radium Dial girls swivelled their knees with the best of them. The luminous glow of the radium on their hair and undulating dresses made those parties even more special. 'Many of the girls,' Catherine Wolfe recalled, 'used to wear their good dresses to the plant so that they would become luminous when they went out to parties later.'

It gave the girls even more reason to invest in high-end fashion: they bought the latest cloche hats, high heels with bows upon them, handbags and strings of pearls. And it wasn't just after work that the good times roared either; in work, too, the girls had a blast. As in Orange, the bosses – Mr and Mrs Reed, and Miss Murray – worked downstairs, giving the girls on the second floor a free rein to have fun. In their lunch hour, the girls would go into the darkroom with the leftover radium paint: they'd had a swell idea for a new game.

'We just figured,' said Marie, 'well, we've got a little radium left in our jars … we gotta get new [material] for starting after lunch … so we started fooling around. We'd paint our faces up and put [on] moustaches. I always did paint by my nostrils and then my eyebrows and a moustache and a chin. We went in the darkroom to make faces at each other.' Charlotte Nevins remembered that they would 'turn the lights off and then [we] could look in the mirror and laugh a lot. [We] glowed in the dark!'

It was a strangely spooky vision. In the darkroom, no daylight shone. There was no light at all – except for the glowing element the girls had painted on their bare skin. They themselves were completely invisible.

'You don't see nothing, no body, all you see is the radium,' recalled Marie. 'All you're looking at is eyebrows and moustaches and your teeth.' But, as she said, it was all 'just for fun'.

More and more girls joined them at Radium Dial. Frances Glacinski, Ella Cruse, Mary Duffy, Ruth Thompson, Sadie Pray, Della Harveston and Inez Corcoran were among them; Inez sat right next to Catherine Wolfe in the studio. 'We were a bunch of happy, vivacious girls,' remembered Charlotte Nevins fondly. 'The brightest of the girlhood of Ottawa. [We had] our own little clique.' That clique worked together, danced together and had outings along the river and at Starved Rock, a local beauty spot.

They were such good, good times. And, as Catherine's nephew later said of those halcyon days, 'They thought it was never gonna end.'

9

Orange, New Jersey
June 1923

It was the Roaring Twenties in Orange, too – but Grace Fryer wasn't in the mood for dancing. It was odd: she had this slight pain in her back and feet; nothing major, but enough to make it uncomfortable for her to walk. Dancing definitely wasn't on the agenda, even though the girls at the bank were still throwing their parties.

She tried to put it to the back of her mind. She'd had a few aches and pains the year before, too, but they came and went; hopefully, when these latest aches cleared up, they would simply go for good. She was just run-down, she reasoned: 'I thought that this was merely a touch of rheumatism and did nothing about it.' Grace had far more important things to think about than an achy foot; she'd been promoted at work and was now the head of her department.

It wasn't just an achy foot troubling her, however. Back in January, Grace had gone to the dentist for a routine check-up; he'd removed two teeth and, although an infection had lingered for two weeks, her trouble had then cleared up. But now, six months on, a hole had appeared at the site of the extraction and was leaking pus profusely. It was painful, and smelly, and tasted disgusting. Grace had health insurance and was prepared to pay to get it sorted; the doctors, she was sure, would be able to fix her trouble.

But had she known what was happening just a few miles away in Newark, she might have had reason to doubt her faith in physicians. Grace's former colleague Irene Rudolph was still paying doctor after doctor to treat her – but without relief. She had by now undergone both operations and blood transfusions, but to no avail. The decay in Irene's jaw was eating her alive, bit by bit.

She could feel herself weakening. Her pulse would pound in her ears as her heart beat faster to try to get more oxygen around her severely anaemic body – but although her heart was drumming faster and faster, it felt to her like her life was inexorably slowing down.

In Orange, for Helen Quinlan, the drumbeat suddenly stopped.

She died on 3 June 1923, at her home on North Jefferson Street; her mother Nellie was with her. Helen was twenty-two years old at the time of her death. The cause of it, according to her death certificate, was Vincent's angina. This is a bacterial disease, 'a progressive painful infection with ulceration, swelling and sloughing off of dead tissue from the mouth and throat due to the spread of infection from the gums'. Her doctor later said he didn't know if the disease was confirmed by laboratory tests, but it was written on her death certificate, nonetheless.

The 'angina' in its name is derived from the Latin *angere*, meaning 'to choke or throttle'. That's what it felt like when the decay in her mouth finally reached her throat. That's how Helen died, this girl who had used to run with the wind in her skirts, making boyfriends gaze and marvel at her zest for life and her freedom. She had lived an impossibly short life, touching the lives of those who knew her; now, suddenly, she was gone.

Six weeks later, Irene Rudolph followed her to the grave. She died on 15 July 1923 at twelve noon, in Newark General Hospital, where she'd been admitted the day before. She was twenty-one. At the time of her death, the necrosis in her jaw was said to be 'complete'. Her death was attributed to her work, but the cause was given as phosphorus poisoning; a diagnosis admitted by the attending physician to be 'not decisive'.

Katherine Schaub, who had watched her cousin suffer through every stage of what she called her 'terrible and mysterious illness', was angry and confused, as well as grief-stricken. She knew Irene had spoken to Dr Allen about her fears that her sickness had been caused by her job, but since then the family had heard nothing. They didn't know the names John Roach or Dr Szamatolski; they knew nothing of the doctor's verdict following his tests. In fact, after reviewing Szamatolski's report and that of the two inspectors, the Department of Labor took no action.

No action whatsoever.

Katherine was an intelligent, determined young woman. If the authorities weren't going to do anything – well then, she would. On 18 July, the Schaubs buried Irene, who had lived such a short, sad life, and the next day, fuelled by sorrow and the senseless waste, Katherine went to the Department of Health on Franklin Street. She had a report she wished to make, she told the official there. And she told him all about Irene, and her tragic death; and how Mollie Maggia had died of the same sort of poisoning a year ago. It was the United States Radium Corporation, she made sure to say, on Alden Street in Orange.

'Still another girl,' she reported, 'is now complaining of trouble.' And she said clearly: 'They have to point the brushes with their lips.' That was the cause of all this trouble, all this agony.

All this death.

Report filed, Katherine left, hoping and assuming that something, now, would be done.

A memo *was* filed about her visit. At the end, it said simply, 'A foreman [at the plant] by the name of Viedt said [her] claims were not true.'

And that was that.

*

Helen and Irene's deaths had not gone unnoticed by their former colleagues, at least. 'Many of the girls I knew and had worked with in the plant,' observed Quinta McDonald, 'began to die off alarmingly fast. They were all young women, in good health. It seemed odd.'

That summer, however, Quinta was caught up in all things family, and had no time to give the situation much more thought than that. On 25 July, she gave birth to her second child, Robert. 'We were all so darned happy together,' she remembered of that time. She and her husband James now had the perfect family: a little boy and a little girl. The kids' Aunt Albina, who was still waiting for the blessing of her first child, doted on them both.

During her pregnancy, like many women Quinta had suffered swollen ankles. Although Helen's birth had been relatively easy, she'd struggled with Robert; it was a difficult labour and forceps had been used. After his birth, she'd assumed she'd get well but instead developed a bad back; and her ankles still bothered her. 'I hobbled around,' she later recalled; she treated the problem with household remedies. And then: 'I went to bed one night,' she remembered, '[and] woke up in the morning with terrible pains in my bones.' She called out a local doctor and he began treating her for rheumatism. His call-out fee was $3 ($40) a pop, and she and James could have done without the additional expense given the new baby, but she just couldn't seem to shake the pains.

By the end of the year, she would have seen the doctor eighty-two times.

As the summer of 1923 drew to a close, Katherine Schaub's complaint from the middle of July was finally investigated by Lenore Young, an Orange health officer. She looked up the dead girls' records – and found that Mollie Maggia had died of syphilis and Helen Quinlan of Vincent's angina.

'I tried to get in touch with Viedt,' she added, 'but he was out of town.' And so she did nothing. 'I let the matter drop. [It] has been neglected ... but not out of my mind entirely.'

Had the dial-painters been privy to her correspondence, those words would have been cold comfort indeed to those continuing to suffer, including Hazel Vincent. Hazel was still being treated for pyorrhea and still having teeth extracted; her teeth old friends dying off, one by one, until her own mouth felt like a stranger. By now she could no longer work as the pain was unbearable.

For her friends and family, it was intolerable to watch. For Theo in particular, who had loved her since they were teenagers, he felt like he was feeling his future disintegrate in his arms. He begged her to let him pay for the doctors and the dentists that she went to, but she was unwilling to accept money from him.

He wasn't going to stand for that. This was the woman he loved. If she wouldn't accept help from him as her boyfriend – would she accept it if she was his wife? And so, even though Hazel was very ill, he married her, because he believed that if she was his wife he would be more able to take care of her. They stood before the altar together and he promised to love her, in sickness and in health ...

The new bride wasn't the only radium girl suffering that fall. In October 1923, Marguerite Carlough, who was still working in the studio, developed a severe toothache that made her face swell up. And then, in November, another young woman fell ill.

'I began to have trouble with my teeth,' wrote Katherine Schaub.

Katherine had seen at first-hand what Irene went through. When her mouth started to ache, it must have shot a bolt of terror right through her. She was brave; she didn't ignore it. Instead, on 17 November, she went to the same dentist who had treated Irene, to see if he could help her where all his efforts had failed for her cousin. Dr Barry removed two of her teeth; he noticed, as he examined them, that they were 'flinty' and broke easily. He added in her file: 'Patient has been employed in radium works in Orange, same place as Miss Rudolph ...' Katherine was told to come back soon.

And she did – again and again. Following the tooth extraction, her gum failed to heal and she returned very frequently to Dr Barry's office: five times within that same month, at a charge of $2 ($27) each visit; the extraction had cost $8 ($111). Katherine wasn't stupid: 'I kept thinking about Irene,' she said anxiously, 'and about the trouble she had had with her jaw ... there was some relationship between Irene's case and mine.' She also realised: '[Irene] had necrosis ... she died.'

Katherine's always-vivid imagination, now fuelled by the knowledge of what she had seen Irene suffer, soon became a constant, flickering cine-reel, silently playing out what must lie before her, over and over. She was 'seriously shocked' and a severe nervousness developed which affected her mental health; a situation that did not improve when, on 16 December 1923, Catherine O'Donnell, another former co-worker, passed away. The doctors said she died of pneumonia and gangrene of lung, but Katherine didn't know for sure. And so Catherine became another ghost girl to haunt her in her head. She was buried in the same cemetery where, six months earlier, Irene had been laid to rest.

So many girls were ailing. As Christmas approached, Grace Fryer was conscious that although her jaw seemed to be getting better, the pain in her back and foot had become worse. 'My foot was stiff; I couldn't bend it,' she remembered. '[When] I walked I had to walk with my foot real flat.' Yet she'd soldiered on throughout the fall and didn't ask for help. 'I said nothing about [my condition] to anyone.'

But she couldn't pull the wool over her parents' eyes. Daniel and Grace Fryer watched their eldest daughter as she went about her life – commuting to the bank, helping out at home, playing with her young nieces and nephews – and they saw that her gait, which had always been confident and unhindered, had changed. She was limping, despite herself. They couldn't let this go on.

'Towards the end of 1923,' their daughter Grace conceded, 'my condition became noticeable, and my parents insisted that I see a doctor.' Dutifully, she made an appointment at the Orthopaedic Hospital in Orange for 5 January 1924.

Before that was Christmas. By Christmas Eve 1923, Marguerite Carlough felt at her wits' end. All fall she'd struggled on, continuing her work at the studio in spite of increasing ill health. Lip-pointing had been stopped in late 1923; Josephine Smith, the forelady, revealed: 'When [the company] warning was given about pointing brushes in [our] mouths, it was explained to the girls [that] this was because the acid in the mouth spoiled the adhesive.'

Marguerite had followed the new orders. Her mind wasn't on the job, though; she couldn't concentrate as she'd used to. She had extreme fatigue, was pale and weak, and her toothache, which had started in October, was driving her insane. Unable to eat, the weight had dropped off her at an alarming rate; the smart tailored clothes she favoured were now hanging off her frame, no longer made to fit her newly scrawny figure.

When she left work that 24 December, she didn't know it, but it would be for the very last time. Because, that same evening, she visited her dentist. There were two teeth that were especially hurting her, and her dentist advised that both should be removed that same day. Marguerite consented to the extraction.

When her dentist pulled the teeth, a piece of decayed jawbone came out too.

She wasn't going back to the studio after that. She went home instead, to her sister Sarah and her niece Marguerite, to her mom and dad, and she tried to tell them what had happened. Christmas Day was a sober, solemn occasion after her gruesome experience – but at least they were all together. Given the absences in other New Jersey homes that winter, that was something to be grateful for.

Unbeknown to the Carloughs, or to any of the dial-painters, that same month the US Public Health Service issued an official report on radium workers. Though it noted that no serious defects had been found among the staff examined, it revealed that there had been two cases of skin erosion and one case of anaemia among the nine technicians studied. As a result, it made a formal recommendation to the nation – to New Jersey and to Illinois; to Connecticut, where the Waterbury Clock Company was painting its own dials; and to all the places where radium was used. Safety precautions, the report said, should most definitely be undertaken by those handling radium.

10

Ottawa, Illinois
1923

The caretaker at Radium Dial wiped his bare hands down his workshirt: he was covered in luminous material, his clothing stiff with it. The only clear spots on his face were where two big drools of chewing tobacco ran down his chin; he liked to chew as he worked – and he wasn't the only one. The dial-painters kept candy on their desks, snacking between dials without washing their hands; a habit that suited the many teenagers employed. As time went on, the current Ottawa high-school students were among them; they would work 'one summer between high-school years from a few to several weeks', just to earn a bit of pocket money.

As in Orange, the girls encouraged friends and family to join them at the studio. The old high school was a lovely building to work in: a grand Victorian brick edifice with huge arched windows and high ceilings. Frances Glacinski was thrilled when her little sister Marguerite, two years younger, came along to work on the second floor with Catherine, Charlotte, Marie, Peg and all the rest. Marguerite was a pretty girl who was described as 'comely'; she and her sister were of Polish heritage. The girls also welcomed fifteen-year-old Helen

Munch, a thin, dark girl who wore scarlet lipstick and painted her nails to match; she was the kind of person who 'wanted to be going all the time'.

The exception to these teenagers was Pearl Payne, a married woman from nearby Utica. Pearl was twenty-three when she started at Radium Dial, a good eight years older than some of her colleagues. She had married Hobart Payne, a tall, slender electrician who wore glasses, in 1922; she described him as a 'fine husband'. He was a man who told jokes and loved children; folks described him as a 'very knowledgeable guy'.

In fact, his wife was full of smarts too. Pearl was the eldest girl of thirteen children, and although she had to leave school at thirteen to earn money for the family, she revealed, 'During my employment [I] attended night school and a private teacher, completing seventh [and] eighth grade and one year of high school.' And her education didn't stop there: during the war she'd gained a nursing diploma and was all set to start a career at a Chicago hospital when her mother was taken ill; Pearl had quit to care for her. Now her mom had recovered, Pearl was returning to work – and dial-painting, which was better-paid than being a nurse, was what she ended up doing.

Pearl and Catherine Wolfe got on especially well. Pearl was a gentle woman; 'never an unkind word from her mouth, ever,' said her nephew Randy. The two women's personalities dovetailed neatly, and their shared experience of nursing relatives – for Catherine took responsibility for her elderly aunt and uncle – brought them closer. Catherine, three years younger than Pearl, described her as a 'dearest friend'. Funnily enough, the two women also looked alike: Pearl had thick dark hair and pale skin too, though she was rounder-faced and more full-figured than Catherine, and her hair was curly.

Pearl overlapped with Charlotte Nevins by only a few months. In the fall of 1923, Charlotte quit her job at Radium Dial to become a seamstress; she had been a dial-painter for only thirteen months. As had been the case in Orange, however, any time a girl left, a dozen more arrived to take her place; Olive West now joined the studio, becoming close with Catherine and Pearl. All were overseen by assistant superintendent Mr Reed, Miss Murray's deputy, with whom the girls often shared a joke. On his occasional forays into their studio, the dial-painters would tease him – and it was an affectionate repartee that cut both ways. One young woman remembered, 'I was [to be] married and [I] remember going to work that morning in the dress and telling the supervisor, Reed, I was quitting and on my way to be married. He joked and said, "Don't come back, you won't have a job!"' But she concluded, 'I was back at work in a couple of weeks.'

As Mr Reed was deaf, the girls would sometimes talk back to him as he couldn't hear them, but it was all good-natured and they enjoyed working with him. 'I never heard of any of them not getting along, never,' said Peg's sister Jean. 'Everybody was generous and good to one another.'

It was such a lovely atmosphere that Peg Looney found herself falling for the job and forgetting all about her ambition to become a schoolteacher. She was extremely conscientious and would even take dials home to paint, carefully tracing the numerals in that cramped house next to the railroad tracks that she shared with her large family.

'She looked after us real good,' remembered her sister Jean of Peg sharing her good fortune with them. 'I remember [Peg] buying me a beautiful blue dress, trimmed in white, for my eighth-grade graduation,' recalled another sister, Jane. 'It was so pretty.' The sisters all agreed: 'She was everything you'd hope a big sister would be.'

Peg not only brought her wages and her work home, but also the games she learned at the studio. 'She entertained the younger siblings with "Let's go in the dark!"' revealed Peg's niece Darlene. And there they would glow, a row of little Looneys with radium moustaches, shining sprites behind the blankets they'd put up for modesty in the tiny bedroom. Peg's sister Catherine – the nearest to her in age – was enamoured of all she saw and longed to join Peg at the studio, though she never did. Everyone wanted to work there.

That was why Pearl Payne was so disappointed when – after only eight months as a dial-painter – she had to leave to nurse her mother again. She was the kind of woman who wouldn't have begrudged that for a second, and so she simply bade farewell to her friends and went back to Utica, where she remained even after her mother's recovery, keeping house. She gave the studio little more thought as she turned her attention to her next dream: having a family with Hobart.

It meant she wasn't there when, later in the 1920s, the Radium Dial bosses took a company photograph. All the girls – there were just over a hundred of them in attendance that day – filed outside to have their picture taken. The company men were there too; just Mr Reed and his caretaking colleagues, not the executives from head office. The men sat cross-legged on the ground in front of the women, Mr Reed donning a white flat hat and his usual dark bow-tie. The girls ranged behind the men, some sitting on benches, others standing on the steps of the old high school: three rows of dial-painters, as jolly a bunch of girls as ever there was. Many had their hair bobbed short in the latest flapper style. They wore drop-waisted dresses embellished with long scarves and strings of pearls. 'We used to wear our street dresses to the plant,' Catherine Wolfe said – but what street dresses they were.

Catherine sat on the front row, in the centre of the picture, just to the right of Mr Reed and Miss Murray. It was perhaps a sign of her seniority; as one of the longest-serving employees, she was now a trusted worker who would on occasion assume duties above and beyond dial-painting. That day, she wore a dark, mid-calf-length dress with a long necklace of black beads; her feet and hands, as they often were, were folded neatly together. She wasn't like Marie Becker, who would broadly gesticulate as she made one joke or another.

Now, all the girls – the jokers and the quiet ones, the conscientious and the unconcerned – sat still for the photographer. Some hugged each other, or interlinked their arms. They sat close together, staring at the camera. And as the shutter closed, it captured them all together, frozen in time for just one moment. The girls of Radium Dial, outside their studio: forever young and happy and well.

On the photographic film, at least.

11

Newark, New Jersey
1924

D r Barry had never had such a busy January. Patient after patient came through his door, pale hands clutched to thin cheeks, discomfort obvious in the women's questioning eyes as they asked him what was wrong.

Perhaps worst of all was Marguerite Carlough, who had first come to him on 2 January with evidence of a recent tooth extraction that had begun the process of the jaw necrosis he was seeing in so many girls. Katherine Schaub was back again; the newly married Hazel Kuser was attended by Barry's partner Dr Davidson; Josephine Smith, the Orange plant's forelady, and her sister Genevieve also sought treatment. Genevieve was best friends with Marguerite Carlough and was extremely anxious for her.

In all, to varying degrees, the dentists saw the same mottled condition of the bone. In all, they saw an illness that they knew not how to treat, although they never let the girls see their perplexity; the dial-painters would never have had the audacity to question them anyway. 'I felt [Dr Barry] knew what he was doing,' Katherine later said, 'I couldn't ask him [why my condition didn't improve].' Katherine's nerves were still very bad; it was as much as she could do to get through the day, let alone ponder on complex medical matters.

For Barry, the sheer number of cases now proved his previous argument that the problem was occupational. He truly believed that phosphorus in the paint was to blame; the symptoms were so like those of phossy jaw that it had to be the issue.

Despite their aching jaws, the Smith sisters were still working in the studio that January. Barry now gave them an ultimatum: quit their jobs or he would refuse to treat them.

Josephine Smith ignored him. Yet, on seeing her friends' conditions, she did take some precautions at work. When she weighed out the material for her team, she tied a handkerchief over her mouth and nose to avoid inhaling the dust.

Probably because some of the afflicted women were still working in the plant, rumours of Barry's threats soon reached the ears of the USRC managers – somewhat to their annoyance. Business was going well: President Arthur Roeder's company had contracts with the US Navy and Army Air Corps, as well as many hospitals and physicians; Undark was by now considered the standard material for government use. Evidently, the firm wanted nothing to get in the way of all these business opportunities. Thus, on hearing of Barry's gossip-mongering, as they probably saw it, they were moved to write to their insurance company in January 1924 to reassure them of the situation: 'There recently have been rumours and comments made by individuals, particularly dentists,' they wrote, 'in which they claim work in our application department is hazardous and has caused injury and poor health to a former operator of ours [probably Marguerite Carlough] and they are advising that other of our operators should discontinue being in our employ.'

It may seem striking that the deaths of Mollie Maggia, Helen Quinlan, Irene Rudolph and Catherine O'Donnell do not feature in this correspondence. But all four women had quit their jobs at the plant well before their deaths, some several years before, and it seems the firm was unconcerned about and possibly ignorant of their deaths. If it had chanced to hear of them, it was only Irene's case that had been attributed to her work, and as the doctors thought it phossy jaw and the firm knew no phosphorus was in the paint, it could rest easy that the suspicions were unfounded. From their point of view, Irene was an orphan, anyway, whose parents had died young; with a genetic inheritance like that, she was probably never long for this world.

As for the others, if anyone at the firm had investigated the mysterious deaths of their former employees, officially Catherine had died of pneumonia, Helen of Vincent's angina, and as for Mollie Maggia – well, everyone knew she had died from syphilis. The firm had employed over 1,000 women during its lifetime; four deaths from such a number was probably to be expected. The company therefore concluded confidently: 'We do not recognise that there is any such hazard in the occupation.'

But, by this time, their former dial-painters disagreed. On 19 January, there was a meeting held in Dr Barry's office with at least Katherine Schaub, the Smith sisters and Marguerite Carlough present. The girls talked over their identical conditions with their increasingly concerned dentist. 'We discussed employment at the radium plant,' Katherine remembered. 'There [was] some talk of industrial disease.' The girls agreed 'there was something going on about this thing'.

Yet ... what could they do about it? Katherine had already complained to the authorities and nothing had come of it. Even though the evidence pointed to some problem at the plant, no one really knew what the cause was. And much more pressing for the women than the cause, anyway, was searching for a cure – or at least some relief. Their health was their primary concern. Hazel Kuser was by now almost constantly on palliative drugs because her pain was agonising. Marguerite Carlough had come to Barry hoping for treatment of her jaw – but she was to be disappointed. 'I refused to operate [on] the girl,' Barry later said, 'for the reason that previous experience [with Irene Rudolph and Katherine Schaub] taught me that the moment there was any operative procedure attempted, the case would flare up and would be much worse than it was at the time I saw her.' And so, although the girls were wracked by pain from their teeth, he refused to remove them. All he could offer the panicked women was to keep them under observation.

He couldn't see what else he could do. He did ask others for help, consulting a highly skilled Newark physician, Dr Harrison Martland. But when Martland examined the girls, he too was puzzled. 'After seeing several girls in the dental office,' Martland later wrote, 'I lost interest in the matter.'

The girls were on their own.

Just down the road at the Orange Orthopaedic Hospital, Grace Fryer wasn't having much more luck. Just as she'd promised her parents, she had kept her appointment with Dr Robert Humphries to have her painful back and foot examined. Humphries was the head doctor at the hospital, an 'exceedingly high-grade man'. A Canadian in his forties, Humphries listened carefully to Grace's complaints and then diagnosed muscle-bound feet and chronic arthritis. He strapped her up for several weeks but noted with concern that there was very little improvement.

Humphries was treating another young woman that spring by the name of Jennie Stocker. He didn't connect her with Grace Fryer, who worked in a bank, but Jennie had been a dial-painter until 1922 and she and Grace had worked together during the war. She had 'a very peculiar condition of the knee' that had been mystifying Humphries ever since he had taken her case.

So many doctors across New Jersey were confused that first month of 1924 – but they didn't share notes and so each case was viewed in isolation. As January drew to a close, Theo and Hazel Kuser decided that they would look elsewhere for treatment. New Jersey was just a short distance from New York City, where some of the best doctors and dentists in the world had their practices. On 25 January, Hazel, bravely swallowing down her pain, made the journey into the Big Apple for treatment at the office of Dr Theodore Blum.

Blum was one of America's first oral surgeons, a prestigious specialist who had pioneered the use of X-rays for dental diagnosis. His fees were extortionate, but Theo insisted that they visit him anyway. He could borrow money on their furniture to pay the bills, he reasoned. If it eased Hazel's pain, if Dr Blum could stop this endless decay in her mouth, then it would all be worth it.

As a mechanic, Theo Kuser was not wealthy, and nor was his family; his father, also called Theo, was a postman. Theo Snr. had saved up money to buy a house for his old age, but he now offered some of his savings to his son for Hazel's treatment. He took it gratefully and the appointment was duly attended.

Blum was a balding man with a neatly trimmed moustache, spectacles and a high forehead. As he introduced himself to Hazel and began his examination, he quickly realised that he had never seen a condition like hers before. Her face was swollen with 'pus bags', but it was the condition of her jawbone that was most perplexing: it seemed almost 'moth-eaten'. It literally had holes in it.

But what, Dr Blum now pondered, could have caused it?

Blum was worth his money. Later, he would try to find out the exact chemicals in the luminous paint, although to no avail. For now, he took a medical and employment history from Hazel and made a provisional diagnosis: she was suffering from 'poisoning by a radioactive substance'. He admitted her to the Flower Hospital in New York to operate on her jaw. It would be the first, but not the last, of such procedures Hazel had to endure.

Yet although Blum had offered a diagnosis, and swift and specialist treatment, he didn't offer the one thing that Theo had been yearning for: hope. That was all he really wanted, to know that there was light at the end of the tunnel; that they could get through this and come out the other side into a shining day, and another one, and another day after that.

Instead, Blum told him 'there is little chance of recovery'.

All the money in the world couldn't save his wife now.

The radium girls' agony hadn't gone unnoticed in the community. That same month, a civic-minded resident wrote to the Department of Labor to raise concerns about the Orange plant. This time, it was John Roach's boss, Commissioner Andrew McBride, who stepped in, grilling health officer Lenore Young on what she'd found out the previous summer. She apologised for seeming 'negligent', interviewed the affected girls and then recommended that the Public Health Service be called in.

Yet McBride felt there was not sufficient evidence to warrant doing so. His reasoning may have been political, for the Department of Labor was pro-business. Under state law, it had no authority to stop an industrial process even if it was harmful. As a result of these factors, the department now gave the plant a clean bill of health – and completely stopped looking into the dial-painters' illnesses. They took this decision even though more and more women were suffering the same symptoms.

It was a stalemate. No diagnosis. No clue as to the cause. No one lifting a finger to find out what was really going on in that radium studio in Orange.

But then the stalemate was ended by an unexpected source: the United States Radium Corporation itself.

As more and more girls fell ill, the company found that – in a stark contrast to the glory days in the war – they were encountering 'considerable difficulty' recruiting staff: a number of the girls had quit and no one wanted to replace them; production was now being held up. When Genevieve Smith – shocked into action by her best friend Marguerite's decline – also handed in her resignation on 20 February 1924, it was the straw that broke the camel's back. Viedt, the vice president, was ordered to find out why Genevieve was leaving and she cited Dr Barry's ultimatum; the dentist was persisting with his outlandish claims.

The lack of operators was a big concern to the company, but

there was another worrying development about the same time that really made them sit up and take note of what was happening to their former employees. For more than three years, Grace Vincent, Hazel's mother, had been watching her daughter suffer. Hazel was in constant agony; no mother could bear it. Dr Blum had said there was no hope now, and Mrs Vincent had nothing to lose. She went down to the studio in Orange and left a letter there. In it, she told the firm 'she was about to make [a] claim for compensation on account of [her daughter's] illness'.

That got their attention.

At once, Viedt reported these developments to the New York headquarters. Not long after, USRC executives decided to launch an investigation to determine if there was anything dangerous in the work. For too long there had been rumour and suspicion; it couldn't continue. After all – now, it was bad for business.

12

In a sign of how seriously the company was taking the downturn in its operations, President Arthur Roeder himself took charge of the investigation. In March 1924, he approached Dr Cecil K. Drinker, professor of physiology at the Harvard School of Public Health, and asked him if he would conduct a study at the Orange plant. Drinker was a qualified MD as well as a recognised authority in occupational disease; Roeder was taking no chances but bringing in the very best. To Drinker, Roeder wrote: 'We must determine definitely and finally if the material is [in] any way harmful.'

To Roeder's delight, Drinker found his letter 'very interesting' and offered to meet him in April for further discussion. Roeder told him of two cases, one fatal – probably Irene Rudolph – and one 'very much improved'; Roeder emphasised knowingly of the latter: 'I have been informed that her family have had considerable tubercular trouble.'

In response Drinker advised that 'We are inclined to feel' – Drinker worked with his equally brilliant wife, Dr Katherine Drinker, and another doctor called Dr Castle – 'that the two occurrences which you mention have been coincidences.'

However, he added, 'At the same time we are agreed that it is not at all safe to permit any conviction upon that point without a rather complete examination.' In April 1924 the study would begin.

It is not wholly clear who Roeder meant by his 'very much improved' case. It was probably Marguerite Carlough, since she was the woman who had left his employ most recently (although, in fact, she was still greatly suffering), but it could also have been Grace Fryer, who was finally reaping the benefits of the expensive medical care she was paying for. Dr Humphries was still examining her once a week to see how her strapped back and feet were doing; now, at last, he was happy to note she was improving.

Humphries, however, had care only of Grace's body – and it was her jaw that was now becoming her major source of pain. The same month Roeder wrote to Drinker, Grace was admitted to hospital in New York for a week-long stay; her latest round of X-rays showed a 'chronic infectious process in the jaw' and she sought treatment from Dr Francis McCaffrey, a specialist, who operated on her, excising some of her jawbone. As Knef and Barry had found, however, once an operation had been carried out, another was needed, and then another after that. 'I have been compelled to go to the hospital so often,' Grace later said, 'that it seems like a second home.'

Grace – as with so many of her former colleagues – now became trapped in a vicious circle with each operation incurring yet another bill. Before too long, she had to swallow her pride and ask her parents for money; but the rising medical costs desecrated both her savings and the family's bank account.

That spring, USRC, too, was concerned about money. April seemed rather distant for Drinker's investigation to begin, given the pressing delay in production at the plant. Though Viedt had managed to hire an extra six girls, it wasn't enough; the executives still had to address the 'psychological and hysterical situation' that was now unfolding in the studio.

So, while it waited for Drinker to begin his study, the firm organised its own examinations of the current team of dial-painters, conducted by the Life Extension Institute. The girls were tested confidentially – but the reports were shared with the firm. 'The individuals concerned,' Viedt wrote Roeder, 'do not know that we have these copies ... information given in them is very confidential and they might object to our having them.' Though the institute found some girls had infected teeth, it concluded that their ailments 'did not reflect any specific occupational influence'. Roeder wrote contentedly to Viedt that the results were 'just as I anticipated'.

Viedt, however, who was more involved in operations at the studio, was not so reassured. 'I do not feel quite as optimistic about this matter as you do,' he wrote to his boss. 'While the Life Extension Institute have made a report, I do not believe that this will satisfy our various operators and that we must wait for Dr Drinker's final report in order to really convince them that there is no injurious element.'

Roeder then added his own two cents. 'We should create an atmosphere in the plant of competence,' he wrote decidedly to Viedt. 'An atmosphere of confidence is just as contagious as one of alarm and doubt.' And he advised that, in his view, 'the most important action is to see Barry and perhaps others who have been making statements [having] jumped at conclusions apparently without thought or knowledge'.

Viedt knew an instruction when he heard one: in late March 1924, he duly paid a visit to Barry and Davidson.

The dentists received him coolly. They had no doubt that the agonising condition they were seeing in their patients was due to the girls' former employment at USRC. During Viedt's visit, they became outraged by what they considered a cold-blooded attitude.

'You ought to close down the plant,' Davidson told Viedt angrily. 'You've made $5 million. Why go on killing people for more money?'

Viedt had no answer.

'If I could have my way,' Davidson told him bitterly, 'I would close your plant.'

The dentists weren't the only ones whose interaction with the girls had brought them to the point of frustration. Lenore Young, the Orange health officer, having seen the Department of Labor do nothing with her recommendation that the Public Health Service be called in, now discreetly took matters into her own hands. On 4 April 1924, she wrote confidentially to Katherine Wiley, the executive secretary of the Consumers League, a national organisation that fought for better working conditions for women. 'The authorities are hesitating,' Young confided in Wiley. '[The Consumers League] must keep after them to see that *something* happens.'

Wiley was a smart and enterprising woman, in charge of the New Jersey branch of the League. A somewhat plain, dark-haired lady in her early thirties, with features too small for her face, she was a tenacious and driven person. When Young asked for her help, Wiley responded immediately. She was assisted by John Roach from the Department of Labor who – unknown to his boss McBride – gave Wiley a list of the women affected, so she could conduct her own investigation.

It came not a moment too soon. For on 15 April 1924, another young woman lost her life. Jennie Stocker – who Dr Humphries had been trying to treat for her peculiar knee condition, without avail – died suddenly after a short illness at the age of twenty.

The day after she died, Roeder kept his appointment with the Drinkers. He showed them around the plant and then they went up to the studio and spoke to several of the women, including Marguerite Carlough. It is surprising that she was in the studio, and not simply because she no longer worked there; since Christmas Eve 1923 she had been confined to her home, except for her visits to Dr Barry. It is possible that the company had asked her to come in specially to meet the Drinkers, being determined to lay to rest the rumours that Marguerite's work had made her sick.

She was accompanied by her sister Sarah Maillefer, who now walked with a cane. Sarah was still working at the plant as a dial-painter; the Carlough family was poor, and with Marguerite no longer able to work and the medical bills mounting, they needed every penny they could get. Sarah's trouble, of course, was very different from that of Marguerite. Evidently, her lame leg was not connected to the awful disease that was plaguing Marguerite's mouth.

Dr Cecil Drinker was a handsome man with abundant fair hair. He introduced himself to Marguerite with immediate concern for her well-being. Her thin face was very pale and she clutched a bandage to her seeping cheek; she complained of 'pains in the bones of her face'. It was obvious that she was seriously ill.

Katherine Drinker turned to Roeder and told him that the day's tour could not be considered an adequate survey. It was imperative, she said, that the Drinkers returned to Orange to make a comprehensive study of the plant and its employees. And so, over two days, from 7 to 8 May 1924, the full Drinker study took place. The scientists, now fully read up on all the radium literature, returned to the plant with their colleague Dr Castle and conducted a detailed investigation. Together, the three doctors inspected all the different facets of the operation, accompanied on their tour by Vice President Viedt.

They met the chief chemist, Dr Edwin Leman, and noted that he had 'serious lesions' on his hands. Yet, when they mentioned them, he 'scoffed at the possibility of future damage'. Perhaps he was mindful of his president's suggestion that an atmosphere of confidence should be promulgated by the plant's top men.

Such an unconcerned attitude, the Drinkers soon realised, was 'characteristic of those in authority throughout the plant'. 'There seemed to be,' Cecil Drinker later wrote, 'an utter lack of realisation of the dangers inherent in the material which was being manufactured.' Roeder even told him that 'no malignant growths ever developed on the basis of radium lesions; a statement so easy to disprove as to be ridiculous'.

Up in the studio, the doctors carried out thorough medical examinations of the workers. Twenty-five employees were selected as a representative number; one by one, the nominated dial-painters knocked nervously on the door of the restroom, where the women's tests were held, before being summoned in.

Sarah Maillefer was one of those chosen. As the doctors requested, she opened her mouth wide so they could prod at her teeth; kept still as they probed firmly around her nose and throat; offered the vulnerable inside of her arm so they could take a vial of blood. The exam then transferred to the darkroom; here, Katherine Drinker 'examined a number of these women, some quite intimately, to determine to what degree they were luminous when sufficiently in the dark'.

Oh, that luminosity. That *glow*. Katherine Drinker was stunned by it. As the women undressed in the darkroom, she witnessed the dust lingering on their breasts, their undergarments, the inside of their thighs. It scattered everywhere, as intimate as a lover's kiss, leaving its trace as it wound around the women's limbs, across their cheeks, down the backs of their necks and around their waists ... Every inch of them was marked by it, by its feather-light dance that touched their soft and unseen skin. It was spectacular – and tenacious, once it had infiltrated the women's clothing. The Drinkers noted that it 'persisted in the skin' even after vigorous washing.

The Drinkers didn't limit their study to the plant: they visited Dr Barry and also met some of the dial-painters who were now exhibiting such similar symptoms, including Grace Fryer. Thanks to the attentions of the expert Dr McCaffrey in New York, however, Grace was the exception to the rule; the Drinkers were pleased to note she had 'recovered satisfactorily' from her illnesses.

The same could not be said for Marguerite Carlough. Finding

no relief with Barry, she had started consulting Dr Knef, who'd treated Mollie Maggia. Marguerite's appearance – she who had once favoured dramatic feathered hats and glossy fashion – was by now very bad. Yet the worst was not on the outside, despite her deathly pale skin and emaciated body, but on the inside, 'from the [constant] discharge of foul pus in her mouth'. She was suffering excruciatingly.

Knef attended her as best he could. 'At least once a day I'd go up there,' he remembered. It was a fifteen- to twenty-mile drive from his office to the Carlough home on Main Street, Orange, but on occasion he would attend her from two to six times a day. Sometimes, he recalled, 'I have been with her as high as three days and three nights on a stretch.' Such close attention was well beyond the budget of the Carloughs, so Knef was essentially working for free. It was good of him, but it didn't necessarily mean that Marguerite was receiving the best-qualified attention.

Yet Knef knew more than most about the disease, even if he didn't, at that time, understand the full implications of his growing knowledge.

Dr Knef was a medical man through and through. When Mollie Maggia's jawbone had so shockingly broken against his fingers, he had been fascinated by it – so he had kept it, this oddly moth-eaten, misshapen piece of bone. Every now and again, after her death, he had examined it, turning it over in his hands, but he was none the wiser; anyway, she had died of syphilis, whatever the strangeness of her bones. He'd therefore popped the fragment into his desk drawer, where he kept his X-ray negatives, and eventually it slipped his mind.

And then, one day, his duties had required him to dig through that crowded desk drawer for the X-ray films. He had scrambled through the bits and pieces he kept in there, searching for them. To his astonishment, when he finally pulled them out, the films were no longer ebony black. Instead, they were 'fogged', as though something had been emanating onto them.

But there was nothing in that drawer but old files and forgotten scraps of bone.

He turned the X-ray film this way and that. The spoiling of the film was undeniable. It was a message, though little did Knef know it, but its meaning was unclear.

Mollie Maggia was still voiceless, even after all this time.

13

T hough Knef did not know why the films were fogged, Marguerite Carlough appreciated his considerate attentions nonetheless. With a spark of hope, she said that 'of late she had felt a little better'. But visiting doctors noted that was 'a statement not in accord with her appearance'. When Katherine Wiley, from the Consumers League, met her in May 1924 as part of her independent investigation, she was shocked, calling her 'this poor sick young thing who looked fairly transparent'. The level of sheer suffering was difficult to witness. Wiley later wrote: 'After seeing one of the victims, I can never rest until I have seen something done whereby I am assured it will not happen again.' She resolved 'to stick to this thing until there is some action somewhere'.

And stick to it she did. She interviewed more of the girls, including Josephine Smith, but found her 'unwilling to discuss the subject while continuing to be employed by the Corporation'. Wiley left no stone unturned, also visiting Katherine Schaub and Edith Mead, who had nursed Mollie Maggia through her terrible illness. The nurse had not forgotten her former boarder. 'Miss Mead wishes,' Wiley wrote, 'to do anything that she can to prevent such a tragedy happening to anyone else.'

And Wiley felt exactly the same. Hearing of Hazel's mother's desire to claim compensation, she consulted a local judge to get his advice on how the families could take legal action. But here she learned that New Jersey law was set against the women. The state, in fact, had somewhat pioneering legislation; a new law had come in only that January that made industrial diseases compensable. But – and it was a big but – only nine diseases were on the permitted list and there was a five-month statute of limitations, meaning any legal claim had to be filed within five months of the point of injury. Not only was 'radium poisoning' – if indeed that was what the girls were suffering from – not on the list, but most of the girls had not been employed by USRC for years, let alone five months. The judge told Wiley frankly, 'When radium poisoning is made a compensable disease, if ever, it would not be retroactive; so that, as far as these girls were concerned, nothing could be done.'

The families had hit the same brick wall. Stone-broke and desperate, Marguerite Carlough was also now considering legal action, in order to get some money to pay for her treatment, but neither she nor Hazel Kuser's family had been able to find a lawyer who would take the case without getting cash upfront. As Wiley noted grimly, 'They have none.'

On 19 May 1924, Wiley returned to the Department of Labor with the results of her investigation. She took them straight to the top, to Commissioner Andrew McBride, but he was 'furious' to find that the Consumers League had stuck its nose into the matter. When he learned that his deputy Roach had supplied Wiley with the women's names, he hit the roof. McBride summoned Roach to their meeting and, Wiley said, 'rebuked him in my presence [in a] severe calling down'. Wiley wasn't daunted one bit by McBride's ferocity, however: she continued to argue with him. McBride, frustrated with this nagging woman, asked her what she wanted to happen.

'An investigation by the US Public Health Service,' she said immediately.

'Put it in writing,' he replied wearily. She did – at once.

Even as Wiley was continuing to champion the women's cause, developments were afoot at the centre of all the trouble: the United States Radium Corporation. The Drinkers had been busy assessing the results of all their tests and now, on 3 June 1924, they delivered their full and final report to the firm.

Fifteen days later, on 18 June, Viedt wrote to Roach at the Department of Labor to share the doctors' verdict. He didn't send the full report, which was lengthy, simply a table of the medical-test results of the workers, which showed the employees' blood to be 'practically normal'. 'I do not believe,' wrote Viedt confidently, 'that this table shows a condition any different than a similar examination would show of the average industrial worker.' The department agreed: the table showed that 'every girl is in perfect condition'.

The company was in the clear. President Roeder wasted no time in spreading the news. 'He tells everyone,' an observer commented, 'he is absolutely safe because he has a report exonerating him from any possible responsibility in the illness of the girls.' Immediately, just as Roeder had hoped, the situation improved at the plant: 'Rumours quieted considerably,' noted an internal memo with satisfaction.

It was, therefore, unfortunate timing that at this moment Dr Theodore Blum begged the company to help his patient Hazel Kuser. Since she'd first visited Blum in January, her condition had deteriorated rapidly, despite many operations, two blood transfusions and multiple hospital stays. The bills were coming in faster than Theo Kuser and his father could pay them. Though Hazel's wealthy physicians, intrigued by her strange condition, were providing her with a significant amount of free treatment, the bills ran into the thousands. Theo was mortgaging everything he owned, while his father's life savings hurtled into a financial black hole, swallowed up as soon as father and son withdrew them from the bank.

Her family could not afford the care Hazel urgently needed,

so Blum appealed directly to the company. He made sure to say that he wasn't trying to lay the blame at the firm's door – even though, by this time, Blum was willing to state that the disease was undoubtedly caused by the material used in dial-painting. 'It is not a question of whether or not your firm is responsible,' he wrote carefully, 'but I feel that if you have the money to spare you should let them have it in some way.' He wasn't interested in culpability: this was a matter of life and death.

The response from USRC was swift. Full of the confidence given them by the Drinkers' report, the company refused to help in any way whatsoever; to do so would establish 'a precedent which we do not consider wise'. Five years before, the firm had been stung when they'd offered $5 compensation for some ruined laundry; they were not going to make the same mistake again. Instead, they gloried in the conclusions of the recent study: 'The results of the very thorough investigation which we [made] upon this condition which you claimed was caused by her employment in our plant have shown that there was nothing in our work which might be considered the cause.' The letter concluded, somewhat insincerely, 'We are sorry that we cannot help you in this way.'

Blum was stunned. 'I was only appealing to the humane sense of the officers of your corporation to find out what you would do to help this poor creature,' he wrote back. 'I must admit that I am surprised that you failed to see the humane side of the question.'

But the corporation cared not a jot for his jibes. They were guiltless – and they had a report that proved it.

14

Katherine Schaub couldn't wait for her summer holiday. It had been a horrible twelve months: her cousin Irene dying last July, almost a year ago now, and then Katherine's own trouble with her teeth starting in November. She knew she was now called a 'nervous case' by her doctors, but as much as she tried not to think about her situation, it was very hard not to. She had recently taken work in an office with the idea that it would help take her mind off things.

As it turned out, Katherine had become a bit of a flibbertigibbet in her career, flipping from one company to the next, leaving due to ill health, or her nerves, or because she was looking for the next much-needed distraction. She would move from the roller-bearing company to an insurance company to a motorcar firm and back again, never staying in one place especially long, always having to leave for one reason or another. Anyway, wherever it was she was working, most of her earnings had to go on medical treatment.

Her state of mind worried her father, William, she knew. He was so good to her, always trying to lift her spirits or paying one of her doctor's bills from his own wages. He didn't earn much – he was a janitor in a factory, and the family lived in a dingy third-floor flat – but he was happy to give all he could to his daughter if it would make her well.

This summer, Katherine planned to take a much-needed rest. She was only twenty-two – an age Irene had never seen, she realised sadly – and she needed to remember what it was to feel young. All this worry was dragging her down.

Yet when July 1924 arrived, Katherine noted: 'I could not go away. The condition in my jaw was causing me considerable anxiety and I decided to consult a skilled dental surgeon in New York City; I had to use my vacation money for a new set of X-rays.'

By chance – although perhaps not, given his standing in his field – she chose to consult Dr Blum, who was also treating Hazel Kuser. Back in May, Katherine had had another tooth pulled by a different dentist; as was now becoming the distressing norm, the socket had not healed. The infection was agonising: 'The pain [I have] suffered,' she said, 'could only be compared with the pain caused by a dentist drilling on a live nerve hour after hour, day after day, month after month.' When Blum examined her in July 1924, he 'advised work to be done when she is in a physical condition to have it done'; until then, Katherine had to return home unaided.

It was the not knowing what was wrong that was the worst thing, she thought. 'I had stopped at nothing in an effort to regain my lost health,' she mused dejectedly, 'but so far I had failed. No one was able to help me.'

She was at Blum's office again and again over the summer; not quite the vacation she had planned. Once, she was compelled to obtain an emergency appointment after suffering agonies in the entire right side of her head. She pulled her blonde hair back from her thin face in his office, trying to demonstrate to Blum where the pain was, all down the right side of her skull.

Blum gently probed at her swollen jaw. And, upon pressure, pus discharged from the tooth socket. Katherine felt it burst into her mouth, and felt sick. 'Why should I be so afflicted?' she would later ask. 'I have never harmed a living thing. What have I done to be so punished?'

On one of her visits to Blum, she ran into Hazel, who was there for treatment too. She was unrecognisable; this mysterious new condition, in some patients, led to grotesque facial swellings, literal footballs of fluid sprouting from their jaws, and it seems Hazel may have been afflicted in this way. Accompanied by her mother, she was in no condition to talk. It was Grace Vincent, having to be her daughter's voice, who told Katherine that Hazel had been seeing Blum for the past six months.

It could not have been a good advertisement for the dentist's skills. Before the summer was out, Hazel would be rushed to hospital in New York, where she would stay for three months, far from her family and Theo in Newark. To pay for her hospital treatment, her husband mortgaged their home to the hilt.

And Hazel and Katherine weren't the only ones seeing doctors. Back in Orange, Quinta McDonald was finding it harder and harder to look after her little ones. Her daughter Helen was now four years old; baby Robert had just turned one. It was the pain in her hip that was the problem, shooting all down her right leg. She was hobbling about now with more than just a noticeable limp – it was more like a lurch as she stumbled from one foot to the next. It was the strangest thing, but, she said, 'It seemed to me that one leg was shorter than the other.'

She must be imagining it. All the twenty-four years of her life her limbs had measured up straight; why would that suddenly change now?

Nevertheless, it was debilitating, especially with Robert crawling about the house at 100 mph and her increasingly unable to keep up with him. She made an appointment with Dr Humphries at the Orange Orthopaedic Hospital, perhaps recommended to her by Grace Fryer. In August 1924, Humphries took an X-ray and perused it for analysis. He'd noted in his physical examination of Quinta that she 'could not move her hip to complete function', so he was looking especially for a problem around her hip joint.

Ah. There it was. But *what* was it?

On the X-ray, there was a 'white shadow', as Humphries called it. It was peculiar, showing 'a white mottling throughout the bone'. He had never seen anything quite like it. As John Roach later wrote of the bewildering maladies: 'The whole situation is baffling and perplexing ... this strange and destructive [force] is an unknown quantity to medical and surgical science.'

In fact, there was one person who had realised exactly what the problem was – one person, that is, beyond the chemist Dr Szamatolski, who had long ago identified that, 'Such trouble as may have been caused is due to the radium.' In September 1924, Dr Blum, having now treated Hazel Kuser for eight months, made an address to the American Dental Association about jaw necrosis. He referenced only Hazel's case, and merely in a brief footnote, but it was he who made the first-ever mention in medical literature of what he now termed 'radium jaw'. He didn't believe the company's protestations of innocence; in fact, fuelled by their cold-hearted response when he had begged them to help his patient, he now promised Hazel 'all necessary assistance should court action be brought versus the company'.

One might have thought that this new term – radium jaw – and the dentist's ground-breaking diagnosis would have captured the imagination of the medical community. But in fact it went entirely unnoticed – by other dentists; by the dial-painters, who were not privy to medical publications; and by physicians, like Dr Humphries in Orange.

Standing before Quinta McDonald's X-ray in that summer of 1924, completely at a loss, Humphries nevertheless had to offer a diagnosis to his patient. Quinta remembered, 'They told me that I had an arthritic hip.'

Humphries duly strapped her leg for a month but, unlike with Grace Fryer, there was no improvement. And so, that summer, Quinta McDonald was encased in plaster, from her diaphragm down to her knees, to keep her body absolutely still in the hope that it would mend her troubles. 'I could still hobble around,' she said, 'with a cane.'

But hobbling around wasn't much good for the mother of two young children. It was even harder trying to care for Robert and Helen after that. It is likely that Quinta's sister Albina – who was still without her own family – helped out; the two sisters now lived some fifteen minutes' walk away from each other in Orange.

To Quinta's relief, the dramatic treatment seemed to bear fruit: 'That cast eased the pain and helped a little,' she remembered. She tried not to think of what was happening beneath the cast, of what she'd started to suspect, that 'one leg was beginning to shrivel up and become shorter than the other'. The cast stayed on for nine long months. As summer turned into fall and she felt some improvement she gave thanks that Dr Humphries's treatment appeared to have helped her.

It was a time for thanks. On Thanksgiving itself, 27 November, Hazel Kuser was finally released from her New York hospital and allowed to return to Newark to be with Theo and her mother Grace. As the family gathered together, they tried to feel the blessing of the fact that at least she was home.

But she wasn't the same person anymore. She had 'suffered so frightfully that her mind seemed affected'. Her priest, Karl Quimby, who was attending the family to offer spiritual comfort, said, 'She suffered excruciating agony.'

It was perhaps, then – when they tried to think of Hazel and put her first – the biggest blessing of all when, on Tuesday 9 December 1924, she finally passed away. She died at 3 a.m., at home, with her husband and mother by her side. She was twenty-five. By the time she died, her body was in such a distressing condition that the family would not allow her friends to see it at the funeral.

It was Theo who notified the authorities of her passing; Theo who organised the embalming of her battered body and her burial, on 11 December, in Rosedale Cemetery. These were the final things he could do for her, for the woman he had loved since he was a boy.

He didn't want to think about the future; about the fact that the mortgage on their home had been foreclosed; about the fact that his father had impoverished himself in helping him and Hazel with their bills. By the time she died, Theo Snr. had spent all of his life savings. The family's bills – for hospitals, X-rays, ambulances, physicians, house calls, medicine and transport to New York – ran to almost $9,000 ($125,000). They had ruined themselves, and it was all for nothing.

Katherine Wiley, of the Consumers League, who had stayed in touch with the family as she continued to support the dial-painters' cause, found the situation unbearable. Frustrated that nothing had been done by the authorities, she now pursued two leads. First she wrote to Dr Alice Hamilton, a brilliant scientist who was considered the founder of industrial toxicology and who always championed the victims of occupational disease; Hamilton was the first-ever female faculty member of Harvard University and her department chair happened to be one Cecil Drinker.

Hamilton knew nothing of Drinker's report on the Orange plant, for although Roeder was using it to quash the fears of his employees and to justify the company's refusal to help the afflicted women, Drinker had not yet submitted it for any official publication. Thus, on receipt of Wiley's letter, not aware of any conflict of interest, Hamilton enthusiastically expressed the desire that the Consumers League should take up the cases 'vigorously – with whatever cooperation I can give you'. She wrote, 'From what I can hear of the attitude of the company, it is pretty callous.' She proposed that perhaps she could undertake her own study as a 'special investigator'.

Wiley's second line of attack was to reach out to Dr Frederick Hoffman, a fifty-nine-year-old statistician who specialised in industrial diseases and worked for the Prudential Insurance Company. After reading Wiley's letter Hoffman began making enquiries; his first port of call, at Wiley's urging, was to visit Marguerite Carlough.

It was now almost a year since Marguerite had made that fateful Christmas Eve trip to the dentist. By the time Hoffman visited her in December 1924, he found her 'a lamentable case which is lingering between life and death, with apparently no hopeful outlook for the future'. He couldn't help but be moved. Before the year was out, Hoffman, a recognised authority on occupational hazards, had sent a strongly worded letter to President Roeder at USRC. 'If the disease in question were compensable, I seriously doubt if your company would escape liability,' he wrote pointedly. And he added: 'That it will be made compensable in [the] course of time if further cases should arise is self-evident.'

A warning shot had been fired – and the Orange dial-painters were determined that this would be only the start of it. Marguerite in particular could not stop thinking that she had given her all to that company – and this was how they repaid her. Out in the cold; not a cent to spare to ease her suffering. And not just her; her friends, too.

Though it had been a long time since she had felt like herself, Marguerite could dimly remember how she used to be: a dynamic young woman in sleekly fitted tailored clothes with fabulous millinery. That winter, as the calendar pages turned and the New Year began, she gathered all her courage and what little strength she had left. She asked her family for help, being too weak now to do what she needed. But this was important. This she would do, even if it was her last act on earth.

Against all the odds, Marguerite Carlough found a lawyer to take her case. And on 5 February 1925, she filed suit against the United States Radium Corporation for $75,000 ($1 million).

The dial-painters' fightback had begun.

15

Ottawa, Illinois
1925

Marguerite's legal case made the local news in Newark. It's unlikely that the Radium Dial girls in Ottawa got to hear of it – but their employers certainly did. The radium industry was a small pond, and Radium Dial was one of the biggest fishes of them all.

By 1925, the studio in Ottawa had become the largest dial-painting plant in the country, supplying 4,300 dials a day. Business was booming – and Radium Dial wanted to take no chances of having a hold-up in their operations, such as their fellow radium corporation had suffered when the rumours first started in New Jersey.

Radium Dial now conceived a master plan to avoid the same problem. They opened a second dial-painting studio in Streator, sixteen miles south of Ottawa, where less was known about radium. Both plants ran simultaneously for nine months, but once it was apparent the Ottawa workers hadn't heard the rumours from the east and weren't going to quit, the firm shut down the second studio; some employees transferred to Ottawa, others simply lost their jobs.

The company also decided, just as USRC had done before

them, to have their workers medically tested later in the year; the exams were conducted by a company doctor in Mr Reed's home on Post Street. Not all the women were tested; Catherine Wolfe was not among them. That was a shame, because just lately she had not been feeling too well. After two years of working at Radium Dial, she later remembered, 'I began to feel pains in my left ankle, which spread up to my hip.' She'd started to limp just a little, every now and again, when that ache made itself known.

Another dial-painter who wasn't tested was Della Harveston, who had been part of the original clique with Catherine, Charlotte and Mary Vicini, Ella Cruse and Inez Corcoran. She had died the previous year of tuberculosis.

Red-haired Peg Looney, however, *was* summoned by Mr Reed to his home for a test. Yet when her colleagues asked her how she'd got on, she had to tell them she hadn't a clue. In Orange, the medical-exam results had been secretly shared with the company behind the women's backs; in Ottawa, they went straight to the corporation and cut out the middle woman entirely. Neither Peg nor any of her tested co-workers were told the results. Peg settled back at her desk in the studio without worry, however, picking up her brush and licking her lips in preparation for painting. She wasn't at all concerned; the company, she was sure, would tell her if anything was wrong.

All the girls in Ottawa still lip-pointed, little knowing that 800 miles away the practice had been banned. Yet behind the scenes at the head office of Radium Dial, its executives, mindful of the New Jersey lawsuit, now started putting some thought into finding an alternative method of applying the paint – just in case. Chamois was trialled, but found too absorbent; rubber sponges were employed, but they didn't work right. Radium Dial's vice president Rufus Fordyce admitted, however, that their endeavours were somewhat half-hearted: 'No strenuous effort,' he later acknowledged, 'ha[s] been made to find and provide any suitable manner to eliminate the procedure.'

The company eventually tasked Mr Reed with the job of finding an alternative method. Soon, he would start tinkering with the idea of a glass pen, such as was employed by Swiss dial-painters, and began to work up various designs. In the meantime, the Ottawa girls kept on. *Lip . . . Dip . . . Paint.*

Their fun times kept right on too. These days, many of them had a man on their arm as the young women started courting. Back in high school, Peg Looney's favourite song had been the independent-minded 'I Ain't Nobody's Darling', but now she had changed her tune: she was stepping out with a bright young man called Chuck. Anyone with half a brain could see that any one of these days he was going to propose.

Chuck was his nickname; his full title was the much more distinguished-sounding Charles Hackensmith. He was a gorgeous, well-muscled, broad-shouldered and tall young man with curling fair hair; his high-school yearbook said the phrase that defined him was: 'And the cold marble athlete leapt to life.' Yet to be the beau of clever Peg Looney, you couldn't be all brawn and no brain, and Chuck was as smart as could be: he made the Senior Honor Roll at high school and was a college guy. 'He was big educated,' said Peg's sister Jean. 'He was everything. He was just elegant, really. Awful good.' He grew up living just around the block from Peg and her large family and although he was now away at college, he came home on weekends – and that was when the young couple really let their hair down.

Chuck had a shack at his house, where he would throw parties and play records on his beat-up old gramophone. As spectators clapped along and drank illicit home-brewed root beer, the dancing would begin. Whenever Chuck hugged Peg to him, he left not an inch between them: two bodies pressed close as they danced to the latest jazz tunes. Chuck was flirtatious; this girl, he knew, was something special.

Everybody would go down the Shack; Marie Becker would have a hoot there. She would zip about between friends if a party was taking place, encouraging everyone to attend. Marie was courting Patrick Rossiter, a labourer who had a large nose and big features, whom she'd met at the National Guard Armory while skating. He was 'a devil', said his family. 'He used to like to have fun.' Catherine Wolfe, too, would attend as a good friend of Peg; she was single at that time. And all the Looneys would be there as well – 'The whole family!' exclaimed Jean. 'And there were ten of us!'

There was so much happening in Ottawa in that spring of 1925 that the visit from the government inspector to the studio barely registered with the women. But that, of course, was just what Radium Dial wanted. In the wake of the New Jersey cases, a national investigation into industrial poisons had been launched by the Bureau of Labor Statistics, which was based in the capital, Washington, DC. The bureau was run by Ethelbert Stewart; his agent on the ground was a man called Swen Kjaer. And when Kjaer met with Rufus Fordyce, Radium Dial's vice president, ahead of coming to inspect the Ottawa studio, he was 'requested to handle the subject carefully, so as not to cause an alarm among the workers'. Maybe as a result, only three girls would even be questioned.

Kjaer began his study in April 1925. He went first to the Chicago office of Radium Dial, where he interviewed Fordyce and some laboratory workers; Kjaer noticed the latter had lesions on their fingers. The lab workers acknowledged that radium was a dangerous material to handle 'unless proper safeguards are provided'. Consequently, the men in Radium Dial's laboratories were provided with them: Kjaer noted that operators were 'well-protected by lead screens' and also given vacations from work to limit their exposure.

On 20 April, Kjaer arrived in the little town of Ottawa for the studio inspection. His first port of call was to speak with Miss Murray, the superintendent.

'Why,' she told him, '[I] never heard of any illness which might in the slightest manner be caused by the work.' In fact, she went on, 'Instead of proving detrimental to the health of the girls, [I] know of several who had seemingly derived benefit from it and showed decidedly physical improvement.'

Kjaer asked her about lip-pointing. She told him that the girls 'had been admonished not to tip the brushes in the mouths without washing them carefully first in the water provided for such a purpose'. But she conceded, 'Tipping in the mouth is constantly practiced.'

Kjaer could see that for himself as he toured the studio the same day. Every single girl there was lip-pointing; yet they were all, he noted, 'healthy and vigorous'. On the day he toured the plant, he observed that the girls did have water on their desks in which they were cleaning their brushes – but later, when Fordyce supplied him with a photograph of the studio taken at a different time, Kjaer noticed that water was not in evidence on the tables.

As part of his inspection, Kjaer also interviewed Ottawa's dentists to find out if they'd come across any extraordinary conditions of the mouth in their patients. In New Jersey, it had been Drs Barry and Davidson who had first raised the alarm; should there be a problem in Ottawa too, it seemed logical that its dentists might be the first to know of it. And so he called on three different dentists on that April afternoon, including one who had the largest dental practice in town. That dentist took care of a number of the girls employed at the factory; he told Kjaer there had been 'no evidence of malignant disorder'. He promised to notify the bureau promptly in case anything should turn up. The other dentists, too, gave the girls a clean bill of health. They took pains to state, in fact, that 'there seemed to be very little dental trouble among those workers'.

Kjaer spent only three weeks on his national study – an incredibly short time given the size of the country and the potential gravity of the situation – before it was suddenly stopped. Kjaer's boss, Ethelbert Stewart, later said of the decision: 'Radium paints came to our attention in connection with our campaign against white phosphorus; phosphorus then was our chief interest and we found that it was not used in the elements which go into luminous paint.' The investigation had been but an offshoot of a wider study into industrial poison.

Yet there was another reason, too. Stewart later confessed, 'I abandoned the inquiry not because I was convinced that no problem existed outside of the United States Radium Corporation, but because the expense of follow-up made it impossible for the Bureau to continue.'

In those three short weeks, however, Kjaer had reached a conclusion. Radium, he determined, *was* dangerous.

It was just that nobody told the girls ...

United States Radium Corporation HQ
30 Church Street, New York
1925

Arthur Roeder was having a very bad day. Ever since the Carlough girl had brought her lawsuit, every day seemed to be a bad day. The publicity had been horrendous – his company's name dragged through the mud, as this little upstart charged that the firm had made her 'totally incapacitated for work' and had 'seriously injured' her. The coverage was affecting business; there were now only a few dial-painters left.

Roeder didn't necessarily know it, but the scandal had also impacted on the dial-painting studio his firm had helped set up at the Waterbury Clock Company; following local news reports of the Carlough case, the watch firm had banned lip-pointing.

In fact, there might have been another reason for that, though the clock company would never admit it. In February 1925, a dial-painter there by the name of Frances Splettstocher had died, just a few weeks after falling ill in agonising pain; she'd had a jaw necrosis that had bored a hole right through to her cheek. Her death was not formally linked to her job, but some of her colleagues made the connection. One Waterbury girl said she 'became frightened when Frances died and would not work in the dial-painting department again for any money'.

Frances's father also worked for the firm. Though he was 'sure' that Frances's job had killed her, he 'did not dare make any kick about it' for fear of being sacked.

Oh, for such obedient employees.

Roeder was fighting the Carlough case via the highly skilled (and highly expensive) USRC company lawyers. They'd immediately filed a motion to strike out the girl's complaint, arguing that the case should be presented to the Workmen's Compensation Bureau, where it would fail as the girl wasn't suffering from one of the nine compensable diseases. So far, however, their legal manoeuvring wasn't working – the judge had directed that a jury should decide the case.

The situation, from Roeder's perspective, grew worse by the day. The family of Hazel Kuser had joined the lawsuit, with a claim of $15,000 ($203,000). The ambulance-chasing lawyers had been after Helen Quinlan's mother Nellie too – but she, believing what the doctors told her of her daughter's death, saw no reason to go to them. It was a small mercy.

It was just as well, Roeder thought, that Miss Carlough's sister, Sarah Maillefer, had quit her job in the dial-painting studio when the lawsuit was brought; there was no way she could have continued in their employ. He mused on Mrs Maillefer for a moment. Viedt had told him what a sickly woman she was – lame for three years, walking with a cane; and all the while the company had assisted her so she could stay in her job. Well, in Roeder's opinion, apples didn't fall far from trees – and if one sister was sickly, the chances were it ran in the family.

He blamed the 'women's clubs' for all this bother. Katherine Wiley had been writing to him since the start of the year; she had, he thought disapprovingly, an 'unusual interest' in the matter. He'd done his best to put her off, but it hadn't worked. Even when Roeder had flattered her, saying he thought it 'perfectly proper that your League should interest itself in reports of this type', she had failed to come onside. Increasingly, she was becoming more than an annoyance.

And then there was the investigating statistician Dr Hoffman. Although he wrote Roeder that 'nothing could be further from [his] mind than to raise a controversy for no purpose', his correspondence was exceedingly critical of the company. He had written to Roeder again about Marguerite Carlough, saying she was in a 'truly pitiable condition'. He had urged Roeder or a company representative to visit her in person, but that was not going to happen.

Roeder could handle such begging letters – the company had easily seen off Blum's request for money before – but it was Hoffman's investigation that was really troubling him. The man was planning to publish his report at the end of it – probably before the influential American Medical Association – but Roeder failed to see how Hoffman, who was not a physician and had no specialised radium knowledge, could be permitted to do so. Roeder had always thought 'that a presentation of any subject before an important medical convention was based on extensive research or investigation or both'. In his opinion, 'such an investigation should at least cover the United States, and would hardly be complete without including Switzerland and parts of Germany and France'. What was Hoffman thinking of, coming up with conclusions based only on his very brief studies in a few parts of the US? (Hoffman had also visited the Radium Dial studio in Ottawa and some dial-painting plants in Long Island as part of his research.) If Hoffman wanted to examine the matter fully, Roeder thought, surely he should commit to several more years' hard work and extended international study before presenting his conclusions.

But, instead, Hoffman had limited himself to sending questionnaires to the doctors and dentists who had attended the women, and conducting interviews with those affected. Hoffman later noted: 'I heard the same story from all of them. They did the same identical work, under identical conditions ...

and consequently it was the same consequence.' Despite the brevity of his research, he seemed determined to publish.

Why, Roeder thought in frustration, he hadn't even visited the factory; though, to be fair, that was perhaps because Roeder had tried to stymie his investigation – the firm had offered no assistance. Roeder had tried to appease Hoffman, writing, 'We sincerely believe that the infection you refer to is not caused by radium. If there is a common cause, I think it lies outside our plant.' Yet Hoffman's study had continued. Roeder couldn't understand his tenacity.

Unbeknown to the company president, it was perhaps partly driven by the fact that even the paint's inventor now acknowledged that the girls' trouble was due to their work. In February 1925, Sabin von Sochocky had written to Hoffman to say that 'the disease in question is, without doubt, an occupational disease'.

Roeder sighed and turned back to his desk to read his correspondence, smoothing down his dark hair – flattened, as it always was, with pomade – and self-consciously adjusting his elegant bow-tie. Yet his heart sank further when he saw what was before him: another letter from Miss Wiley.

'My dear Mr Roeder,' she wrote easily. 'I [have] learned that Dr Drinker made an investigation [last spring]. I have heard nothing of the result, but have been looking with great interest to the time when it would be published . . .'

A troubled look crossed Arthur Roeder's affluently rounded face. The Drinker investigation: that was another thorn in his side. He'd so looked forward to the delivery of the doctors' report last June – here, finally, would be the scientific proof, the unchallengeable confirmation, of what he knew to be the truth: that these grim illnesses and deaths had absolutely nothing to do with his firm.

He had been stunned when he read the covering letter Drinker had enclosed with the report. 'We believe that the trouble which has occurred is due to radium,' Drinker had written almost a year ago, on 3 June 1924. 'It would, in our opinion, be unjustifiable for you to deal with the situation through any other method of attack.'

Well, that was ... unexpected. The Drinkers *had* delivered a provisional opinion on 29 April, following their initial academic research, that 'it would seem that radium is the probable cause of the trouble'. But that was before they'd even returned to the plant, and Roeder had been certain that further study would prove them wrong.

Yet the final report hadn't made for better reading. 'In our opinion, so great an incidence among these employees of this unusual disease ... cannot be a coincidence but must be dependent on some type of bone damage occasioned by the employment.'

The Drinkers had methodically gone through the paint's ingredients, dismissing each one in turn as non-toxic, but at radium they declared there was 'ample evidence' of the dangers of over-exposure. 'The only constituent of the luminous material which can do harm,' the Drinkers concluded, 'must be the radium.'

They even gave a detailed hypothesis of what they thought was happening inside the women as a result of their exposure. Radium, they noted, had a 'similar chemical nature' to calcium. Thus radium 'if absorbed, might have a preference for bone as a final point of fixation'. Radium was what one might call a bone-seeker, just like calcium; and the human body is programmed to deliver calcium straight to the bones to make them stronger ... Essentially, radium had masked itself as calcium and, fooled, the girls' bodies had deposited it inside their bones. Radium was a silent stalker, hiding behind that mask, using its disguise to burrow deep into the women's jaws and teeth.

As Drinker had read in scientific literature, radium, since the beginning of the century, had been known to cause serious flesh wounds. It was why workers exposed to large amounts of radium dressed themselves in heavy lead aprons and wielded ivory-tipped tongs; why lab workers at Radium Dial were restricted in the amount of time they could spend in its presence. It was why Dr von Sochocky didn't have the tip of his left index finger anymore; why Dr Leman, the chief chemist in his former company, had lesions all over his hands; why von Sochocky's partner Willis no longer had a thumb. The impact it had externally could easily kill a man, as Pierre Curie had noted back in 1903.

That was the effect it had on the outside. Now imagine the impact of it, once it had craftily concealed its way inside your bones.

'Radium, once deposited in bone,' wrote Drinker in his report, 'would be in a position to produce peculiarly effective damage, many thousand times greater than the same amount outside.'

It was radium, lurking in Mollie Maggia's bones, that had caused her jaw to splinter. It was radium, making itself at home in Hazel Kuser, that had eaten away at her skull until her jawbones had holes riddled right through them. It was radium, shooting out its constant rays, that was battering Marguerite Carlough's mouth, even at this moment.

It was radium that had killed Irene, and Helen, and so many more . . .

It was radium, the Drinkers said, that was the problem.

The doctors enclosed a table of their test results of the workers and, crucially, analysed them. 'No blood [from the USRC employees],' they wrote, 'was entirely normal. These same findings were noted in previous reports by the Life Extension Institute, but the Institute did not appear to have been aware of their meaning.' While some employees registered marked changes in their blood, other results were noted to be 'practically normal'. But not one worker had wholly normal blood; not even a woman who had been with the firm for only two weeks.

The Drinkers commented specifically on the case of Marguerite Carlough, whom they had interviewed on their very first visit to the studio: the case at the root of all Roeder's present woes. And here, for a moment, they dropped the detached tone that characterised the rest of the technical report. 'It seems to us important to express our opinion,' they wrote, 'that Miss Carlough's present serious condition is the result of her years of employment in your plant.' They wanted, they said, 'to call your attention to the fact that this girl needs the best of medical attention if she is to survive'.

Almost a year on, the company had not lifted a finger to help her.

The report finished with various safety recommendations, 'precautions you should take at once'. Ever since this thing had blown up in Roeder's face, there seemed to be nothing but safety recommendations. He had recently instructed Viedt to put some of them into practice: 'This is much more economical,' he'd told his deputy in a memo, 'than paying $75,000 lawsuits.'

Roeder had been aghast at the Drinkers' report by the time he'd finished reading. Surely it couldn't be true. He had taken a few days to collect his thoughts and then, over the course of several weeks in June 1924, he had exchanged further correspondence with Dr Drinker. Seeming to forget the doctor's undoubted brilliance – the very attribute that had driven Roeder to recruit him in the first place – Roeder now pronounced himself 'mystified' by the doctor's conclusions and longed to 'reconcile in my own mind the situation that you have found'. Yet, perhaps anticipating an offer from Drinker to discuss it further, Roeder stressed that he was far too busy to meet him; so much so that he was 'contemplating giving up my Saturdays which I usually spend at the seashore during the summer' to spend more time at work.

On 18 June 1924, the day that Harold Viedt wrote to the Department of Labor to share the company's sleight-of-hand summary of the Drinker report, Roeder and Drinker were still in debate by letter. In the company president's correspondence of that date, he wrote dismissively to Drinker, 'Your preliminary report is rather a discussion, with tentative conclusions, based on evidence which is circumstantial.'

Of course, the doctor had responded. 'I am sorry our report impressed you as preliminary and circumstantial and fear that reiteration can do little to alter such an impression.' Yet he stated once again: 'We found blood changes in many of your employees which could be explained on no other grounds.'

The two had then got into a heated dispute, with letters flying back and forth. Roeder was adamant: 'I still feel that we have to find the cause.'

Privately, Drinker was surprisingly understanding about the president's position. He wrote to an associate: 'The unfortunate economic situation in which he finds himself makes it very hard for him to take any stand save one in regard to radium, namely that it is a harmless, beneficent substance which we all ought to have around as much as possible.' He added: 'It does not seem to me that [the company can] be blamed' for what had happened to the girls.

The doctor's standpoint may in part have been due to the discipline in which he worked: industrial hygiene. Until 1922, Drinker's department at Harvard was wholly funded by business; even in 1924, commercial firms contributed money for special projects. To offend such a prestigious institution as USRC would not be wise. As one industrial physician put it: 'Are we in industry to help carry out some soft, silly, social plan? Are we in industry to buy the goodwill of the employees? No. We are in industry because it is good business.'

Thus, after a final exchange of views between Roeder and Drinker, during which – perhaps to keep the doctor at bay – Roeder made sure to mention 'the almost complete closing down of our application plant for lack of business', it had all gone quiet. The full report had never been published; the Department of Labor had been satisfied with the company's version of events; the current dial-painters were no longer listening to hysterical rumours and were back at work; and Arthur Roeder had been able to get on with business as usual.

Until now.

Until Katherine Wiley had stuck her nose in where it wasn't wanted.

Unknown to Roeder, Wiley and the female doctor she had previously asked for help, Dr Alice Hamilton (who worked in the same department as Drinker), were stirring things up with his erstwhile investigators. Hamilton had learned that the reason the Drinkers' report had not yet been published was because Cecil Drinker believed Roeder should first give consent, which naturally was not forthcoming as the company was concealing the true results. Wiley thought Drinker's position 'showed a very unethical spirit'; she called him 'dishonest'.

The two women had thus come up with a master plan. Little knowing that USRC had already given a misleading precis of the report to the Department of Labor, they conspired to ask John Roach to *request* the results from Roeder. Such a move, they judged, would force Roeder's hand and bring the report into the light, as he could hardly refuse Roach in his official capacity.

Therefore, when Roach revealed to Wiley that he had in fact already seen the Drinkers' report – and that it put the company in the clear – she was taken aback. Wiley immediately told Hamilton; and Hamilton, who not only knew the Drinkers personally but perceived that they would be perturbed by this misrepresentation of their data, wrote at once to Katherine Drinker.

'Do you suppose,' she wrote in mock-innocence, 'Roeder could do such a thing as to issue a forged report in your name?'

Katherine Drinker responded immediately; she and her husband were 'very indignant' at the idea that Roeder might have distorted their findings; 'he has proved,' Katherine concluded savagely, 'a real villain.' Encouraged by his wife, Cecil Drinker wrote to Roeder – still, it has to be said, using language that flattered and appeased the president – to suggest the publication of the full study, urging that 'it can only be to your interest to see the publication ... your strongest position is one which must convince the public that you have done everything humanly possible to get to the bottom of the trouble in your plant.'

Wheels thus set in motion, Hamilton wrote to Wiley that she now believed the situation almost resolved. Surely, she said, Arthur Roeder would not be 'stupid enough [as] to refuse to let Dr Drinker publish the report'.

But she had underestimated the audacity of the president.

17

A rthur Roeder had not become the head of the United States
Radium Corporation without being an astute and wily businessman. He was an expert negotiator, skilled in manipulating
situations to his advantage. It was always wise, he considered,
to keep your friends close – but one should always keep one's
enemies closer.

On 2 April 1925, he invited Frederick Hoffman to the Orange
plant.

The statistician, in fact, visited two or three times, noting in
particular the lack of warning signs about lip-pointing. And
perhaps Roeder observed his notes, or maybe what happened
next was simply part of the ongoing safety precautions the
president had instructed Viedt to put in place. For on Hoffman's
final visit, on Good Friday 1925, Roeder called his attention to
new notices in the studio which commanded employees not to
put the brushes in their mouths. Hoffman approved: 'They had
impressed me,' he later said, 'with improved conditions.'

Roeder knew what he was doing. With relations between
the two men cordial, he pressed home his advantage. 'I wish
that I could persuade you,' Roeder wrote Hoffman, 'to defer

publishing a paper on the subject of "radium necrosis".' He wanted Hoffman, he said, to have the 'opportunity to thoroughly investigate the subject'.

Hoffman responded genially: 'I wish to express to you my sincere appreciation of the courtesy extended to me during my visit and of my own sympathy for the trying position in which you find yourself.' However, Roeder was too late. 'I find on looking over my file that an abstract of [my paper] was furnished to the [American Medical] Association some time ago for inclusion in the Handbook which has gone to the printer ... The paper is now out of my hands.' Hoffman added that he had agreed to supply the Bureau of Labor Statistics – the government agency led by Ethelbert Stewart – with a copy of his report.

One can only imagine Roeder's reaction to that news; although he had smoothly tried to allay the concerns of the bureau as well. When Swen Kjaer had interviewed Roeder that spring about Marguerite Carlough, Roeder told him frankly that he 'thought that the ailment was not due to any cause in the factory; in fact, was probably an attempt to palm off something on [the firm]'.

At least the Carlough girl had given him an excuse to put off John Roach. When Roach had heard that the report supplied by the firm was a whitewash, he had immediately requested a full copy of the study. But Roeder had replied that, because of the Carlough suit, 'The matter has been placed in the hands of our New Jersey attorneys, Lindabury, Depue & Faulks, and I am consequently referring your request to Mr Stryker of that firm.' He said the same in response to Drinker's plea for him to publish the full study: 'In view of the [legal] situation, I have taken no action in regard to submitting your paper for publication; we are not issuing any reports now except on advice of counsel.'

Now, however, the situation started to spiral out of Roeder's hands. Drinker began to lose patience with the president's continued stalling and wrote directly to Roach to find out exactly what the company had said of his study. Roach duly sent him Viedt's letter of 18 June 1924 – and Drinker was stunned to find that, just as Hamilton had told his wife, the company had lied. 'We have [both] been deceived,' he declared to Roach, 'in our dealings with the United States Radium Corporation.' He was so shocked by the firm's behaviour that he arranged a face-to-face meeting with Roeder in New York to confront him.

Roeder was still trying to calm the troubled waters. When Drinker told him sharply that he 'felt the conduct of his firm in this affair had not been very creditable', Roeder 'assured [him] that their desire had been quite the reverse and he would at once see to it that [Roach] received a complete copy of the original report'. Though somewhat reassured, Drinker was not wholly placated. He therefore made a deal with the company president. As long as Roeder kept his word, Drinker promised him, 'I will do nothing about publication.'

It was a good deal for Roeder: the game was now up with Roach, after all, and the lack of wider publication meant that the litigating Marguerite Carlough would have no access to the expert report that directly linked her illness with her employment. Yet it was also an ultimatum – and the powerful Arthur Roeder was not the sort to kowtow to pressure from those he had employed.

In fact, he did not seem at all perturbed by the doctor's attempted negotiations; he simply passed on Drinker's demands to Stryker, the company lawyer. Roeder was paying Stryker good money; he would trust him to deal with these latest developments. In the meantime, Roeder had an ace up his sleeve. Drinker, he thought, wasn't the only expert in town.

Enter Dr Frederick Flinn.

Dr Flinn specialised in industrial hygiene, just like Drinker. He was the assistant professor of physiology at the Institute of Public Health at Columbia University; previously, he had been a director for a handful of mining companies. He was a serious man in his late forties with thinning hair and wire-framed glasses. Within a day or so of being asked to undertake research into the harmful effects of radioactive paint, he had met with Roeder, who agreed to furnish money for his study.

This was not Flinn's first interaction with USRC; he had been involved with the firm the previous year, part of its defence against a lawsuit for damages regarding the fumes from the Orange factory, about which residents were still complaining. The company was also likely familiar with Flinn's work with the Ethyl Corporation in early 1925, when the doctor had been hired to find evidence that leaded gas was safe.

Flinn began work the next morning with a tour of the Orange plant – but his remit did not end there. Through the contacts of USRC, Flinn gained access to the dial-painters of other firms, including the Waterbury Clock Company, giving them physical check-ups. To begin with, Flinn said, 'I made my first examinations without any cost to the companies.' But, later, he was paid by the firms employing the girls.

One of the radium companies he worked for was the Luminite Corporation in Newark, where he now encountered Edna Bolz Hussman, the 'Dresden Doll' beauty who had worked at the Orange plant during the war. Since her marriage to Louis in September 1922, Edna had worked for Luminite only intermittently, just to keep some housekeeping money coming in to boost Louis's wages as a plumber. They didn't need much though; they had no children. Instead, they shared their home with a small white terrier.

Edna was employed at Luminite one day when Dr Flinn asked if he could examine her. Although Edna later said that she did 'not know directly in whose behalf the examination was had' and that it 'did not take place at my request', it did, nevertheless, take place. Flinn examined her elegant body carefully and took some blood.

At that time, Edna had slight knee pains, but she was paying them no attention and it is not known if she mentioned them to him. She had probably heard the rumours about the Carlough lawsuit, however, so it must have come as a huge relief when Flinn gave his verdict following the tests. '[He] told me,' she later said, 'that my health was perfect.'

If only her former colleagues were as fortunate. Katherine Schaub was having a dreadful time. It had been, she later wrote, 'a very depressing winter'. Her stomach was now troubling her; so much so that she could not retain solid food and had endured an abdominal operation. She felt like she was being passed from pillar to post, dentist to doctor, and no one offered any answers. 'Since [my] first visit [to a doctor], it has been nothing but doctors, Doctors, DOCTORS,' she wrote in frustration. 'To be under the care of a skilled physician and yet not show any sign of improvement was most discouraging.' Her illness was affecting her whole life, for though she tried to work, her ailments now made it impossible for her to be engaged in any form of employment.

Grace Fryer, however, was still maintaining her job at the bank. Thanks to Dr McCaffrey's attentions, the infection in her jaw seemed to have cleared up, but she was very apprehensive that it might come back. And although her mouth was all right, her back still plagued her. Dr Humphries's strapping treatments no longer had any effect; 'I have been to every doctor of any note in all New York and New Jersey,' she said – but not one of them could determine the cause of her ailments; often, they made things worse. Grace's chiropractic treatments, in the end, 'became so painful that I was compelled to stop taking them'.

In Orange, Grace's friend Quinta McDonald was having no better luck. In April 1925, she was finally removed from the constricting plaster cast that had encased her body for nine months. Yet despite the doctors' best efforts, her condition declined. Now, she could walk only with the greatest difficulty. By the end of the year, her family doctor had been called out ninety times: a bill of some $270 ($3,660).

It was awful timing. She found herself unable to manage the fifteen-minute walk to her sister Albina's house just at the time when she most wanted to be with her. Highland Avenue sloped sharply down towards the railway station on the way to her sister's home, and Quinta simply couldn't get down the hill anymore, even with a stick, let alone back up it. Albina Larice, to the whole family's delight, was pregnant, after almost four years of trying. It was such good news, and there was little enough of that to go around at the moment.

While the Maggia family at least had a reason to celebrate that spring, just down the road on Main Street, the Carloughs were really struggling. They were still spending money they didn't have on Marguerite's care; by May 1925 the medical bills ran to $1,312 (almost $18,000). Sarah Maillefer was distraught by her little sister's condition. She tried to keep talking to her, soothing words or jokes to lift her spirits, but Marguerite's hearing was greatly impaired in both ears because of her infected facial bones and she struggled to hear what Sarah said. The pain was awful: her lower jaw was fractured on the right side of her face and most of her teeth were missing. Her head, essentially, was 'extremely rotten' – with all the putrefaction that implies. But she was alive, still. Her whole head was rotting, but she was still alive.

Her condition was so bad that, at last, it prompted Josephine Smith to quit her job. Nobody could see what had happened to Marguerite and not be moved. Frederick Hoffman and Dr Knef were still fighting her corner too. Seeing her rapid decline, they now sought aid from a perhaps unlikely source – USRC founder Sabin von Sochocky.

Von Sochocky was no longer part of his company. He had no ties to the corporation and, if anything, may even have felt bitter about the way he had been ousted. Perhaps, too, he felt some responsibility. One of the girls' allies later wrote of him, 'I feel absolutely satisfied that there is no prejudice but every desire to assist in a useful way.'

And that is what von Sochocky now did. Together with Drs Hoffman and Knef, the trio admitted Marguerite to St Mary's Hospital in Orange to find out what was the matter with her. On admittance, she was anaemic and weighed 90 pounds (six-and-a-half stone); her pulse was 'small, rapid and irregular'. She was hanging on, but barely.

A week or so after she was admitted, which was partly thanks to Hoffman's intervention, the statistician did the dial-painters his biggest service yet: he read his paper on their problems before the American Medical Association – the first major study to connect the women's illnesses to their work; the first, that is, to be made public. And his opinion was thus: 'The women were slowly poisoned as a result of introducing into the system minute quantities of radioactive substance.'

That 'minute' was important, for the company – *all* the radium companies – believed dial-painting to be safe because there was such a tiny amount of radium in the paint. But Hoffman had realised that it wasn't the amount that was the problem, it was the cumulative effect of the women taking the paint into their body day in and day out, dial after dial. The amount of radium in the paint may have been small, but by the time you had been swallowing it every single day for three or four or five years in a row, there was enough there to cause you damage – particularly when, as the Drinkers had already realised, radium was even more potent internally, and headed straight for your bones.

As early as 1914, specialists knew that radium could deposit in bone and cause changes in the blood. The radium clinics researching such effects thought that the radium stimulated the bone marrow to produce extra red blood cells, which was a good thing for the body. In a way, they were right – that was exactly what happened. Ironically, the radium did, at first, boost the health of those it had infiltrated; there *were* more red blood cells, something that gave an illusion of excellent health.

But it was an illusion only. That stimulation of the bone

marrow, by which the red blood cells were produced, soon became *over*stimulation. The body couldn't keep up. In the end, Hoffman said, 'The cumulative effect was disastrous, destroying the red blood cells, causing anaemia and other ailments, including necrosis.' He concluded emphatically, 'We are dealing with an entirely new occupational affection demanding the utmost attention,' and then – perhaps thinking of Marguerite's lawsuit, which was dragging sluggishly through the legal system – added that the disease should be brought under the workmen's compensation laws.

Katherine Wiley was, in fact, attempting to do just that through her work with the Consumers League, campaigning to have radium necrosis added to the list of compensable diseases. In the meantime, Marguerite's only hope for justice was the federal court – but her case was unlikely to be heard before the fall. As Alice Hamilton noted with dismay, 'Miss Carlough may not live till then.'

Hoffman continued to present his discoveries. He noted that although he had looked for cases of radium poisoning in other studios across the United States, 'There was none affected outside of this plant.' Unwittingly, Hoffman now revealed exactly why that was, but he didn't grasp the relevance of his statement. 'The most sinister aspect of the affliction,' he wrote, 'is that the disease is apparently latent for several years before it manifests its destructive tendencies.'

Several years. The Radium Dial studio in Ottawa had been running for less than three.

Both Hoffman and von Sochocky, whom he'd consulted for his paper, were struck by the lack of other cases. For USRC, it was clear evidence of why the girls' illnesses could not possibly be occupational. Hoffman and von Sochocky, however, who were convinced that dial-painting *was* the cause of the girls' sickness, did what any scientists would do: they looked for a reason. And when von Sochocky gave Hoffman the top-secret paint formula, they believed they had found it. '[Von Sochocky] gave me to understand,' Hoffman later said, 'that the difference between the paste used in the plant in Orange and the paste used elsewhere was mesothorium.'

Mesothorium – radium-228 – and not radium; at least, not the radium-226 that people used in their tonics and pills. That had to be the answer. And so Hoffman, building on Dr Blum's work, commented in his paper: 'It has seemed to me more appropriate to use the term "radium (mesothorium) necrosis".'

In conclusion, it wasn't – not exactly – *radium* that was to blame.

However, when news of Hoffman's report hit the headlines, the radium industry fought back. Radium was still the wonder element and new products were being launched all the time – one such right there in Orange. A highly radioactive tonic called Radithor, produced by William Bailey of Bailey Radium Laboratories – a client of USRC – had been launched in early 1925. He and others spoke out publicly against the attempts to link radium to the dial-painters' deaths: 'It is a pity,' Bailey said, 'that the public [are being] turned against this splendid curative agency by unfounded statements.'

But while the radium men were quick to strike back, Hoffman's paper, although it attracted some publicity, was a rather niche, specialist publication. Not many people subscribed to the *Journal of the American Medical Association*. And who was Frederick Hoffman anyway? Not a physician, who might *really* know about these things. Even the women's allies were aware of his lack of authority. 'It seems to me unfortunate,' wrote Alice Hamilton to Wiley, 'that Dr Hoffman is the man to make this situation public. He does not command the confidence of physicians and the work that he does will not be thorough nor proof against attack.'

What the women needed was a champion. A medical mastermind – someone who could not only command authority, but also, perhaps, find a way of definitively diagnosing their disease. Blum had had his suspicions, Barry too, but neither of them had actually proved that radium was the cause. Most importantly of all, they needed a doctor who wasn't in the pocket of the company.

Sometimes, the Lord works in mysterious ways. On 21 May 1925, a Newark trolley car was trammelling along its tracks on Market Street when there was a commotion on board. The commuters, making their way home in the evening rush hour, made room for the passenger who had suddenly collapsed to the floor. They called out to give him some air, for the trolley car to stop; a kindly passer-by no doubt bent to mop his brow.

It was all in vain. Only a few minutes after the man had first been stricken, he died. His name was George L. Warren. In life, he had been the county physician for Essex County – a senior medical figure with responsibility for the welfare of all residents within the county's borders, which included those in Newark as well as Orange: the locations where former dial-painters were now dying unstoppably.

With Warren's passing, his position became vacant. The role of county physician – what would become the powerfully titled Chief Medical Examiner – was now open. Whoever filled it would make or break the case.

18

It was a unanimous appointment. The board congratulated the new county physician with firm handshakes and much approving nodding of heads.

Dr Harrison Martland, please step up.

Martland had already shown an interest in the dial-painters' cases, having met briefly some of Barry's patients. Although, not being able to determine the cause of the problem, he had by his own admission 'lost interest', the cases had remained on his mind. Reportedly, when Hazel Kuser died, he had endeavoured to arrange an autopsy to determine the cause of death, but Theo had been so attentive in making the final arrangements for his beloved wife that her body had been buried before Martland could contact the proper authorities.

Martland had perhaps also been hampered by territorial politics. Previously he'd had authority to investigate issues in Newark alone; since the plant and many of the victims had been located in Orange, it wasn't necessarily the done thing for him to examine the matter further. Now, however, with the broader remit given him by his new role, he had the power to get to the bottom of it.

Martland was a man of extraordinary talents, who had studied at the College of Physicians and Surgeons in New York; he ran his own laboratory at the Newark City Hospital where he was chief pathologist. Though he had a wife and two children, he was in many ways married to the job – he made 'no difference between weekdays and Sundays' and worked late most nights. He was forty-one years old, a 'heavy but distinguished-looking' man with jowls. His hair, which was light brown and greying at the temples, lay flat on his scalp; he wore circular spectacles. He was the kind of man who worked in his shirt-sleeves, '*sans* tie', a colourful personality who drove open-top automobiles and 'did his exercises to the loud phonograph music of Scots bagpipes' every morning. Everyone called him Mart or Marty, never Harrison and certainly never Harry. As chance would have it, he was also a Sherlock Holmes enthusiast.

The Case of the Radium Girls was a mystery to challenge even the greatest of medical detectives.

Martland took his new responsibilities seriously. As he himself said, 'One of the main functions of a medical examiner is to prevent wastage of human life in industry.' The cynical would say, however, that this proclamation had absolutely nothing to do with why he took an interest in the radium cases at that moment. The cynical would say there was only one reason a high-profile specialist finally took up the cause.

On 7 June 1925, the first male employee of the United States Radium Corporation died.

'The first case that was called to my attention,' Martland later remarked, 'was a Dr Leman.'

The chief chemist of USRC, he who had 'scoffed' at the Drinkers when they'd expressed concern about the blackened lesions on his hands the year before, was dead. He died aged thirty-six of pernicious anaemia, after an illness of only a few weeks. His death had occurred much too rapidly for a normal case of anaemia so Martland was called in to conduct an autopsy.

He suspected radium poisoning, but the chemical analyses he carried out on Leman's body failed to show any sign of the element; specialist testing would clearly be required. Martland, as Drs Knef and Hoffman had done a short time before, now turned to Sabin von Sochocky, an authority on radium, for assistance. And he asked someone else for help too. Where could he possibly find the best-qualified radium expert in town? Surely the United States Radium Corporation knew a little bit about it?

Together, Martland, von Sochocky and USRC's Howard Barker tested Leman's tissues and bones in the radium-factory laboratory. In exchange for its help, USRC asked Martland to promise that he would keep his conclusions secret.

The tests were a success. The doctors reduced Leman's bones to ashes, then tested the ash with an instrument called an electrometer. In so doing, they made medical history in measuring radioactivity in a human body for the first time. In so doing, they determined that Leman had died from radium poisoning; his remains were saturated with radioactivity.

As Martland and von Sochocky worked together, von Sochocky asked the medical examiner to help the dial-painters; Knef made a similar appeal. And so, only a day or so after Leman had died, Martland found himself in St Mary's Hospital, meeting a brave young woman called Marguerite Carlough.

She lay weakly in her hospital bed, her shockingly pale face surrounded by limp dark hair. At this time, 'her palate had so eroded that it opened into her nasal passages'. Also visiting Marguerite was her sister Sarah Maillefer.

Sarah was no longer quite as matronly in figure as she had once been; she'd been losing weight for the past year or so. It was the worry, she thought. Worry for Marguerite, who was so badly ill; worry for her daughter, who was now fourteen years old. Like most mothers, she rarely worried about herself.

A week ago she'd noticed that she'd started to bruise easily. And it was more than that, if she was honest with herself: large black-and-blue spots had broken out all over her body. She'd come to see Marguerite anyway, not wanting to miss the visit, limping up the stairs with her walking cane, even though she felt very weak. Her teeth were aching, too, but you had to put things into perspective: look at her sister, she was far worse off. Even when her gums started to bleed Sarah thought only of her sister, who was so close to death.

As Martland met the Carlough girls, he observed that although Marguerite was more ill than Sarah, Sarah was also not well. When he asked her, she confessed that the black-and-blue spots were causing her intense pain.

Martland ran tests and found Sarah to be very anaemic. He told her the results; spoke with her about her jaw trouble. And then Sarah, perhaps finally worried over what it might mean, 'went bad quite rapidly' and had to be admitted to hospital. But at least she wasn't alone. She and Marguerite shared a hospital room: two sisters together, facing whatever might lie ahead.

The hospital doctors examined Sarah closely, concerned at her decline. Her face was swollen on the left side, her glands hot and tender. She was running a temperature of 39 degrees – increasing up to 41 degrees in the evenings – and by now had marked lesions in her mouth. She was, it appeared, 'profoundly toxic'.

Martland wanted to test the two women to see if radium was the cause of their illnesses – but the only tests he knew, those he had conducted with von Sochocky and Barker, required burning bone to ash. You couldn't very well do that with living patients.

It was von Sochocky who came up with the answer. If the women were radioactive, all they had to do was devise some tests to prove it. These tests, which would be honed and largely invented by Drs Martland and von Sochocky, were created specifically to test the dial-painters' bodies. No physician had ever attempted to test living patients in this way before. Later, Martland would discover that a specialist had done something similar before him, but in June 1925, with the clock running down on Marguerite Carlough, he innovated the tests knowing nothing of the other scientist's work. He was, indeed, a man of extraordinary talents.

The pair devised two methods: the gamma-ray test, which involved sitting the patient before an electroscope to read the gamma radiation coming from the skeleton, and the expired-air method, whereby the patient blew through a series of bottles into an electroscope so that the amount of radon could be measured. This latter was born from the idea that, as radium decayed into the gas radon, if radium was present in the girls' jawbones, the toxic gas might be exhaled as they breathed out.

The doctors took their equipment to the hospital to try on Marguerite. But, when they got there, it was Sarah Maillefer they decided to test first.

Being in hospital had not helped her. Despite being given a blood transfusion on 14 June, Sarah had become so ill she'd had to be removed from the room she shared with her sister. When Marguerite asked where she was, the nurses told her Sarah had been 'removed to receive special treatments'.

That was true, in a way. The tests Sarah was about to have *were* special, for she was the first dial-painter ever to be tested for the presence of radium. The first who would prove whether or not all that conjecture was correct.

This was the moment of truth.

In a hospital room in St Mary's, Martland and von Sochocky set up the equipment. They first tested Sarah's body. As she lay weakly on the bed, Martland held the electrometer eighteen inches above her chest, to test her bones. A 'normal leak' would be 10 subdivisions in 60 minutes: Sarah's body was leaking 14 subdivisions in the same time. *Radium.*

Next, they tested her breath; the normal result they were looking for was 5 subdivisions in 30 minutes. This test wasn't as easy as simply holding the measuring device over Sarah's prone body, though. This test, she had to help with.

It was very hard for her to do, because she was so poorly. 'The patient was in a dying, almost moribund condition,' remembered Martland. Sarah found it difficult to breathe properly. 'She couldn't for five minutes' time.'

Sarah was a fighter. It's not clear if she knew what the tests were for; whether she had the capacity at that stage even to know what was going on around her. But when Martland asked her to breathe into his machine, she tried so very, very hard for him. *In ... out ... In ... out.* She kept it going, even as her pulse raced and her gums bled and her gammy leg ached and ached. *In ... out ... In ... out.* Sarah Maillefer breathed. She lay back on the pillows, exhausted, spent, and the doctors checked the results.

The subdivisions were 15.4. With every breath she gave, the radium was there, carried on the very air, slipping out through her painful mouth, passing by her aching teeth, moving like a whisper across her tongue. *Radium.*

Sarah Maillefer was a fighter. But there are some fights that you cannot win. The doctors left her in the hospital that day, on 16 June 1925. They didn't see as her septic condition increased; as new bruises bloomed on her body, blood vessels bursting under her skin. Her mouth would not stop bleeding; pus oozed from her gums. Her bad leg was a constant source of pain. *Everything* was a constant source of pain. She couldn't take it anymore; she became 'delirious' and lost her mind.

But it didn't take too long, not after that. In the early hours of 18 June, only a week after she'd been admitted to hospital, Sarah Maillefer died.

The same day, Martland conducted an autopsy; the results would take some weeks to come back. He was bound by no promises of secrecy this time. He spoke to the media on the day Sarah died as they gathered to hear of this latest death. 'I have nothing more than my suspicions now,' he told them. 'We are going to take the bones and some of the organs of Mrs Maillefer's body, reduce them to ashes and make extensive laboratory tests with the most delicate instruments available for radioactive substances.' And then he continued, probably striking fear into the hearts of Sarah's former employers: 'This poisoning, if my suspicions are correct, is so insidious, and sometimes takes so long to manifest itself, that I think it possible it has been going on for some time throughout the country without being discovered.' That time was now at an end, although Martland wasn't rushing into anything: 'We have nothing more definite than a theory at present,' he said. 'I will not make the statement that commercial "radium poisoning" exists until we can prove it.' But, the implication was, *once he could* . . .

The press were all over it; Sarah's death even made the front page of the *New York Times*. Yet while the whole world knew of her passing, there was someone who didn't.

Her little sister, Marguerite. She hadn't seen Sarah since 15 June, the night she'd been taken from their shared room. She'd enquired several times as to how her sister was. Even though Marguerite had seen Sarah decline, she must have had hope. Sarah had always been the strong one once Marguerite had sickened, and she had only been seriously ill for a few days.

The nurses put her off when she asked after her big sister. But on 18 June, when the papers were filled with the news of Sarah's death, Marguerite had innocently asked to see a newspaper.

'No,' the nurses said, wanting to spare her.

'Why?' asked Marguerite. Of course Marguerite Carlough would ask why.

And so the nurses told her of her sister's death. 'She is said to have borne the news bravely – and expressed regret that she could not be present at the funeral.' She was far too ill to go.

It was Sarah's father, Stephen, who told the authorities of his daughter's death; who arranged her funeral; who looked after her teenage daughter, Marguerite. It was he who watched her coffin being lowered into the ground at Laurel Grove Cemetery, shortly after 2 p.m. on Saturday 20 June.

Sarah may have been thirty-five, but it was his little girl who'd gone.

S arah wasn't even in her grave before her former company was denying it was to blame.

Viedt gave a statement to the press. There was 'small possibility', he said, 'of the existence of a "radium poison" menace'. Speaking of the newly appointed Dr Flinn, the USRC-employed company doctor, he revealed, 'We have engaged persons of the greatest reliability and reputation to conduct an investigation.' He told the press that Sarah had been examined by the Life Extension Institute while working for the company and, perpetuating the position taken back in June 1924 when the firm had chosen to ignore the unpublished Drinker report, revealed that 'nothing was found in our plant not found among average industrial employees'. It was, he said, 'absurd to think the same condition could have caused the deaths of Dr Leman and Sarah Maillefer. The latter could not have handled in one hundred years of her work half the amount of radium Dr Leman handled in one year. The amounts handled by [Sarah] were so infinitesimal that in the opinion of company officials the work could not be considered as hazardous.'

Yet those infinitesimal amounts still left a trace – something that Martland was discovering. Sarah's autopsy was conducted nine hours after her death. She was the first-ever dial-painter to be autopsied; the first radium girl to have an expert examine every inch of her body for clues as to what could have caused her mysterious downfall.

The medical detective made notes as he moved down her silent corpse, working from head to toe. He stretched her mouth wide, looked inside. It was 'filled with old, dark, clotted blood'. He inspected her left leg, the one she'd been limping on for three years; it was, the doctor noted, 4 centimetres shorter than her right.

He weighed and measured her internal organs; stripped out her bones to run his tests. He looked inside those bones, into the bone marrow where the blood-producing centres lay. In a healthy adult, the bone marrow is usually yellow and fatty; Sarah had 'dark-red marrow throughout [the] entire shaft'.

Martland was a medical man. He had seen for himself the application of radium to treat cancers in hospital, and he knew how that worked. Radium had three types of rays that it constantly emitted: alpha, beta and gamma rays. The alpha ray was a very short ray and could be cut off by a thin layer of paper. A beta ray, which had a little greater penetrating power, could be cut off by a sheet of lead. (Modern science says a sheet of aluminium.) The gamma ray was very penetrating, and it was 'by its gamma ray' a radium expert said, 'you might say it is magic', for the gamma radiation was what gave radium its medicinal value, being able to travel through the body and be directed at a tumour. It was the gamma and beta rays the lab workers protected themselves against with their lead aprons; they didn't need to worry about the alpha rays as they could do no harm, being unable to penetrate skin. That was just as well, for alpha rays, which formed 95 per cent of the total rays, were 'physiologically and biologically intensely more irritating than beta or gamma' rays. In other words: the worst kind of radiation.

In Sarah Maillefer's body, Martland now realised, the alpha

rays had not been blocked by a thin sheet of paper or by skin – they were not blocked by anything. The radium was in the very heart of her bones, in close proximity to her bone marrow, which was constantly bombarded by rays from the radioactive deposits. 'The distance,' Martland later said, 'is approximately one-hundredth of an inch from the blood-forming centres.'

There was no escaping this most dangerous poison.

Given the extreme power of the alpha rays – those 'whirling, powerful, invisible forces we do not yet understand', as von Sochocky had once written – Martland now realised that it didn't matter that the amount of radium Sarah had worked with was 'infinitesimal'. From the tests, the doctor estimated that her body contained 180 micrograms of radium; a tiny amount. But it was enough. It was 'a type of radiation never before known to have occurred in human beings'.

He continued his tests. And now he discovered something that no one had ever appreciated before. For he didn't just test Sarah's affected jaw and teeth for radioactivity – the site of all the dial-painters' necroses – he tested her organs, he tested her bones.

They were all radioactive.

Her spleen was radioactive; her liver; her gammy left leg. He found it all over her, but chiefly in her bones, with her legs and jaw having 'considerable radioactivity' – they were the parts most affected, just as her symptoms had shown.

It was an extremely important discovery. Dr Humphries in Orange had never connected the cases he had seen because the women presented different complaints – why would he have thought that Grace Fryer's aching back might be connected to Jennie Stocker's peculiar knee or Quinta McDonald's arthritic hip? But it was the same thing affecting all the girls. It was radium, heading straight for their bones – yet, on its way, seeming to decide, almost on a whim, where to settle in the greatest degree. And so some women felt the pain first in their feet; in others, it was in their jaw; in others still their spine. It had totally foxed their doctors. But it was the same cause in all of them. In all of them, it was the radium.

There was one final test that Martland now conducted. 'I then took from Mrs Maillefer,' he remembered, 'portions of the femur and other bones and placed dental films over them. [I] strapped [the films] all over [her bones] at various places and left them in a dark room in a box.' When he'd tried this experiment on normal bones, leaving the films in place for three or four months, he had not got the slightest photographic impression.

Within sixty hours, Sarah's bones caused exposure on the film: white fog-like patches against the ebony black. Just as the girls' glow had once done, as they walked home through the streets of Orange after work, her bones had made a picture: an eerie, shining light against the dark.

And from that strange white fog Martland now understood another critical concept. Sarah was dead – but her bones seemed very much alive: making impressions on photographic plates; carelessly emitting measurable radioactivity. It was all due, of course, to the radium. Sarah's own life may have been cut short, but the radium inside her had a half-life of 1,600 years. It would be shooting out its rays from Sarah's bones for centuries, long after she was gone. Even though it had killed her, it kept on bombarding her body 'every day, every week, month after month, year after year'.

It is bombarding her body to this day.

Martland paused in his work, thinking hard. Thinking not just of Sarah, but of her sister Marguerite, and of all the other girls he had seen in Barry's office. Thinking of the fact that, as he later said, 'There is nothing known to science that will eliminate, change or neutralise these [radium] deposits.'

'Radium is indestructible,' Dr Knef concurred. 'You can subject it to fire for days, weeks or months without it being affected in the least.' He went on to make the damning connection. 'If this is the case ... how can we expect to get it out of the human body?'

For years the girls had been searching for a diagnosis, for someone to tell them what was wrong. Once they had that, they believed faithfully, then the doctors would be able to cure them.

But radium poisoning, Martland now knew, was utterly incurable.

Following the results of his tests, Martland shared the proven cause of Sarah's death. 'There is not the slightest doubt,' he wrote, 'that she died of an acute anaemia, following the ingestion of luminous paint.'

As hers was the first properly tested case, it became of considerable interest to medical men. Dr Flinn, the company doctor, immediately wrote to Martland: 'Would it be possible for me to get a section of [Mrs Maillefer's] tissues, so I can compare them with those of my [laboratory] animals which I am expecting to kill some time in the next few weeks?' Dr Drinker, too, followed the progress of the case with great interest. He had not finished his fight with USRC – because Arthur Roeder had not kept his word.

It was the company lawyer, Josiah Stryker, who had handled the delicate matter of the Drinker report and the Department of Labor. He *had* taken the report to Roach – but refused to let him retain a copy. 'It will be available to [you],' he said airily to Roach, 'at any time [in my office].' Stryker had left with the report in hand. He added as he went, 'If the Department insist on having a copy in their own file, [I will] provide one.'

Well, the department had insisted; but the company sent it to McBride, Roach's boss – the man who had been 'furious' when tenacious Katherine Wiley had intervened in the dial-painters' cases and rebuked Roach as a result – and not to Roach himself.

When Drinker found out, *he* was furious. On the day Sarah Maillefer died, he wrote to Roeder: 'I am arranging for the immediate publication of [my] report.' He was, as the saying goes, planning to publish and be damned. But Stryker was quick with his response: publish *and be sued*.

Yet if Roeder and Stryker thought they had Drinker pegged, they had thought wrong. One of Drinker's brothers happened to be a good corporation lawyer. The doctor asked him what he thought of the firm's threat, and the brother said to 'tell 'em to sue and be damned!' So Drinker called their bluff.

The Drinker report – first filed on 3 June 1924 – would finally be published in August 1925, with a press date of 25 May, five days before Hoffman had first read his study, in order to give Drinker precedence on the discovery of the link between the girls' illnesses and the radioactive paint. Whatever date they put on it, it was published well over a year after being submitted to USRC. Commentators on the case later said, 'This report by the Harvard investigators was a scientific document of the greatest importance, not only to remedy conditions in this plant, but to acquaint other manufacturers, using the same radium formula, with its toxicity and potentially lethal effects. Science and humanity alike demanded immediate publication of this report . . . but [it] was resolutely suppressed.'

The company had tried to keep everyone in the dark – the Department of Labor, the medical community, the women they had doomed to die. But the light, finally, was flooding in now. The momentum was building for the women's cause, even as radium's supporters tried to throw it off-track – and Martland, the girls' illustrious medical champion, was first in the firing line as pro-radiumites sought to undermine his credibility. William Bailey, the man behind the Radithor tonic, remarked cuttingly, 'Doctors, who have never had the slightest experience with radium and know no more about it than a schoolboy, have been trying to garner some publicity by claiming harmful effects. Their statements are perfectly ridiculous!' Bailey added that he would happily 'take in one dose all the radium used in the factory in one month'.

USRC, too, was quick to step in, with a spokesman saying

dismissively, 'Radium, because of the mystery which surrounds much of its activities, is a topic which stimulates the imagination and it is probably this rather than actual fact that has caused the outcry.' Roeder weighed in to the debate, claiming publicly that many of the women were 'unfit' when they started dial-painting and as an excuse had unfairly impugned the firm. It wasn't even the female victims alone that USRC attacked. A spokesman said that Leman, the chief chemist who had died, 'was not in robust health when he began with radium work'.

Yet the momentum that had begun and was now building – first with Hoffman's report, then Sarah's sacrifice, and now the Drinker report too – was unstoppable. Even Andrew McBride, who had previously seemed reluctant to intervene, now beat the drum of change. He made a personal visit to the Orange studio and asked why the Drinkers' safety recommendations had not been put into effect; he was informed that the firm 'did not agree with them all, many of them had already been followed and some were impractical'.

McBride wasn't swayed. Now he said that he believed that 'human life is far too important to be neglected or wasted if it is possible to conserve it'. Consequently, he declared that if the firm did not carry out the Drinkers' suggestions, 'I would issue orders to close their factory ... I would compel them to comply or close, no matter at what cost.'

For those who had long supported the girls, it was a complete turnaround. Karl Quimby, the priest who had offered spiritual comfort to Hazel Kuser, was relieved to observe that, at last, someone in authority was finally taking note. Seeing Dr Martland's findings reported widely in the eastern press, he felt moved to write to him: 'I can scarcely express to you how gratified I am that you are doing this splendid thing. [I am] wishing you every success and assuring you of the appreciation of a goodly number of people.'

But it was, of course, the dial-painters themselves for whom the biggest difference was made. Soon after Sarah passed away,

Martland carried his testing equipment back into St Mary's. It was Marguerite Carlough's turn to be measured for the radium that the doctor believed was lurking in her bones.

She was in 'a terrible condition' on the day he ran the tests, with her mouth, as it had always been, the thing that was most agonising. The alpha rays of the radium, Martland now believed, were slowly drilling those holes into her jawbone. Despite the pain, Marguerite put the breathing tube into her mouth and blew. Just like her sister had done before her, she breathed, as steadily as she possibly could. *In ... out.* On the day Martland ran her tests, the normal leak was 8.5 subdivisions in 50 minutes. (The normal number changed depending on the humidity and other factors.) When he checked Marguerite's results, they showed 99.7 subdivisions in the same time.

At least, she thought, it would help her legal case.

She had more reason than ever to want to win now: following her sister's death, the Carloughs had added Sarah's claim to the litigation. USRC was now fighting three cases: for Marguerite, Hazel and Sarah. Marguerite was the only one of those three left alive. And so she wanted to do everything she could to help the case; not only for herself, but for her sister. That was something to live for, to strive for, to battle through the pain for. While she was in St Mary's, her lawyer, Isidor Kalitsch of Kalitsch & Kalitsch, interviewed her even as she lay in bed, taking her formal testimony so that – whatever happened – he had it to fight the girls' case.

Yet Hazel, Sarah and Marguerite weren't the only girls afflicted. That was something Martland knew – but what he didn't know was how to contact the others, to get more girls to come forward. Some were ultimately connected to him by their dentists and physicians, but others came to him via a young woman by the name of Katherine Wiley.

'In the midst of my difficulties in the summer of 1925,' Katherine Schaub later recalled, 'Miss Wiley again called at our home. This time, she was interested in my own case, for she had heard that I had been ill. [She] suggested that I consult the county medical examiner for a diagnosis.'

Katherine had been troubled by her ailments for a long time by that stage. She had seen what had happened to Irene; she had read what had happened to Sarah. She wasn't stupid. She knew why Miss Wiley had called on her, and what Dr Martland thought he would find. To her sister Josephine, she said slowly, 'It must be that I have radium poisoning.'

She tried it out in her mind, like slipping on a new dress. It clung to her, skin-tight: nowhere to hide. It felt most peculiar; not least because Katherine was doing well that summer. She did not appear ill anymore. Her jaw wasn't troubling her; all the infections in her mouth had cleared up. Her stomach, following her operation, was much improved. 'Her general physique was good.' She couldn't have what the others had, she couldn't; for they had all died, and here she was, still living. Yet there was only one way to be sure. There was only one way to *know*. Katherine Schaub duly made an appointment with the county physician.

She wasn't the only one. Quinta McDonald had become increasingly concerned about her own condition of late: her teeth, which she had once considered her best feature, had started loosening in her mouth and then falling out spontaneously, straight into her hand. Ironically, her daughter Helen was losing her milk teeth at the same time. 'I can stand the pain,' Quinta later said, 'but I do hate to lose my teeth. The upper ones are so loose they merely hang.'

Following her new trouble, Quinta had started consulting Dr Knef, the kind dentist who had treated her sister Mollie. Knef had been working with Martland to treat Marguerite; so it was Knef who arranged for Quinta to have Martland's special tests. And with her came her old friend Grace Fryer, who now had no jaw trouble at all, who was apparently in good health – but whose back hurt worse with every passing day.

One by one they came. Katherine. Quinta. Grace. They

weren't ill like Sarah, or Marguerite, or Dr Leman. They weren't at death's door. They stayed still as Martland scanned their bodies with his electrometer; asked them to breathe into a tube; tested them for the tell-tale anaemia that would betray what was happening inside their bones.

To each he said the same. 'He told me,' Grace remembered, 'that my system showed the presence of radioactive substances.' 'He told me,' Quinta said, 'that my trouble was all due to the presence of [radium].'

He told them that there was no cure.

A deep breath was needed for that news. *In . . . out.*

'When I first found out what I had,' Grace remembered, 'and learned that it was incurable . . .' She tailed off, but eventually continued: 'I was horror-stricken . . . I would look at people I knew and I would say to myself, "Well, I'll never see you again."'

They all had that same thought. Quinta, heading home to her children: *I'll never see you again.* Katherine, breaking the news to her father: *I'll never see you again.*

For Katherine, though, the diagnosis brought relief too. 'When the doctors [finally] told me that [my tests] showed positive radioactivity,' she remembered, 'I was not as frightened as I thought I would be. At least there was no groping in the dark now.'

Instead, there was light. Glowing, glorious light. Shining, stunning light. Light that led their way into the future. 'The medical examiner's diagnosis,' commented Katherine Schaub with characteristic acumen, 'furnished perfect legal evidence for a lawsuit.'

For too long the women had waited for the truth. The scales, at last, were tipping against the company. The girls had been given a death sentence; yet they had also been given the tools to fight their cause – to fight for justice.

The diagnosis, Katherine Schaub now said, 'gave me *hope*'.

PART TWO

Power

20

There was much to be done. Even before the summer was out Dr Martland had lent his voice to Katherine Wiley's campaign to amend the industrial-compensation law. Yet the legal change was only part of it. For the girls, who now understood how unforgivably careless the company had been with their lives, the real question was how the firm's executives could have considered them expendable. Why didn't their basic humanity compel them to end the practice of lip-pointing?

Grace Fryer, for one, was filled with anger as her clever mind raked over what had happened. For she now recalled all too well a fleeting moment in her memory that sealed the company's guilt.

'Do not do that,' Sabin von Sochocky had once said to her. *'You will get sick.'*

Seven years later . . . here she was in Newark City Hospital.

Now she realised: von Sochocky had *known*. He had known all along. But if he had, why had he let them slowly kill themselves with every dial they painted?

Grace had an opportunity to put that question to the man himself – immediately. For when Martland tested her and

Quinta for radioactivity in July 1925, he wasn't the only doctor present. Von Sochocky sat quietly beside the technical equipment as the girls were told that they were going to die. And as Grace listened to the words fall from Martland's mouth – 'all your trouble ... presence of radioactive substances' – the memory of that warning came rushing into her mind.

Still reeling from the news, Grace nonetheless jutted out her chin with archetypal resolve and looked levelly at her former boss.

'Why didn't you tell us?' she asked simply.

Von Sochocky must have bowed his head. He stuttered out something about being 'aware of these dangers' and said he had 'warned other members of the corporation without avail'. Earlier that year he'd told Hoffman that he 'endeavoured to remedy the situation but was opposed by members of the corporation who had charge of the personnel'.

He now said to Grace: 'The matter was not in [my] jurisdiction but Mr Roeder's. Since the matter was under his supervision, [I] could do nothing.'

Well, there was nothing the girls could do now about their fatal illness, that was for sure – and there was nothing von Sochocky could do about his. For he also blew into the machines that he and Martland had devised that summer; perhaps just out of interest, or perhaps with deep suspicion, for he had not been well. Von Sochocky's breath, as it turned out, contained more radiation than anyone they had tested so far.

From the very start, Grace bore her diagnosis bravely. She had a courageous spirit and refused to let Martland's prognosis affect her life. She had always loved her life and, if anything, she now valued it even more highly. So she tucked the diagnosis away in her mind and then carried on. She didn't stop work; she didn't change her habits: she kept swimming, she kept socialising with her friends and she kept on going to the theatre. 'I don't believe in giving up,' was what she said.

As for Quinta, like her friend Grace she was said to have

borne the news with a 'brave and smiling' demeanour. Far worse than her own diagnosis, in the mind of a kind-hearted woman like Quinta, was seeing her friends suffer. 'She often worried,' remembered her sister-in-law Ethel Brelitz, 'because the others were similarly afflicted.' At least she had the steadfast Dr Knef to help her with her treatment; Quinta's teeth grew worse over the summer and she relied on Knef's attentions more and more.

Almost as soon as they'd received the news, Grace, Quinta and Katherine Schaub hoped to bring a lawsuit against USRC in order to get some help with their crippling medical bills. Knowing that Marguerite Carlough had successfully done so earlier that year, they hoped that it would be straightforward. Isidor Kalitsch, Marguerite's lawyer, was the obvious place to start their fight for justice; Quinta made the first appointment to see him. With some trepidation – for she had never done anything like this before – she limped into his office and outlined her case. He listened carefully, and then broke bad news: her action was barred by the statute of limitations.

The new girls had run into an old problem. The Workmen's Compensation Bureau – where the company wanted the existing cases heard – had a five-month statute of limitations in New Jersey; Marguerite, who had filed suit about thirteen months after she'd left USRC, was thus going through the federal court, which had a more generous two-year statute. That was well matched for Marguerite, for she'd stayed with the company long after the other girls had left, so when she first became ill she was still an employee. But Quinta hadn't worked for the firm since February 1919. She was now trying to start a lawsuit more than six years later; four years too late, according to the law, even though her symptoms hadn't started until 1923 and she hadn't received a diagnosis of radium poisoning until a few weeks ago.

But the law cared nothing for the fact that this brand-new disease took several years to manifest. The law was the law – and it said neither Quinta, nor Grace, nor Katherine had any recourse to justice; or, at least, that was the interpretation of Isidor Kalitsch. It fell to Quinta to tell the others what he'd said: 'Nothing could be done.'

It was galling news for them all. 'When I realise,' said Grace Fryer, 'that I am paying for something [that] someone else is to blame for . . .' Grace now tried another lawyer, Henry Gottfried, with whom she'd already had some dealings, but Gottfried told her that the case would take 'considerable money to develop'. He said he could do nothing for her unless she gave him cash upfront. '[But] I had no money!' remembered Grace in frustration. '[For] I was compelled constantly to attend doctors. I felt very badly [yet] lawyers did not seem to be interested in the matter without a fee.'

Part of the reluctance of attorneys to take the case was undoubtedly due to the power of the United States Radium Corporation. For not only were the legal issues potentially insurmountable, but the girls' opponent in court would be a hugely wealthy, well-connected company, with government contacts and the financial resources to eke out the fight for as long as it took. Said Katherine Schaub: 'Each of the attorneys to whom I appealed felt that it was hopeless to try to collect damages from the radium company.'

Another problem, too, was just how new the disease was; given the longevity of the radium-therapeutics industry, could it really be true that radium had hurt the girls? Perhaps, as Roeder had said, the girls were trying to 'palm off something' on the firm.

Now the effects of the company's suppression of the Drinker report really made themselves felt. Thanks to that concealment, published studies on the link between radium and the women's illnesses had been available for only a matter of weeks. None of the lawyers had ever heard of radium poisoning. No one knew anything about it – no one, that is, except for Harrison Martland.

Martland was in direct contact with the girls over that summer, offering what assistance he could, and one day Katherine Schaub came to his lab to discuss something very important. She had always wanted to write – well, now she and Martland wrote something together, though its topic was macabre. In time, it would come to have its own name.

The List of the Doomed.

Martland wrote it out on the back of a blank autopsy report. He sketched out a series of pencilled lines to create a neat chart and then picked up his fountain pen and wrote in flowing black ink, on Katherine's direction:

1. Helen Quinlan
2. Miss Molly Magia [sic]
3. Miss Irene Rudolph
4. Mrs Hazel Kuser
5. Mrs Maillefer
6. Miss Marguerite Carlough ...

The list went on and on. Slowly, methodically, Katherine supplied him with as many names as she could recall: those girls she knew were ill or had died, as well as those who weren't yet sick. She recalled some fifty former co-workers, whose names she gave to Martland.

In the years to come, the doctor was said to retrieve the list from his files whenever he heard of the death of a dial-painter. With chilling prescience, he would find her name on the list, written there back in the summer of 1925, and meticulously write a neat red D beside the woman's name.

D is for Death.

Katherine was at that time in fair health. But as her formal diagnosis sank in, she found that she could not stop thinking about that prediction. *D is for Death.* She had already been made nervous by Irene's passing; now, every ache became a symptom that could lead to her own sudden death. 'I know I am going to die,' she said. She stressed it, trying it on for size: 'Die. DIE. It doesn't seem right.' When she looked in the mirror these days, it wasn't the same Katherine who stared back at her anymore. 'Her face, once usually pretty,' a newspaper wrote of her at this time, 'is now pinched and drawn with suffering. The suspense and worry have undermined her spirit.'

That was the thing. The *worry*. It put her 'in a very precarious mental condition'. Her former company, keeping tabs on her, put it more harshly – they called her 'mentally deranged'.

'When you're sick and can't get around much,' Katherine herself said, 'things are different. Your friends aren't the same to you. They're nice to you and all that, but you're not *one* of them. I get so discouraged sometimes that I wish . . . well, I don't wish pleasant things.'

She became 'very badly ill' and consulted a nerve specialist countless times. But Dr Beling couldn't stop her spiral of thoughts, nor halt that flickering cine-reel of ghost girls still playing in her head. Katherine had always been lively and sociable before, but now, her sister said, 'She is not the same girl at all. She has completely changed in temperament.'

Katherine's periods stopped; she couldn't eat; her features themselves seemed almost to change, with her eyes becoming larger and more bug-like, as though sticking out on stalks. That was what happened when you stared your own death in the face. She murmured: 'Night and rainy days are the worst times.'

Before the year was out, Katherine Schaub would be confined in a hospital for nervous disorders. It was little wonder, given the trauma she could see her friends enduring; the surprise was that more dial-painters weren't similarly afflicted.

These days visitors to Marguerite Carlough in St Mary's Hospital found her much the same. Her blood was almost white and her blood count only 20 per cent. But it was her head, her face . . . her X-rays now showed that the radium had eaten away her lower jaw 'to a mere stump'. Just as he'd experienced with Mollie Maggia, Knef found himself helpless to stop the rot.

Another patient at St Mary's that August of 1925 was Albina Maggia Larice – but for much happier reasons. Her stomach bloomed with pregnancy; her cheeks flushed with pride. For nearly four years she and James had been trying for a child. Each month that had passed without the good news she yearned for had left a bitter taste in her mouth, as her body betrayed her time and again. Next month, she would tell herself ... but then next month had always brought the same acidic disappointment.

Not anymore. At long last, Albina thought contentedly, rubbing a loving hand across her swollen belly, she was becoming a mother – she would cradle her child in her arms, tuck it into bed at night, keep it safe from harm ...

When the pains had started, she'd made her way to St Mary's. Albina was clutching her stomach, trying to stop from crying out. It was strange but, somehow, even though she didn't know how this was *supposed* to feel – somehow, in some way, something felt wrong. It just felt wrong.

The doctors put her in a room, laid her down on a bed. She pushed and pushed when they told her to. She felt the baby move through her, felt it as her baby came. Her son. She felt him, but Albina never heard him cry.

Her baby was born dead.

Albina Larice didn't suffer from the same aching pains as her sister Quinta: the arthritic hips, the loose teeth. She'd once had a rheumatic knee, shortly before she'd married James, but, she said, 'I got rid of that; it never bothered me anymore.' Yet just two weeks after her baby was stillborn at St Mary's, as though her body was breaking to match her heart, severe pain appeared in her limbs and her left leg began to shorten. In October 1925, as her family doctor's treatment had brought no relief, Albina consulted Dr Humphries at the Orthopaedic Hospital. It was there, as she overheard the doctors talking about her, that she heard one of them remark that she was a radium case.

It was shock after shock, trouble after trouble. 'I am,' Albina later said, 'so unhappy.'

As the doctors had done with Quinta, they encased her in plaster for four months, hoping it would help her improve. But Albina didn't feel any benefit. 'I know,' she murmured dejectedly, 'I'm getting weaker, weaker, weaker . . .'

Along the corridor from her in the hospital was another former dial-painter. Edna Hussman, the Dresden Doll as was, had been seeing doctors since September 1925, apparently for rheumatism; when their treatments didn't help, she'd sought out Humphries.

Her trouble had begun back in July. 'I first started,' she later said, 'with these here pains in my hip. When I would be walking, I would get these sharp pains, and I would stumble. [It happened] nearly every time I would walk. I just limped along, held on to different things around the house to get around; that is the only way I could get around.'

Humphries, who noted that Edna's left leg was an inch shorter than her right, took an X-ray. Edna had walked to the hospital with the help of her husband, Louis, so he didn't think she could be too badly hurt. But when he assessed the picture, he had to think again: her leg was broken. She'd broken her leg when she stumbled yet, since the stumble was a slight stumble and not a fall, she had not realised she was so badly injured.

Humphries remembered of Edna's case: 'She had a spontaneous fracture of the neck of the femur [thigh bone] – and that doesn't occur in young people as a rule. I have never seen a young woman with [such] a fracture occurring spontaneously.'

Never – until now.

'By that time,' Humphries went on, 'we knew she was working in the radium plant and we were beginning to get wise to something unusual happening to these cases. [But] her X-rays did not show any white shadow or anything but a fracture.'

It was not radium poisoning. It backed up what Dr Flinn had told Edna when he'd examined her. While she might not be able to walk anymore, Flinn had assured her just a short time ago that she was in perfect health. So she must be OK.

Due to her X-ray, Humphries simply treated her for a broken leg. 'They put me in a plaster cast,' Edna remembered, 'and I was laid up for a whole year in this here plaster cast.' Louis took her back to their little bungalow and their small white dog, and life went on.

Flinn, too, continued on with his work. He had come across

a treasure trove of information, supplied to him unwittingly by Katherine Wiley. 'I went to see Dr Flinn,' Wiley later recalled, 'and found him most interested. He said that he would be glad to have the names and addresses of all the sick girls that I knew.'

Wiley wasn't aware that Flinn was working for USRC, for he did not disclose that information. Nor did she know that the firm had 'asked Dr Flinn to see these girls and to give medical advice'.

And so, with Flinn now having her home address, on 7 December 1925 Katherine Schaub received a letter.

'My dear Miss Schaub,' Dr Flinn wrote, on the headed paper of the College of Physicians and Surgeons, 'I wonder if you would be kind enough to either come to my office or if you prefer to my house in South Orange and let me give you an unbias [sic] opinion . . .'

But Katherine Schaub, in 'a terrible nervous state', was in no condition to see Dr Flinn. 'I was ill when I received the letter,' she remembered, 'I was in bed and I could not go out.'

She wrote back to explain her predicament, as Flinn recalled: 'I never replied to [her] letter,' he said, 'as I told my technician that if she wasn't willing to come either to my home or my office I certainly wasn't going to put myself out; a girl of that class didn't appreciate it when you did try to aid her.'

Flinn wasn't bothered about being unable to examine Katherine, for he had plenty of other avenues to explore; he later boastfully said, 'I [have] examined practically every girl now working in this industry.' With his contacts at USRC, Luminite and the Waterbury Clock Company – to name a few of the firms who had engaged him – he now had unprecedented access to the dial-painters. However, despite his boast, he doesn't seem to have examined many ex-employees.

If he had, he might have discovered that a second Waterbury girl, Elizabeth Dunn, had recently been taken ill. She'd left her dial-painting job earlier in 1925 (whether before or after Flinn started his study is not clear) after a simple slip on a dance floor had left her with a broken leg; what might be called a spontaneous fracture. Had Flinn discovered her case – or the death of her former colleague, Frances Splettstocher – it would have been crucial evidence that the dial-painters' illnesses went beyond the Orange plant and *were* caused by their occupation.

Flinn was also busy discrediting the work of Dr Martland. In December 1925, Martland, another doctor called Conlon and the girls' dentist Dr Knef published a joint medical study based on their work with the women that year. Their conclusion was that this was 'a hitherto unrecognised form of occupational poisoning'. The article became, in time, a classic example of a medical mystery solved.

In 1925, however, with this being such a pioneering statement, no such respect was shown. Martland's conclusions were so radical that they were disputed fiercely; and not just by Flinn. A radium-medicine specialist, Dr James Ewing, commented drily at a meeting of the New York Pathological Society: 'We are a long way from speaking of the ill effects of radium therapy.'

He might have been – but Martland certainly wasn't. In fact, Martland singled out the medicinal use of injected or ingested radium as dangerous and stated that 'none of the known radioactive substances produce any curative results'.

That was a red rag to the bullish radium men. This wasn't just about some dial-painters dying; Martland was now attacking a hugely lucrative industry. 'The original [study was] ridiculed by most authorities on radium,' Martland remembered later. 'I have been under constant attack for my efforts to protect the public and to secure some compensation for disabled, death-facing girls. Manufacturers of radium have been particularly active and insulting in their efforts to discredit me.'

It was with good reason, as far as the radium companies were concerned. A letter from the Radium Ore Revigator Company told the doctor that his article had 'automatically reduced our sales [to] less than half our previous quarter'.

Yet it wasn't just those with a financial stake in the benefi-
cence of radium who had their doubts. Even the American
Medical Association – which in 1914 had formally included
radium on its list of 'New and Nonofficial Remedies' – was sceptical. It all made the girls' claims look increasingly suspicious to
the lawyers they contacted for help.

This public reception to Martland's work could not have
pleased USRC more. Soon they would be fighting back with
medical studies of their own; Vice President Barker wrote in a
memo with almost undisguised glee, 'Our friend Martland [is]
still maintaining that we are killing [the dial-painters] off by the
[dozen]; [his] article is some of their propaganda. [Yet] I understand Flinn's report is to be published soon. *His* findings have
been entirely negative and I think his report represents a very
good piece of work.' He added, 'I am rather inclined to think
that he will be given a certain amount of money to continue his
work.'

Flinn was just the ticket as far as the company was concerned.
USRC, no doubt, would have been aghast to learn what Flinn
had written to their former investigator, Dr Drinker. 'Though I
am not saying it out loud,' Flinn wrote, 'I cannot but feel that the
paint is to blame for the girls' conditions.'

But while the scientists publicly fought over the cause of the
girls' disease, there was one woman who was in its clutches, still
fighting for all she was worth. Marguerite Carlough had been
'half dead' for weeks. In Hoffman's opinion, hers was 'the most
tragic [case] on record'. With her immune system dangerously
weakened, she contracted pneumonia on top of everything else.
But she managed to make it home for Christmas, to spend it
with her niece and mom and dad. It was two years since that
Christmas Eve when she'd had her tooth pulled, and all this
trouble had begun. It was six months since her sister Sarah had
passed away.

In the early hours of Boxing Day 1925, at the age of twenty-
four, Marguerite followed her sister to that undiscovered

country. She died at home on Main Street at 3 a.m. Her bones, Martland later said, showed 'beautiful concentrations' on the X-ray films that he wrapped around her in death.

Two days later, for the second time in six months, her parents laid to rest a daughter in the peaceful quiet of Laurel Grove Cemetery. Yet Marguerite had not died quietly: as the first dial-painter to file suit – the first to show it was even possible to fight back against the corporation that killed her – she went out with a roar.

It was a sound that would echo long afterwards: long after she died; long after she was buried; long after her parents made their slow way home from her funeral, and closed their door against the world.

All she wanted, Grace Fryer thought, as she flicked through the local paper, was a bit of good news. There had been only one piece of it so far in 1926. Miss Wiley's new law, to her and the dial-painters' delight, had been signed into being: radium necrosis was now formally a compensable disease. In many ways, it had been a lot easier to get it through than Wiley had anticipated.

Other than that, however, it had not been a great spring. Grace's jaw problems had started back up again – she had now lost all but three teeth in her lower jaw and had to see Dr McCaffrey three times a week – while her back was incredibly painful. She hadn't had it looked at by a doctor in a while though; it was too expensive. Nonetheless, despite her troubles, Grace still commuted daily to her office. She commented simply: 'I feel better when I am working.' Indeed, she was said to meet people cheerfully in the bank.

Yet there was another reason to keep on with her job. Quinta remarked that Grace was working 'so she won't be a burden in her family'. Grace had incurred medical bills of some $2,000 ($26,800) and her parents certainly couldn't cover them. Yet even if Grace put *all* her earnings, which were about $20 ($268) a week, towards her medical treatment, it would take her two years to pay what she owed. She had no idea where she might find the money ... well, no idea but one. By now she had spent almost a year pursuing different attorneys; pretty much on her own, too. Faced with lawyer after lawyer turning down the case, the other girls seemed to have given up.

Albina was not at all well; she saw only close friends and was unable to leave the house due to her locked hips. James Larice did his best to put a smile on his wife's face – 'He cheers me up,' Albina said, 'and says I'm a "good sport"' – but it didn't help. 'I'm such a burden,' she cried despondently. Although her sister Quinta carried on resignedly, her disability was progressing too: the 'white shadow' now showed in both her legs, while Knef could do nothing to save her teeth.

As for Katherine Schaub, nobody even saw her anymore: she stayed at home and refused to go out. 'While other girls are going to dances and the theatres and courting and marrying for love,' Katherine said mournfully, 'I have to remain here and watch painful death approach. I am so lonely.' She left the house only to attend church. While Katherine had not been especially religious before, she now pronounced, 'You don't know what a consolation I obtain from going to mass.' As she was now unable to work, her medical bills fell to her family. Her father, William, who was in his mid-sixties, did his utmost to help, but Katherine's sister confided, 'It's pretty hard on Dad. He can't work like he used to.'

As time went on, despite these crippling bills, the girls started to doubt that a lawsuit was the right way to go. Perhaps it wasn't fair to blame the company? For Katherine had eventually consulted Dr Flinn – and his 'unbias opinion' was that 'radium could not and had not harmed her'. Katherine, naturally, told the other women this; and that got them confused. As Albina put it, '[We] all thought it significant that of the several doctors who treated [us], only one doctor, Dr Martland, had informed [us] that [our] illnesses were due to radioactive substances.' With the women in bad health and a question now raised over the company's culpability, a lawsuit was the last thing on their minds.

The last thing on the other girls' minds, maybe – but it remained a high priority for Grace Fryer. Still reading the local newspaper, she turned the pages slowly, deep in thought. And then, to her astonishment, she noticed a small piece buried within the paper. Scarcely able to believe her eyes, she read: SUITS ARE SETTLED IN RADIUM DEATHS.

What? She quickly read on – and found the headline did not lie. USRC had settled out of court the suits of Marguerite Carlough, Sarah Maillefer and Hazel Kuser. The women had beaten the corporation – got money for what the firm had done to them. Grace could barely believe it. Surely that was an admission of guilt? Surely this opened the door for her and her friends to bring a lawsuit? She read on in excitement: 'Mr Carlough [the girls' father] received $9,000 [$120,679] for the death of Marguerite Carlough and $3,000 [$40,226] for the death of Mrs Maillefer, and Mr Kuser received $1,000 [$13,408] for the death of his [wife].'

It was hardly big money. Theo's settlement, in particular, barely dented the $8,904 (almost $120,000) debts that he and his father had incurred for Hazel's care – especially once Mr Kalitsch took his 45 per cent cut. This was a higher-than-usual split with the attorney, but the families did not have much choice but to agree, for he had been the only lawyer to take the case. In the end, Theo was left with just $550 ($7,300), but it was better than nothing.

Grace wondered what on earth had happened to make the company pay out, when it had fought the women for almost eighteen months with no sign that they would concede a single cent. In fact, behind the scenes at USRC, there were probably several reasons – not least of which was that the women, especially the Carlough sisters, had strong cases; the company might well have lost before a sympathetic jury. Even looked at from a basic legal perspective, the cases were promising: the girls had filed within the two-year statute; there was Katherine Wiley's new law supporting the girls' claims that they had been killed by such a thing as radium necrosis; and there was also the issue of the Drinker report. Sarah had still been employed by USRC at the time the firm had chosen to suppress it; if it came out that the company had received information that could have saved her – or at the very least mitigated the harm caused – and hadn't acted upon it, it would look very bad indeed.

Grace was galvanised into action: *this* was the good news she'd been waiting for. She got back in touch with lawyer Henry Gottfried and just two days after she'd read about the settlements the wheels of her own claim were set in motion. On 6 May 1926, USRC received the following message from Gottfried: 'Gentlemen, unless you communicate with me relative to [Miss Fryer's] claim for damages on or before Monday, May 10th, 1926, I will be compelled to institute suit.'

USRC, like clockwork, immediately referred the matter to their attorney Stryker, who seemingly asked Gottfried to name a number. On 8 June, Gottfried wrote to say that Grace would be willing to settle for $5,000 ($67,000).

It wasn't a gigantic sum of money; it would cover the extensive medical bills that Grace had already run up, plus provide a nest egg to pay the future expenses that would undoubtedly be required. Grace wasn't a greedy person, and she didn't really want to start a big lawsuit. If the company would simply make her a fair offer, she was prepared to take that compensation and be done.

USRC took just one week to respond. 'I have your letter of the 8th,' Stryker replied on 15 June, 'and note your suggestion. I cannot advise my client to adopt it.' The company 'refused to do anything for Miss Fryer without suit'.

Grace's heart must have sunk when she heard. She must have been confused, too. For not only had the company agreed to pay out damages to her former colleagues just the month before, but

Miss Wiley had put that new law in place. Didn't that change anything?

But, it now transpired, it did not. And here it became clear just why Wiley had had such an easy ride getting the radium-necrosis bill passed. For a start, the bill could not be applied retroactively, so no one injured before 1926 could claim. Also, as the new amendment became part of the existing law, it automatically had a five-month statute of limitations attached, a length of time that would never be long enough for radium poisoning to become evident in any dial-painter. Finally, and most crucially of all, it covered *only* radium necrosis – specifically jaw necrosis, of the aggressive kind suffered by Mollie Maggia and Marguerite Carlough. None of the other medical conditions arising from the women's poisoning – their life-sapping anaemia, bad backs, locked hips, broken thighs, even simply their loosening teeth – was compensable. Wiley had found that the new bill was 'not unpopular' with the state Manufacturers' Association; now, suddenly, it became obvious why. The law, as written, was designed so that no one would ever collect compensation.

Wiley soon realised her mistake. With renewed fervour, the Consumers League began campaigning to get radium *poisoning* into the law books. Tellingly, however, this fight would take them much, much longer before any changes were made – far too long, and far too late, to help Grace Fryer, sitting despondently at home in Orange in June 1926.

There may have been another reason the company wasn't inclined to settle: there is some evidence to suggest that the firm was not doing quite as well as it once had financially; one executive referred to the situation as being 'hand to mouth'. Part of the problem was finding staff; its remaining employees were 'jumpy and nervous' and new workers scarce. Before the year was out, USRC would cut its losses and shut down the Orange plant, putting the site up for sale. Even so, it wasn't down-and-out completely. The firm simply transferred operations to New York.

Grace Fryer wasn't down-and-out either, although USRC's

response led to a double blow as, learning of the firm's refusal to settle, Gottfried dropped her case. Yet she felt more determined than ever to battle on; she was her father's daughter and this child of a union delegate would not back down so easily from a fight against a guilty firm. 'I feel we girls should not give up all hope,' she said.

She went on to consult at least two other lawyers – but, frustratingly, without success. Part of her problem was that her former company, just as it had planned, was now beginning to benefit from expert publications that said radium poisoning was *not* to blame for the girls' illnesses. The most high profile of these, which would be published in December 1926, was authored by one Dr Flinn.

'An industrial hazard does not exist in the painting of luminous dials,' he wrote plainly. He said the girls' problems were due to a bacterial infection. Hoffman dubbed the report 'more bias than science'.

Yet it went way beyond bias: Flinn was lying through his teeth. For not only did his published conclusions contradict his stated opinion to Dr Drinker – 'I cannot but feel that the paint is to blame for the girls' conditions' – but in June 1926, six months *before* his study was published, Flinn finally discovered two radium-poisoning cases at the Waterbury Clock Company. These proved, once and for all, that it wasn't a bacterial infection that had been passed around a single studio: it was the women's profession that had killed them.

Despite knowing of the cases for so long, Flinn didn't correct or withdraw his report and allowed it to be printed, giving USRC published expert evidence to draw on in their continued denial of responsibility. Later, Flinn did say that he regretted that decision. But from his future behaviour, one can infer, not *too* much …

Flinn wasn't finished with the Orange girls, despite his proclamation that their illnesses were only an infection. In July 1926, the month after USRC turned down Grace's attempt at a settlement, he finagled his way into examining Grace herself;

the timing was probably a coincidence. Flinn – accompanied by another man who Grace did not know – took her blood and an X-ray. And when the results were in, Flinn pronounced with a smile, 'Your blood picture [is] better than mine!'

'He told me,' Grace later remembered, 'I was in better health than he was and there was nothing wrong with me.'

But that was not what Grace's body told her.

The girls were all in dire straits that summer, notwithstanding Flinn's assertions of good health to Grace, Katherine and Edna Hussman. Quinta McDonald was still attending Dr Knef for her loosening teeth; and it was Knef, now, who chose to make an appointment with the radium company. One morning during that summer of 1926, he met with the board of directors, including President Roeder and an up-and-coming vice president called Clarence B. Lee, at the USRC headquarters in New York. Knef was at his wits' end from trying to treat the women, and he now made the radium firm an offer he hoped they couldn't refuse.

'If you will play ball with me,' Knef told the gathered executives, 'I will play ball with you people. Get [me] a list, the names of the girls, [and] I will keep my mouth shut as long as I can. Quite a few cases will just die a natural death. I can hold these girls off for four or five years ... There are my cards on the table. I have to be compensated somewhere.'

Knef wasn't at his wits' end in sympathy for his patients, mind you, but because he wanted to be paid. Perhaps it was the Carlough settlements that had set him off; he was itching to be remunerated for all the free treatment he had given: 'Took it all out of my own pocket!' he now exclaimed in vexation. His appeal to the company for payment might have been fair enough – after all, their paint had caused the sicknesses for which he had treated the women – but this scheme he was now presenting, of lying to the girls, of letting them die in ignorance to protect the firm, went way beyond being paid what he was owed. All loyalty to the girls had vanished.

'What is your proposition?' asked the executives, seemingly intrigued.

'I told Mr Roeder I ask $10,000 [$134,000]. I don't think I ask one cent too much.'

USRC gave due consideration to his offer. 'Are you sure that all these girls will come to you?'

'I believe now that the majority will come to me,' Knef replied, confident that the women saw him as a friend.

'Would you tell them that [your services were] being paid for by the company?'

'I won't tell them that I have any connection with you people,' Knef said with a smile.

Perhaps encouraged by the positive way the meeting was unfolding, Knef now made another suggestion. 'If you people want me,' he said, leaning forward on the boardroom desk to stress his point, 'I can get up on the stand and testify ... "Do you believe that this girl is troubled with radioactivity?" I'd have to say no. I can [say] whatever I want to believe; [the] moon is made of blue cheese!'

'You can make it go one way or the other, can you?' asked the executives.

'I could if I wanted to; that is, if I am working for you people. It [is] customary for experts to testify for the people who pa[y] them.'

The money was all-important to Knef. And here, perhaps, he made a fatal error. This dentist from Newark, who had worked his whole life only in dentistry, now tried to play it tough with the big boys of big business. 'I am going to get this one way or the other,' he said threateningly. 'Do you want me as a friend or do you want me as an enemy? If I can't come to an agreement with you people, I am going to sue these people [the girls; then] they will have to sue [you] to get the money. Fair warning: when I fight, I fight ferocious as a lion. I am a *very* valuable man to be with you.'

That had gone well, he must have thought. He must have

smiled with his next line, confident that he had them on a hook. 'I am going to be as reasonable with you as I can. I am not here to gouge you or bleed you or anything else.'

The executives summed up the position: 'Unless we pay you $10,000, you are in a position to make a lot of trouble for us. If we do pay it, you will help us.'

'I can help you, yes,' said the dentist eagerly.

Another director spoke up. 'The future by itself [i.e. Knef getting paid for treating the girls in future, and holding them at bay from filing suit] is not sufficient? You must get the $10,000?'

'As I tell you,' said Knef cockily, 'I must have my compensation.'

He had blown it. He may not have known, either, that the corporation already had Dr Flinn in place, doing such fine work for them. Roeder stood up swiftly, ready to dismiss him. 'Your proposition is immoral,' he declared. 'We will have nothing whatever to do with it.'

'Immoral, is it?' echoed Knef. 'Is that final?'

It seemed it was.

When evidence of his proposition came out, down the line, USRC would take the moral high ground for the fact that they had sent the dentist packing.

The meeting had lasted for fifty-five minutes exactly.

23

Ottawa, Illinois
1926

The bells of St Columba pealed out joyously across Ottawa. A wedding seemed to happen every other week these days as the dial-painters got married; many were bridesmaids for each other. Frances Glacinski married John O'Connell, a labourer; Mary Duffy wed a carpenter called Francis Robinson. Marie Becker got engaged to Patrick Rossiter; Mary Vicini courted Joseph Tonielli; and Peg Looney and Chuck Hackensmith eventually made plans to marry in June 1930. Charlotte Nevins – who hadn't worked for Radium Dial since 1923 – was also one of those falling head over heels; she was still in touch with lots of the girls and told them excitedly of the charms of Albert Purcell. They'd met at the Aragon Ballroom in Chicago dancing; Charlotte knew exactly how to swivel as she showed off her Charleston and in so doing she caught the eye of Al, a labourer from Canada. 'They were best friends,' revealed a close relative; within two short years, Charlotte Nevins became the latest bride walking down the aisle of St Columba.

The church where many of these weddings took place was a white-stone building with a grey slate roof and a beautiful altar that was the envy of the region – it was imitation marble and filled the whole space. St Columba was fairly narrow, but its arched ceiling was so much higher than the building was wide that the effect was breathtaking. One of its few parishioners not caught up in this maelstrom of marriage was Catherine Wolfe. A young man at church *had* caught her eye, however: his name was Thomas Donohue.

He was thirty-one to Catherine's twenty-three. Tom was a diminutive man with bushy eyebrows and a thatch of dark hair; he had a moustache and wire-frame glasses. He did a variety of jobs, including engineer and painter, a rather apt parallel to Catherine, as the dial-painters were listed as 'artists' in the town directory, in a nod to the glamour of their work. Later in life, Tom would labour in a local glass factory, Libbey-Owens, where he worked alongside Al Purcell and Patrick Rossiter.

He was a 'real quiet man who never talked much'. That may have been due to his upbringing, for Tom came from a large Irish-immigrant family. As a relative said: 'He was the sixth of seven kids; he was never gonna get much word in.' The whole family grew up on the Donohue farm, which was based in Wallace township, just north of Ottawa; one of those places where you could see forever across the fertile fields and the sky seemed to swallow you up. Like Catherine, who said a rosary every day with her own set of beads, Tom was highly devout; so much so that he'd attended an all-male Catholic school with the idea that he might join the priesthood, but that didn't come to be.

Tom worshipped at St Columba, just as Catherine did; his grandfather had paid for one of the stained-glass windows when the church was built. The Donohues didn't come into town that often, though: 'People didn't travel as much in them days as they do now,' remarked Tom's nephew James. 'If you went to town more than once a week, you were a big shot.'

Tom Donohue was definitely not a big shot; he was 'not an extrovert in any way'. He was, in fact, just like Catherine. 'They were both very quiet people,' said their niece Mary. 'Very shy people.'

Perhaps that was partly why they wouldn't marry until 1932.

It is possible that Catherine told her deskmate at Radium Dial, Inez Corcoran, about Tom. Inez had her own story to share, for she was engaged to Vincent Lloyd Vallat, the proprietor of a gas-filling station; they were due to marry later that year.

Not all the married dial-painters quit their jobs. It seems the studio didn't want to lose its highly skilled workers, so the company became a pioneer in offering part-time terms to working moms. 'I quit ten, twelve times,' one girl recalled. 'They always took me back; it took too long to train new ones.' The company needed to retain its best girls because business was still booming – and *how*. Westclox hit a new production high of 1.5 million luminous watches in 1926 and Radium Dial painted all of them.

The new husbands noticed something strange in their house-holds when their wives came home from work. One later wrote, 'I remember when we were married and she hung her smock in the bedroom: it would shine like the northern lights. The first time I saw it, it gave me an eerie feeling – like a ghost was bouncing around on the wall.'

Like someone else was in the room with them, just watching and waiting for the right time to strike.

There was no indication that the good times were ever going to end. No Ottawa women had fallen ill; one worker had a 'face [that] broke out in blotches' while another said, 'I quit because I was sick to my stomach,' but these complaints were nothing to do with the work. Another woman, who had left the studio at the end of 1925, did have 'terrible, excruciating pain in my hip socket', but it went away. 'Though we had several doctors,' she recalled, 'it never was diagnosed.' She never went back to Radium Dial but, she said, 'I had friends who worked there for many years with no bad effects.'

Notwithstanding the lack of bad effects, however, the com-pany executives had not forgotten the downturn in business suffered by their competitor in New Jersey – and they doubtless noted with concern the settlements USRC had been forced to pay out. Mr Reed's invention of his glass pen had duly been supplied to the workforce, albeit with no explanation as to why.

The girls knew nothing of what was happening out east. The news piece about Marguerite Carlough's case was hidden inside a local paper 800 miles away. Dr Martland's radical study from the year before was hotly debated ... but only in the specialist medical press. Although his findings were reported by the general media of New York and New Jersey, such findings barely caused a ripple in the Great Lakes of the Midwest. Girls who lived in Ottawa didn't read the *New York Times*.

To be fair to Radium Dial, therefore, they didn't *have* to implement the change. The girls were not laying down their brushes in protest, and it wasn't as though there was any outside pressure to alter the working practices. For despite the national study by Swen Kjaer, with his conclusion that radium dial-painting was hazardous, and despite the medical studies now being published that said the same, no organisation had intervened on the national stage to prevent workers beyond Orange being harmed.

But although Radium Dial introduced glass pens, which were intended to put a stop to lip-pointing, it doesn't appear that they were fit for use. Perhaps the firm rushed them out in a panic. From the girls' perspective, they were not a runaway success; Catherine Wolfe thought them 'awkward' and 'clumsy to handle'. And because the brushes weren't removed when the pens were launched, the dial-painters continued to lip-point to clean up the run-overs; run-overs that were now plentiful due to the clumsiness of the new instruments.

The girls conceded, 'We were watched very closely at first to see that we didn't try to go back to the brush,' but it was a watchfulness that didn't last long: 'Supervisor wasn't too observant,' another girl later said.

Using brushes instead of pens to paint the dials was supposed to be a dismissible offence, but it wasn't a rule that was

upheld. One girl recalled that she and six or seven others got a
little behind in their work because of the inefficiency of the glass
instruments, so one day decided to catch up using brushes.
They were fired by Mr Reed, who saw them do it, but this girl
'went back right away, apologised; she was reinstated, as were
the other girls at a later date'.

Gradually, after just a few months, the glass pens fell out
of use. Catherine Wolfe remarked: 'We had our choice to use
the glass pencil or Jap art brushes, whichever we found the
most efficient to use.' Well, if *that* was the criteria, there was
no contest. Some commentators later criticised the women for
returning to the brush: 'Those who were greedy,' one wrote,
'and profited from it would do it the fastest way and the fastest
way to make perfect numbers was to use the mouth.' But the
girls were paid by piecework, not salary, so the impact of using
the pens made a huge difference financially.

Of course, they were not the only ones to profit from the
choice they made: Radium Dial benefited too. And although
Mr Reed had been tasked with inventing the pen, once it
became clear it didn't work, the company relaxed its rules and
allowed the girls to go back to lip-pointing without further
intervention. After all, with 1926 bringing that new Westclox
production high, it was hardly the ideal time for the firm to
insist on a new method of production, particularly one that
was so ineffectual.

'The company left it up to us whether we used glass brushes
or not,' Catherine Wolfe remembered. 'I preferred the hair
brushes, as the others were awkward. I didn't think there was
any danger in placing the brush in my mouth.'

So she and Inez and Ella Cruse – another of the original
girls – still lipped and dipped all day throughout that year of
1926. Catherine pointed the brush on every single number she
did.

Towards the end of the year, she laid down her brush to say
a special goodbye to her friend Inez, who worked beside her;

it was her colleague's last day on the job before a very special event. On Wednesday 20 October 1926, Inez Corcoran married Vincent Lloyd Vallat. Together, the happy couple stood at the altar and made their solemn vows – vows that would see them through their future: through every dream, every day, every delightful thing to come.

Their voices echoed lightly round the cool church walls.

'Until death do us part ...'

24

Orange, New Jersey
1927

G race Fryer limped into Dr Humphries's office, trying not to cry out with pain. Humphries was shocked at the change in Grace; he hadn't treated her for some time. She had been sent to him by Dr Martland, who said she was 'in rather a serious condition on account of her spine'.

Dr Martland and Dr Hoffman had both tried to help her, Grace mused, as Humphries ushered her straight to radiology for a new X-ray. Hoffman in particular, she thought, had been very kind. Seeing a sharp deterioration in her health, he had written to President Roeder on her behalf, to appeal to him to help Grace 'in a spirit of fairness and justice'.

Hoffman was surprised by USRC's response: 'Mr Roeder is no longer connected with this corporation.'

It seems the firm hadn't appreciated being put in the position of having to settle lawsuits. Roeder's fingerprints were all over the company's questionable handling of the Drinker report, and perhaps it was felt it was best all round if he moved on to pastures new. He had resigned in July 1926. While no longer the public face of the company, he remained a director on the board.

Despite the change at the top, the company's attitude towards its stricken former employees hadn't changed one bit. The incoming president, Clarence B. Lee, immediately declined Hoffman's appeal for assistance. When Hoffman wrote to Grace to let her know, he added: 'You must take legal action at once.'

Well, Grace thought, she was *trying*. Despite her poor health, she had not stopped looking for a lawyer and was even now waiting to hear back from a firm the bank had passed her on to. In the meantime, she had come to see Humphries to find out what was wrong with her back.

It is difficult to imagine how she might have reacted to the news he had to share. 'X-rays taken at that time,' Humphries later said, 'showed a crushing of the vertebrae.'

Grace's very spine had been shattered by the radium. In her foot, meanwhile, there was 'destruction of the entire' affected bone through 'crushing and thinning'. It must have been agonising to endure.

'Radium eats the bone,' Grace later said, 'as steadily and surely as fire burns wood.'

Humphries could do nothing but try to find ways to make life more comfortable for her; ways to help her live her life. And so, on 29 January 1927, he fitted Grace Fryer, then twenty-seven, with a solid steel back brace. It extended from her shoulders to her waist and was held in place by two crossbars of steel; she had to wear it every single day and was permitted to take it off for only two minutes at a time. It was a demanding schedule of treatment, but she had no choice but to follow her doctor's orders. She later confided, 'I can hardly stand up without it.' She wore a brace on her foot, too, and some days she felt the braces were the only things keeping her together, helping her to carry on.

She needed them more than ever when, on 24 March, she finally heard back from the latest set of lawyers: 'We regret to say that in our opinion the Statute of Limitations barred your right of action against [USRC] two years after you left the[ir] employ.'

It was another dead end.

Grace had just one final card to play. '[Dr Martland] agrees with me,' Hoffman had written, 'that it is of the utmost importance that you should take legal steps at once. [H]e suggests that you see the firm [Potter & Berry].'

She had nothing more to lose; she had everything to gain. Grace Fryer, aged twenty-eight, with a broken back and a broken foot and a disintegrating jaw, made an appointment with the firm for Tuesday 3 May 1927. Maybe this Raymond H. Berry would be able to help where no other attorney had.

There was only one way to find out.

Grace dressed carefully for the appointment. This was make-or-break time. She'd had to change her wardrobe after the brace had been fitted. 'It's awfully hard,' she revealed, 'to get clothes that don't make it show. I can't wear the sort of dresses I used to wear at all.'

She styled her short dark hair smartly, then checked her appearance in the mirror. Grace was used to dealing daily with well-to-do clients at the bank; she knew from experience that first impressions count.

And it seems her potential new lawyers appreciated that sentiment too. Potter & Berry, despite being a small law firm, had its offices in the Military Park Building, one of Newark's earliest skyscrapers; it was then the tallest building in the whole of New Jersey and had been completed only the year before. The lawyer she was meeting inside it, Grace soon realised as he introduced himself, was just as fresh-faced as his office.

Raymond Herst Berry was a youthful lawyer, not even in his thirties. Yet his baby-faced good looks – he had blond hair and blue eyes – belied a brain as sharp as a tack. He was not long out of Harvard and had been valedictorian of his class; already, he was a junior partner in the firm. He had served his clerkship at none other than Lindabury, Depue & Faulks, USRC's legal firm, and perhaps that experience gave him some insider knowledge. Berry took a lengthy statement from Grace. And it seems she may have shared her new lead with her friends; for just three days later, Katherine Schaub also called on Berry.

He was not a man to jump into things. As any lawyer should, he first scrutinised the girls' claims. Berry went to Martland's lab, and interviewed von Sochocky; he then summoned Grace and Katherine back to his office on 7 May. He had conducted his initial investigation, he told them, and he had seen enough. And then Raymond Berry took their case. He was a married man with three young daughters – a fourth would be born the following year – and perhaps having so many girls influenced his decision. Berry was also a veteran; and this case, he could see, would be one hell of a war. In his agreed terms with Katherine, Berry contracted to take the then-standard split of one-third of any compensation. With Grace, however, she seems to have negotiated him down to just a quarter.

Berry's tack-sharp brain had been working hard on the statute-of-limitations question. His theory was this: the girls could not possibly have brought a lawsuit until they knew that the company was to blame. As the firm had actively conducted a campaign to mislead the girls, it should not be allowed to rely upon the delay, which it had caused, as a defence. After all, due to the misdirection, the girls' certain knowledge came only with Martland's formal diagnosis in July 1925. In Berry's view, therefore, the two-year clock did not start ticking until that moment.

It was now May 1927. They were just in time.

With not a moment to lose, Berry began preparations for a lawsuit. Grace's case would be the first to be filed; perhaps because she had been the first to call on him, or maybe because she was stronger than Katherine in terms of her mental health. She was also – in Hoffman's words – 'a very estimable person employed by one of [Newark's] largest business corporations'. Berry may well have known that USRC's lawyers would be looking for any chink in their armour, and Grace's good character stood them all in good stead. Thus on 18 May 1927, Grace's formal complaint was filed against the radium firm.

It made for uncomfortable reading – for USRC. Berry charged that they had 'carelessly and negligently' put Grace at risk so that her body 'became impregnated with radioactive substances' which 'continually attack and break down the plaintiff's tissues ... causing great pain and suffering'. And he concluded: 'Plaintiff demands $125,000 [$1.7 million] damages on the first count.'

There were two counts included. In total, Grace was suing her former firm for a cool $250,000 ($3.4 million).

They kind of had it coming.

From the very start, Grace's case attracted heart-rending headlines that supported her cause: HER BODY WASTING, SHE SUES EMPLOYER: WOMAN APPEARS IN COURT WITH STEEL FRAME TO HOLD HER ERECT declared the *Newark Evening News* after Grace's first appearance in court to file the papers. And such coverage – combined with the friendship networks of the girls – soon led to other dial-painters coming forward. Quinta McDonald was one, with her sister Albina beside her.

And with these married women, Berry now launched lawsuits not only for them, but also for their husbands. As Berry wrote in legal papers for Quinta's partner: 'James McDonald lost the services of his wife and will in the future be deprived of the comfort and aid of her society and will be compelled to expend large sums of money in an endeavour to treat and cure his wife. Plaintiff James McDonald demands $25,000 [$341,000].'

Adding the husbands to the lawsuits wasn't an excessive gesture – the truth was that it was increasingly impossible for Quinta to be the wife and mother she wanted to be. She admitted: 'I do what housework I can nowadays. Of course, I can't do much. I can't bend over now.' Given her extreme disability, she and James had recently been forced to hire a housekeeper; another expense.

Her sister Albina was in dire need of aid too. Her left leg was now four inches shorter than her right, leaving her crippled and bedridden. She and James had not given up on their dream of a family, but she had since suffered a miscarriage, leaving her feeling worse than ever. 'Life,' Albina said dully, 'is empty for me and my husband.'

And there was someone else who was suffering. Edna Hussman had been released from her year-long sentence in her plaster cast, but her ailments continued: her left leg shrank by three inches; her right shoulder became so stiff it was impossible for her to use her arm; and her blood tests showed she was anaemic. When her mother had died, in December 1926, her spirits darkened further.

Yet Edna had hope. Hadn't that company doctor, Dr Flinn, told her she was in perfect health? She took the prescribed drugs for her anaemia and followed her doctors' orders. And then, one night in May 1927, as she groped in the dark for her medicine on the bureau, she caught sight of herself in a mirror. At first, she might have wondered if it was her mother Minnie returned from the grave to haunt her. For in the dead of night, in the dead of dark, a ghost girl glowed in the mirror.

Edna screamed and fainted. For she knew exactly what her shining bones foretold, shimmering through her skin. She knew that glow. Only one thing on earth could make that glimmer. *Radium*.

She went back to Dr Humphries and told him what she'd seen; how much pain she was in. And there at the Orange Orthopaedic Hospital, she said, 'I heard Dr Humphries talking with another doctor. He told the doctor I was suffering from radium poisoning. That is the first I knew of it.'

Edna was a 'peaceful and resigned woman'. She later said, 'I'm religious. Perhaps that is why I'm not angry at anyone for what has occurred.' But that didn't mean she didn't feel it was unjust. She went on, '[I] feel that someone should have warned us. None of us knew that paint paste was dangerous; we were only girls: fifteen, seventeen and nineteen years old.' Maybe that innate sense of injustice was why, in June 1927, only a month after receiving her diagnosis, Edna and Louis Hussman made their way to Raymond Berry.

There were five of them now: five girls crying out for justice; five girls fighting for their cause. Grace, Katherine, Quinta, Albina and Edna. The newspapers went mad, inventing memorable monikers to define this new quintet. And so, in the summer of 1927, it became official.

The Case of the Five Women Doomed to Die had now begun.

25

The company executives, it would be fair to say, were taken completely by surprise by the five lawsuits. In fact, they dubbed it a 'conspiracy', cooked up by what they termed 'Berry's outfit'. Their previous unshakeable confidence in rejecting all pleas for clemency had been based upon the fact that the statute of limitations had made them, in their eyes, invincible – but now, with Berry's adroit interpretation of the law, they were left scrambling for their defence.

It was, perhaps, inevitable who they would blame. The plaintiffs, the firm alleged in their reply to the girls' claims, were 'guilty of contributory negligence in failing to exercise due care and precaution for [their] safety'. And the corporation went further: it denied that the girls were ever instructed to lip-point; it denied any woman in the studio did; it denied the radium powder clung to them. On and on, denial after denial, for pages and pages of legalese. The company admitted only one thing: 'it gave no warning'. That was because it 'denie[d] that radium was dangerous'.

Such denials (such outright lies, one might say ...) comprised its written fightback through the law, yet that was only the start of the firm's attempt to take control of the situation. Now faced with the women squaring up to it in court, it fought back behind their backs. Miss Rooney, the former forelady of USRC, who was a good friend of Edna Hussman, was surprised when one of her erstwhile bosses suddenly appeared at Luminite and asked to speak to her. At first, she chatted happily with the visiting executive about the girls who had worked under her management, sharing intimate details of how they were getting on. The executive, thrilled, noted she gave him 'considerable information'.

It seems both the firm and the former forelady totally underestimated the strength of will of Grace Fryer. 'Miss Rooney says that she feels certain that Grace Fryer was urged into the suit by the lawyers,' a company memo said. They little knew that, without Grace's two-year fight to find an attorney, none of this would be happening.

The company, in fact, started to suspect not just the lawyers but a former friend of stabbing it in the back. 'Miss Rooney seems to have reason to believe that Dr [von] Sochocky was in back of all these cases,' the memo noted. 'I certainly think we should get a line on what [von] Sochocky is doing and where he is.'

The executive who conducted these informal chats with Miss Rooney came not once, not twice, but three times to her workplace to grill her about the girls. By the third visit, it seems she had cottoned on to what was happening. 'Miss Rooney claimed to have no further information this morning,' noted the executive's final memo on the subject. 'Have an idea that she is merely shutting up on information for fear of doing her friend harm.'

But it was of no consequence; the company had got what it needed and there were other sources of information easily to hand. For the firm now paid private detectives to follow the five girls, looking for dirt on them that it could present in court. Berry perhaps suspected that such underhand tactics would be employed, given his decision to lead with Grace's case.

Yet despite the lily-white innocence of Grace Fryer, it didn't mean that any dirt kicked up didn't stick. There was that old rumour – not even a rumour, a cold hard fact in black and white, printed on Amelia Maggia's death certificate. *She* had died of syphilis, so who was to say that all these other girls, who had once worked alongside a girl of her sort, weren't touched by the same Cupid's disease? The rumours wafted around the streets of Orange, sidling as close to the girls as a second skin; just as the radium dust had once done. 'You know how in small towns people gossip . . .' a relative of Grace later said.

USRC had not forgotten that Grace had once proposed a $5,000 settlement. When the firm filed its legal response to Berry, there was a final paragraph included: 'Kindly let us have your rock-bottom figure in settlement,' the firm's lawyers wrote. 'Do not let us dicker. Give us your very best offer.'

As it was his job to do, Berry put the idea to Grace. One can imagine her response; Berry duly wrote back to USRC's legal team – which now comprised three separate firms, some of which were representing the company's insurers, who would have to pay out if the girls won – 'So far as settlement is concerned, [Grace] does not desire to make any proposition.' In other words: we will see you in court.

Berry immediately threw himself into building his case. He quickly met the girls' allies – Wiley, Hamilton, Hoffman, Martland, Humphries and von Sochocky – and spent a lot of time reading their notes and interviewing them. Wiley gave him a damning account of all she had learned: 'Although its employees were falling sick,' she said, 'the corporation did nothing. USRC has done everything it could to obscure the issue and render proper relief to its employees impossible.'

Learning of the cover-up of the Drinker report, Berry immediately grasped the impact of such double-dealing for his case – and wrote to Cecil Drinker to request that he give evidence to help the women. But Drinker responded, via his secretary, that: 'He does not care to make testimony.' Berry would spend all summer trying to change his mind but, in the end, he was forced to issue a formal court summons.

Drinker wasn't the only doctor not keen to testify. 'While I thoroughly sympathise with the girls,' wrote Martland, 'I cannot take sides in a civil suit.' Martland disliked lawyers and had no desire to get caught up in a legal battle; as much as he enjoyed a Sherlock Holmes mystery, he was not a fan of court-room drama.

As Martland's testimony was not certain – Berry would not give up trying to persuade him – Berry started hunting for another specialist who could take new breath tests of the girls to prove they were radioactive. But try as he might, he hit dead end after dead end. He found one specialist in Boston who was willing to assist, but the girls were not well enough to travel.

In the other camp, meanwhile, USRC was having no such problems, as Flinn was still acting as its expert; and not just for them, as Berry soon discovered.

Berry had now learned of the Waterbury Clock Company cases of radium poisoning. Essentially, their existence proved that his clients' disease was occupational. Consequently, Berry wrote to the Workmen's Compensation Commission in Connecticut, the state in which the watch firm was based, looking for evidence he could cite in court. But the commission's response was totally unexpected: 'Had anybody in this vicinity suffered an occupa-tional disease,' its officer wrote, 'I should have [had] it brought to my attention, as occupational diseases have been compensable in this state for several years. No claims have been filed with me. I have heard a number of rumours such as you have, but know nothing about them.'

It was a conundrum. Connecticut law had a much more favourable statute of limitations of five years, which was just about long enough for a dial-painter to discover she had radium poisoning before bringing a lawsuit. By now, at least three dial-painters had died in Waterbury, and others were ill. Had not *one* of their families filed suit?

They had not – and there was a very good reason why: Dr Frederick Flinn. Flinn had enjoyed access to the Waterbury girls almost from when they began falling ill. He was given privileged admittance to the workforce and the girls not only knew him but trusted him. When he told them they were in perfect health, they believed him. And once radium poisoning was discovered, Flinn became 'willing to play a two-faced role: to the dial-painters, he presented himself as a concerned medical expert, whereas for the company he persuaded dial-painters to accept settlements that explicitly freed the company from further liability'.

And this was the reason that not one case had been filed with the Compensation Commission – any claims that might have been raised had been quietly settled by the firm. There was one obvious reason for the difference in the Waterbury Clock Company's approach compared with that of USRC – and the clue is in its name. As Waterbury was a clock company, and not an out-and-out radium firm, agreeing to settlements – and in so doing tacitly admitting that the paint had harmed the girls – didn't affect its wider business, for it didn't make money from selling radium. And so, when its employees started dying, the company simply settled any cases that came up, using the gentle mediations of Dr Flinn. 'In these negotiations,' wrote one commentator, 'Flinn held the upper hand. He knew what he wanted and the women he dealt with were invariably young, unsophisticated, vulnerable, and without the benefit of legal counsel.' Had the Waterbury women taken some legal advice, they could have discovered what Berry well knew: that under Connecticut law, many of them might well have found justice thanks to the more generous five-year statute. It is to be noted, however, that the statute was five years only at the time of the discovery of radium poisoning; following the emergence of the girls' cases, the law was rewritten in order to shorten it.

With the good doctor's intervention, the company spent an average of $5,600 ($75,000) per affected woman, but this figure is skewed by a handful of high settlements. Most victims received less than this average; some, insultingly, were offered only two-digit sums, such as – in one shocking case – $43.75 ($606) for a woman's death.

If one squinted at the situation and tried really hard, one could say that Flinn did the Waterbury girls a favour. He certainly saw it that way; intervening to save them the trouble of bringing a lawsuit. But the company and Flinn held all the cards – and Flinn hadn't finished with what Martland called his 'double-dealing and two-faced' shenanigans. For even though Flinn had now been forced to acknowledge the existence of radium poisoning, it didn't mean that *all* the girls who fell sick suffered from it. And so, as Flinn continued to run his tests on the Waterbury women, he continued to find no positive cases of radium poisoning. Not one – not in 1925, not in 1926, not in 1927. Only in the fading months of 1928 would he concede, at last, that five girls *might* be affected. He told one worker, Katherine Moore, on eight separate occasions that there was not a single trace of radium in her body. She later died from radium poisoning.

Berry, hearing back from the commission, knowing nothing of Flinn's work at Waterbury, was completely stumped by the lack of evidence. But his new friend Alice Hamilton quickly realised what had gone on and filled him in. With the cases settled quietly by Flinn, of course there had been no evidence: no publicity; no Department of Labor knocking on the clock company's door; no lawyers involved at all – just a tidy sum passed across a desk, and a grateful recipient who took it. It was all hush hush.

None of it helped Raymond Berry.

From the beginning, Berry was very interested in Dr Flinn. He had learned from the girls about his declarations of their good health – declarations that had confused them and for some girls taken the wind out of their sails when it came to filing suit. Berry thus decided as early as August 1927 to dig a little deeper into Dr Flinn. His enquiries soon unearthed a shocking piece of news.

Dr Flinn had been examining the girls; taking blood; reading their X-rays. He had been arranging medical treatment and writing to the women on the letter-headed paper of the College of Physicians and Surgeons. '[I] understood,' said Grace's physician Dr McCaffrey, who'd arranged her examination with Flinn, 'that Dr Flinn was an MD.'

But now, when Berry asked the authorities to look into exactly who Flinn was, he received the following letter from the New Jersey Board of Medical Examiners: 'Our records do not show the issuance of a license to practice medicine and surgery or any branch of medicine and surgery to Frederick B. Flinn.'

Flinn was not a medical doctor.

His degree was in philosophy.

He was, as the Consumers League put it, 'a fraud of frauds'.

Ottawa, Illinois
August 1927

Ella Cruse slammed the screen door of her house on Clinton Street and made her way down the few steps outside. She called goodbye to her mom Nellie as she went – but her voice was not as spirited as it once had been.

Ella didn't know what was wrong with her. She had always been 'strong and robust' before, but now she felt tired all the time. She started walking to work, taking her bearings, as always, from the spire of St Columba, which was only a block or two from her home. Ella and her family – mom Nellie, dad James and little brother John – regularly attended services at the Catholic church, like most everybody she worked with.

Nellie's reply to her as she said goodbye had been quiet too; but then her mom disapproved of Ella being a dial-painter. 'I never wanted Ella to work there,' she used to say, shaking her head, 'but [it's] a clean place and they're a jolly bunch of girls.'

Clinton Street was only a couple of blocks from the art studio too, so even with her new snail-pace gait, Ella was soon there. She made her way up the school steps with all the other girls arriving for work. There was Catherine Wolfe, walking with that slight limp she'd recently developed; Marie Becker, talking – as ever – nineteen to the dozen; Mary Vicini, Ruth Thompson and Sadie Pray. Peg Looney was already at her desk when Ella entered the studio, as conscientious as always. Ella said hello to them all; she was 'a popular young woman'.

In 1927, Mary Ellen Cruse (as she'd been christened by her parents) was twenty-four, the same age as Catherine Wolfe. Her glossy chestnut hair was cut into a fashionable bob which ended daringly short at her cheekbones; it was finished with a dramatic fringe that swept across her flawless skin. She wore her eyebrows neatly plucked and had a shy smile that brought out a dimple in her left cheek.

She settled at her wooden desk and picked up her brush. *Lip ... Dip ... Paint.* It was a familiar routine by now, as she'd started working there when she was about twenty – she worked twenty-five days a month, eight hours a day, with no paid vacations.

Boy, she felt like taking a holiday now, though. She was tired and run-down and her jaw felt sore. It made no sense; she was in excellent health normally. Ella had started seeing a doctor about six months back, but even though she'd gone to a couple of different physicians, none had been able to help. It was just like Peg Looney; she'd had a tooth pulled recently, but she said her dentist couldn't make it heal.

Ella looked up as she heard Mr Reed come into their room, watching as he paced up and down on one of his infrequent inspections. He had a certain swagger to his walk these days – but why wouldn't he? He now ran the joint: the super-intendent at last, ever since Miss Murray had died of cancer back in July. Ella turned her attention back to her dials. No time to waste.

It was hard work today, though. All that summer she had complained of pains in her hands and legs, and it was tough to keep up with the delicate painting when her knuckles were so sore. She took a breather just for a minute and rested her head in her hands. But that also worried her: there was a hard ridge under her chin. She didn't know what it was or why it had suddenly appeared in the past few weeks, but it felt most peculiar.

Still, at least it was Friday. Ella wondered what the girls would be up to this weekend – maybe Peg's boyfriend Chuck would have people over at the Shack, or there'd be a plan to catch a movie at the Roxy. Her finger absent-mindedly stroked the small pimple that had appeared on her usually perfect skin a day or two earlier; it was on her left cheek, right by her dimple. When it had appeared she'd picked at it and it had started to swell; she could feel the pain and pressure under her fingertips. Hopefully, her skin would clear up before any parties started.

She tried to concentrate on her work all morning, but found it harder and harder. No parties for her this weekend, that was for sure. In fact, she thought suddenly, no work either. She was done in – and she was done for today. She took her tray of dials up to Mr Reed and said she had to go home sick. It was less than ten minutes before she was back on Clinton Street as the St Columba bells tolled the hour of noon. She told her mom she wasn't feeling well and probably went to bed.

'The next day,' remembered her mother, Nellie, 'we went to the doctor.' That little pimple had swelled up on her daughter's face and she wanted to get it checked. But there was nothing serious about her condition and the doctor was in convivial spirits as he chatted with the Cruses. Ella told him how her mom was always scared of her working at Radium Dial and he retorted with a hearty chuckle, 'That's all bunk, there's not a cleaner place.'

And so Ella and Nellie went back to Clinton Street.

Ella may have skipped church on the Sunday; she certainly wasn't well enough for work come Monday morning. On Tuesday 30 August, her mother called out a physician again; he opened the pimple but nothing came out. He then left them to it; it appeared that whatever was causing Ella's sickness was a mystery.

A mystery it may have been, but *not right* was what Ella Cruse knew it to be. That pimple, that little pimple, just kept swelling and swelling. It was incredibly painful. Nothing she or her mom or even the doctor did could halt it; it was an infection that was unstoppable. Her face became badly swollen; she had a fever too.

'The next day,' recalled Nellie, '[the doctor] looked at her face [again] and ordered her to the hospital.'

Ella was admitted to Ottawa City Hospital on 31 August. But still that spot got bigger and bigger – until you couldn't call it a pimple anymore. It wasn't even a boil; it went way beyond that. Ella's neat haircut still sprouted from her head, as fashionable as ever, but the girl below, in just a few short days, became unrecognisable. Septic poisoning set in and her pretty face and head turned black.

'She suffered the awfulest pain . . .' remembered her mother in horror. 'The awfulest pain I ever saw *anyone* suffer.'

Ella was her only daughter. Nellie kept a bedside vigil as long as the doctors would let her, even though the person in the bed didn't look like Ella anymore. But she *was* still Ella: she was still her daughter, and she was alive and she needed her mom.

Midnight, 3 September. Saturday night slipped into Sunday morning, and Ella's condition declined. As she lay in bed, her system septic, her head swollen and black, her face unrecognisable, the poison in her body did its worst. At 4.30 a.m. on Sunday 4 September, her death came suddenly. She had been at work painting dials just the week before; all she'd had was a little spot on her face. How had it come to this?

The doctors filled in her death certificate. 'Streptococcic poisoning,' they wrote as the cause of death. 'Contributory cause: infected face.'

On 6 September, Nellie and James Cruse traced the familiar path to St Columba to bury their daughter. 'Miss Cruse's death,' reported the local paper, 'came as a shock to all of her friends and family.'

It was a shock. It left a hole; a hole in a family that could never be filled. Her parents said, many years later, 'Life has never been the same since she went.'

Ella's obituary mentioned only one other detail as it mourned this youthful Ottawa girl, who had lived there most of her life, who'd had friends, been popular, worn her hair in a bob and lived her too-few days in the shadow of the church spire.

'She had been employed,' the paper noted, 'at the Radium Dial . . .'

27

Newark, New Jersey
1927

The news of Flinn's non-medical degree was shocking to all. Wiley, reeling from being duped, called him 'a real villain'. Hamilton wrote to Flinn urging him to 'consider very seriously the stand you are taking'. But Flinn was nonplussed. He replied to Hamilton: 'What you mean by "my recent conduct" is beyond my ken.' He seems to have been unperturbed by Berry's discovery of his real degree; in his eyes, he was still an expert in industrial hygiene – just as Hoffman, a statistician, could also be called a specialist in that field – and he had done nothing wrong.

Hamilton was frustrated by Flinn's glib reply. He is 'impossible to deal with', she exclaimed. Berry, meanwhile, reported Flinn to the authorities for practising medicine without a licence.

Separately to the Flinn matter, Hamilton now equipped Berry with what would prove to be an all-important secret weapon: a personal connection with Walter Lippmann and the *World*. The *World* was arguably the most powerful newspaper in America at the time. It promised to 'never lack sympathy with the poor [and] always remain devoted to public welfare', so the dial-painters' case was a perfect cause célèbre for the paper to get behind. Lippmann was one of its leading writers; he would become the paper's editor in 1929 and later be deemed by several sources as the most influential journalist of the twentieth century. To have him in the girls' camp was something of a coup.

Immediately, Berry got a taste of just what Lippmann could do. USRC, as was to be expected, had cited the statute of limitations in its defence; the company argued that the cases should be thrown out of court before the firm's guilt could even be examined. But Lippmann was quick to give his own interpretation of that kind of legal trickery in the *World*, calling the attempt by the corporation to take refuge in the statute 'intolerable' and 'despicable'. 'It is scarcely thinkable,' he wrote, 'that the Court will not agree with counsel for the complainant.'

He was right, in a way; the court did not agree with the company. Instead, the girls' cases – which had all been consolidated into one case to avoid duplicate hearings – were transferred to the Court of Chancery, where their cases would be presented and a ruling given on whether Berry's interpretation of the statute held. Assuming he and the girls were triumphant, there would then be a second trial, which would rule on whether the company was at fault. The Court of Chancery was dubbed 'the Court of King's Conscience': it was where pleas for mercy that might be left unanswered through a strict reading of the law were heard. The trial date was set for 12 January 1928.

There was much to do before then. Berry had at last found a specialist who would run the new radioactivity tests on the girls; Elizabeth Hughes was a physicist and former assistant to von Sochocky. The tests were planned for November 1927. Berry knew, however, that whatever Mrs Hughes found, the results would be questioned in court. USRC, in fact, had already said, 'We should also like to have a physical examination of the plaintiff[s] by *our* doctor,' and Berry anticipated that there would be some dispute over the tests. The results could be variable, no doubt; a humid day could skew the readings and even different doctors looking at the same figures could interpret them contrarily.

Berry's problem thus mirrored that of Dr Martland in 1925. How could he *prove* it was radium that was killing the dial-painters? There was only really one way to do that, and it wasn't something Berry could ask of his clients. For the only way to extract radium from a victim's bones – to demonstrate incontrovertibly that radium was present – was to reduce those bones to ash. 'The deposit [of radium],' commented Martland, 'can be removed only by cremating the bone and then boiling the ash in hydro-chloric acid.'

No: this was nothing that Grace, or Edna, or Katherine, or the Maggia sisters could help with. Except . . .

Except for maybe *one* of the Maggia girls.

Mollie.

It was shortly after 9 a.m. that the men came to Rosedale Cemetery on 15 October 1927. They made their way through the rows of memorials until they stopped at one particular grave. They erected a tent over it and removed the headstone. Then they worked to uncover the coffin, heaving sodden earth out of the hole until they unveiled a nondescript wooden box, which held Amelia 'Mollie' Maggia – the girl, so they said, who had died of syphilis. The men ran ropes under it, then attached stronger silver chains. It was raised just slightly, 'to free it from water that had seeped in around it as a result of the recent rains'.

Then they waited for the officials to arrive. Berry had arranged with the radium company that they would all converge at 3.30 p.m. exactly.

At 3 p.m., the specialists from the company arrived at Mollie's grave.

There were six of them, including Vice President Barker and the ubiquitous Dr Flinn. Prudently, Berry had arranged for a special investigator to be present for the morning's activities; he now watched the company men closely as they milled outside the tent. At 3.30 p.m., as specified, Berry walked up to the grave with Mrs Hughes, Dr Martland and a cohort of New York doctors, who would lead on conducting the autopsy. There were thirteen officials in all, gathered together to witness Mollie's exhumation.

Standing awkwardly among the doctors and lawyers were three other men: James McDonald and James Larice, Mollie's sisters' husbands, and her father, Valerio. The family hadn't protested when Berry had put the idea to them. Mollie's body could provide perfect corroborating evidence for the dial-painters' fight in court. Even after all these years, she could still help her sisters.

After the arrival of Berry's team, preparations were made for raising the coffin. Curtains were drawn around and the entire party went inside the tent. The grave workers heaved on the ropes and chains. Slowly, Mollie rose the six feet to the surface. 'The outer box was in bad condition and easily pulled apart; the casket was likewise ready to fall apart.' Despite the dim fall day, the coffin seemed to glow with an unnatural light; there were 'unmistakeable signs of radium – the inside of the coffin was aglow with the soft luminescence of radium compounds'.

Someone lent over the glowing coffin and pulled a silver nameplate from the rotten wood. *Amelia Maggia*, it read. They showed it to Valerio for identification. He nodded: yes, that was the one. That was the one the family had chosen for his child.

As soon as Mollie's identity was confirmed, the top and sides of the casket were removed. And there she was. There was Mollie Maggia, back from the grave, in her white dress and her black leather pumps, just as she had been dressed on the day she was buried in 1922.

'The body,' observers noted, 'was in a good state of preservation.'

They removed her carefully from her coffin, placed her gently in a wooden box, and then took her by automobile to a local undertaking parlour. At 4.50 p.m., her autopsy would begin. At 4.50 p.m., Amelia Maggia would finally have the chance to speak.

There is no dignity in death. The doctors started with her upper jawbones, which were removed in several pieces; they had no need to do the same with her lower jaw, for it was no longer present, having been lifted out in life. They sawed through her spine, her head, her ribs. They scraped her bones with a knife to prepare them for the next steps. And there was, somehow, a kind of ritualistic care in their steady tasks, as they 'washed [her bones] in hot water, dried, and reduced [them] to greyest white ash'. Some bones they put to the X-ray film test; others they ignited to ash and then tested the ash itself for radioactivity.

When they checked the X-ray film, days later, *there* was Mollie's message from beyond the grave. She had been trying to speak for so long – now, at last, there was someone listening.

Her bones had made white pictures on the ebony film. Her vertebrae glowed in vertical white lights, like a regiment of matches slowly burning into black. They looked like rows of shining dial-painters, walking home from work. The pictures of her skull, meanwhile, with her jawbone missing, made her mouth stretch unnaturally wide, as though she was screaming; screaming for justice through all these years. There was a smudge of dark where her eye had once been, as though she was looking out, staring accusingly, setting straight a lie that had blackened her name.

There was, the examining doctors said, 'No evidence of disease, in particular no evidence of syphilis.'

Innocent.

'Each and every portion of tissue and bone tested,' the doctors concluded, 'gave evidence of radioactivity.'

It *wasn't* Cupid's disease, as the gossip-mongers charged. It was *radium*.

*

The doctors' autopsy findings gathered wide publicity; the girls' fight for justice was slowly becoming famous. And it was this publicity that now brought another girl to Berry's office, though she did not sign with him at that time.

Ella Eckert, Mollie Maggia's friend, the fun-loving girl with frizzy blonde hair who had laughed her head off at so many company picnics, now called on the Newark lawyer in the fall of 1927. She was in better health than any of the five women suing, but she nevertheless told Berry, 'I have spent at least $200 [$2,724] for X-rays, blood tests, medicine and medical attention, all to no avail.' She'd had a fall at work, at Bamberger's, the year before and been forced to give up her job as her shoulder had never healed. Indeed, Berry could see that her arm was 'badly swollen, extending from the shoulder down to the hand'. She said she was in severe pain and begged him to help her.

And that help was not just for her. Ella Eckert had taken her fun-loving ways to what were then considered extremes; she'd had a son with a married man who had since disappeared, and now she was bringing up the boy on her own. She couldn't afford to be out of work, or to get sick: her son needed her.

Berry knew their paths would cross again; in the meantime, his pace of work sped up. An important date was 14 November 1927: this was when the first testimony was taken in the girls' trial, as part of a deposition. Berry had issued his formal summons to Dr Drinker – now the reluctant doctor gave his evidence under oath.

It was at this juncture that Berry met his main adversary: Edward A. Markley, the attorney for the radium firm's insurance company, who was leading USRC's defence. Markley was almost six feet tall, with brown hair and eyes, which he framed with glasses. His father had been a judge and he was the eldest son in his family; he had all the suave confidence and self-possession such attributes would give. He was some six years older than Berry, with all the added experience that implied.

From the moment the deposition-taking started, Berry

realised it was not going to be an easy ride. He was trying to admit all the Drinker evidence: the blinkered, blustering letters that Roeder had sent to justify suppressing the Drinker report; the firm's false claims to the Department of Labor. To every single question, every single item of evidence, the USRC lawyers fought back.

'We object to the question,' said Markley, 'on the ground that the purpose is immaterial.'

'We object,' said Stryker, 'to the witness stating what he told Mr Roeder.'

They even shut down Drinker himself.

'I should like to make a statement for the purpose of the record on my side relative to this,' the doctor began calmly.

'Before you do that, we object to it,' jumped in Markley, before Drinker could proceed.

The lawyers deemed the collection of authentic letters 'a scurrilous statement of rumours' and took a clever line of questioning with the pioneering scientist and his colleagues. To each of the three investigators who had authored the Drinker report, they put the question: 'Had you had any experience in investigating radium poisoning?'

The answer from all, of course, was 'None'. The implication being: how could the word of such inexperienced 'experts' be taken seriously? Only Katherine Drinker pointed out the obvious: 'This is the first time the disease was [discovered].'

Berry was not daunted by all this, however. Submitting the Drinkers' report, he said cheekily, 'It is offered as the best evidence we have and will be used in the event Mr Roeder has "mislaid" the original.'

The company lawyers merely responded, 'If the original is used it is of course subject to our objections ...'

January was going to prove a tough battle: that was for sure.

Before they got there, however, an unexpected event took everyone by surprise. Berry had been concerned about the young woman who'd visited him earlier in the year, for Ella

Eckert, he had heard, had been 'near death' in the Orthopaedic Hospital for weeks now. She had the usual symptoms of radium poisoning: the anaemia, the white shadow through her bones. Yet despite these tell-tale signs, Dr Martland commented: 'This case is very puzzling and not as clear-cut as the others.'

On 13 December 1927, Ella Eckert died. Martland traced her name on the List of the Doomed. *D is for Death.*

She'd had an operation earlier that day, on her swollen shoulder. And herein lay the cause of her mystery. For when the doctors cut her open, they found a 'calcareous formation [was] attached and [had] permeated the entire shoulder region'. It was a growth of 'considerable size'. Such a growth was new to Martland, to all the doctors. No dial-painter, so far as they knew, had ever presented such a thing.

Radium was a clever poison. It masked its way inside its victims' bones; it foxed the most experienced physicians. And like the expert serial killer it was, it had now evolved its modus operandi. Ella had developed what was called a sarcoma: a cancerous tumour of the bone. She was the first known dial-painter to die from such a thing – but she would not be the last.

Her death shocked the five girls suing; her decline had been so fast. Yet it also gave them even more inspiration for the fight that lay ahead.

On 12 January 1928, the trial of the decade would begin.

'I could hardly sleep the night before the court hearing,' Katherine Schaub wrote, 'for I had been waiting for ages, it seemed, to see this very day.'

She was not alone. When the five women, on a frosty January day, arrived at the Court of Chancery, they found themselves surrounded. Newspapermen crowded around, flashing cameras in their eyes, and then packed the seats of the gallery inside.

Berry hoped the girls would be ready for what lay ahead. He had prepped them all he could, calling the five women together two days ago to go through their testimony. Yet the women's mental strength was only part of the equation; and anyone could see that their physical health was failing. The past six months had not been kind to them. 'The condition of certain of the girls,' Berry wrote, 'is truly deplorable.'

He was most concerned about Albina Larice. She could not extend her left leg more than four inches; she could not even put on her own shoes and stockings, because she could not bend over. Her medical prognosis, along with Edna Hussman's, was now deemed the worst. Yet it was not the loss of her health that plagued her . . .

'I have lost,' Albina mourned, 'two children because I'm this way.' Only the previous fall, Berry knew from what the doctors had told him, she'd lost a third baby; a baby who, if things had been different, might well have lived. She'd been so delighted when she found out she was pregnant – but Albina's happiness did not last long. For when her doctors discovered her condition, they would not permit development of the child, due to her health. They ordered her to have a 'therapeutically induced' abortion.

'I've been so discouraged at times,' Albina confided, 'that I've thought about taking gas and ending it all.'

Dr Humphries had said that radium poisoning 'destroys [his patients'] will to live'. Berry could only hope that the women, on this day, could find the will to fight.

Edna Hussman was the first to give evidence; Louis almost had to carry his wife to the witness stand. His beautiful blonde Edna, when seen in snapshot, looked as modelesque as ever, striking a pose with one leg casually crossed in front of the other. But appearances were deceptive: she could no longer move her legs apart, for her hips had locked in place at that 'abnormal angle'. She had also lost the use of her right arm; she could not even raise it to take the oath.

The judge overseeing proceedings was Vice Chancellor John Backes, a very experienced man in his mid-sixties. Berry must have been hopeful for a sympathetic hearing, for Backes's own father had died after being injured in a rolling mill. Backes wore a bushy moustache and glasses; he looked kindly on Edna as she prepared to give her testimony.

Berry eased her in slowly, just as they'd rehearsed. Edna concentrated on him, answering simple questions about where she lived and how she was now a housewife; although, as she said outside the court, 'I cannot keep my little home. I do what I can, but my husband does most of the work.'

Edna was tired. 'The worst thing I have to put up with is not being able to sleep at night because of the pain in my hips,' she revealed. So it didn't help when, only eight questions in, as she began to outline the nature of her work at USRC, the lawyers for the corporation cut in with the first of their many objections.

Berry was expecting it. On 4 January, he'd taken another three-hour deposition with the company lawyers present, this time from the Newark dentist Dr Barry; once again, they'd questioned everything. Irene Rudolph's dental file had a note that read: 'Recovery? OK', which Barry explained meant that she had recovered from the anaesthetic; the lawyers, however, said tartly, 'Isn't the "Recovery" here recovery from the treatment?' They asked the same question, in different permutations, at least eight times before moving on.

Edna Hussman, however, was not a professional man like Dr Barry – she was a twenty-six-year-old crippled housewife, and the company lawyers' aggressive tactics did them no favours. As they harangued her to remember dates and how frequently she had stumbled when her pains began, Backes interrupted their incessant questioning. 'Of what importance is it?' he asked pointedly. Sympathy for Edna increased as her testimony continued. 'I suffer,' she told the court, 'all the time.'

Berry's inexperience in court sometimes showed itself. Despite his brilliant mind, he was still in the early stages of his trial career – but he found the judge was willing to help him out. When Hoffman took the stand after Edna, Backes assisted Berry by helping him to phrase his questions ('What did he do to get the information and what did he learn?' he prompted) and even stepping in to help when he anticipated an objection.

On their cross-examination of Hoffman, the company lawyers tried the same tactic they'd used with the Drinkers.

'Is this the first time that you had occasion to consider the question of radium necrosis?' Markley asked the statistician, his tall frame pacing the courtroom as he fired off questions.

'Yes, sir; entirely new venture.'

'... You had no knowledge, did you?'

'Or nobody else . . .' pointed out Hoffman.

'I am asking you,' Markley said sternly, 'to speak for yourself. [Was this] the first time you ever had anything to do with the subject?'

'Yes, sir,' Hoffman had to agree.

Markley then tried to get Hoffman's evidence dismissed entirely. 'I submit, Your Honour,' he said with a condescending sneer, 'that a mere statistician is not qualified in court to pass judgement.'

Yet Markley found that Backes was not playing ball.

'I think he is a little more than that,' retorted the judge. 'I think you curbed him down some.'

All this time, the five women watched the drama unfold. They were flanked by the witnesses for the company too; the 'chameleon-hued' Dr Flinn sat across from them in the court-room. Grace felt calm inside, knowing she was up next. 'Grace is so accustomed to talking of disease and decay,' wrote a journal-ist of Miss Fryer, 'that she can tell you of these deaths without flickering an eyelash.'

Still, there must have been a few butterflies as the court sergeant-at-arms tenderly assisted her to the witness stand. This was it, Grace thought. This was her chance to tell her story.

She sat somewhat awkwardly in the chair: her metal back brace chafed her skin and a fresh bandage clung to her jaw, following a recent operation. Yet the slim young woman with neat dark hair and intelligent eyes now composed herself as she began her testimony. 'We were instructed to point the brush with our lips,' she said.

'Did [all the girls] do it that way?' asked Backes.

'All I ever saw do it,' answered Grace.

'Were you ever told at any time not to put the brush in your mouth?' queried Berry, cutting to the heart of the matter.

'Only on one occasion,' she said. 'Dr von Sochocky was pass-ing through and when he saw me put the brush to my lips he told me not to do it.'

'What else did he say?'

'He said it would make me sick.'

Her answers were concise and informative. She and Berry had an instant repartee, with question and answer flicking back and forth slickly, just as they had planned. Yet Berry gave her room, too, to describe her suffering, so that everyone could hear what the company had done.

'I have had my jaws curetted seventeen times,' said Grace simply, 'with pieces of the jawbone removed. Most of my teeth have been removed. [My] spine [is] decaying and one bone in [my] foot [is] totally destroyed.'

It was horrifying to listen to; many in the courtroom were in tears. No wonder, when Markley made some smart comment, the judge snapped back at him. 'If I find you guilty, I think you will be sorry,' he said tartly.

Given the warning, Markley approached Grace's cross-examination with some caution. He could doubtless see, too, that she was not going to be a pushover. And she certainly was not.

Critical to the USRC lawyers' arguments, particularly in this Court of Chancery, was the statute of limitations and what the girls knew when. If they'd had information prior to July 1925 that their work had made them sick, they should have brought a lawsuit at that time. So Markley tried to push Grace into saying that she had known her work was to blame earlier.

'Did [your dentist] tell you he thought it was your work that was affecting you?' the lawyer asked as he stalked around the court.

'No, sir.'

The question was repeated.

'Why no,' said Grace smartly. 'I was working for the Fidelity Union Trust Company when I saw him.'

They also quizzed her on all the different lawyers she had seen. And when they came to Berry, they asked her, '[Was he] the first one you had?'

'No, not the first one,' Grace replied, locking eyes with her young lawyer. 'The only one that ever brought suit.'

Katherine Schaub watched proceedings eagerly. 'Everything was going along splendidly, I thought,' she later wrote. She watched Quinta limp to the stand; the judge, Katherine was gratified to note, was immediately concerned. 'I notice you are very lame,' Backes said to Quinta, before Berry had asked a single question. 'What is the trouble?'

'Trouble with my hip – both hips in fact,' Quinta replied. 'As to my ankles, I cannot wear a shoe very long; I [have] terrible pains in my knees, one arm and shoulder.'

Katherine listened attentively. 'Tomorrow there would be another court, and the day after, still another,' she wrote, 'and so on until the entire case was heard. And then – the court would give its verdict. Then perhaps I could get away from everything and forget.' Still half-listening to Quinta, she started to picture her life afterwards, how happy she hoped she would be. Just a few more days of these January hearings, she thought, and then it would all be over – one way or the other.

But it was not to be. 'I was awakened from my dreaming,' she later said, 'by the sound of the vice chancellor's gavel hitting the desk. The vice chancellor was speaking. The next court day, he said, would be April 25. I could have given way to tears, but tears would not do any good, I knew.

'I must summon all the courage I had – and fight.'

Though the delay was galling, the time, in the end, passed quickly. Berry, who was concerned that little was being done for the girls medically, persuaded some New York doctors to admit the women to hospital and all five spent a month in their care. The physicians believed there might be some treatment devised that would eliminate the radium in the girls' bones.

'A Russian doctor,' Grace recalled, 'thought he could help us with some sort of lead treatment [a treatment used in lead-poisoning cases], but it didn't seem to take the radium out of our systems. I guess nothing ever will.' Perhaps grasping the hope-lessness of her situation, Grace summoned Berry and formally drafted her will, even though she didn't have much to leave to her family.

Despite the failure of the treatment, many of the girls remained positive. 'I face the inevitable unflinchingly,' said Quinta. 'What else can I do? I don't know when I will die. I try not to think of the death that is creeping closer, all the time.' Death seemed further from Quinta than some of the others, though, as her condition was progressing more slowly than, for example, Albina's; consequently, it was her habit 'to turn aside pity for herself by commiserating her sister's plight'.

Many of the women found just being out of Newark in the quiet calm of a hospital made a big difference to their outlook. 'I haven't had anything yet but a bath,' Katherine wrote when they first arrived. 'I enjoyed that because someone helped me to take it. A maid is a fine thing to have when you are sick.'

There was one other bonus to being in New York. As Katherine wrote, they were at last 'safe from intrusion [and] safe from the prying eyes of unwelcome advisers'.

For the omnipresent unwelcome adviser, Dr Flinn, had not stopped trying to get to them, even though Berry had found him out. Flinn had recently told – and convinced – Dr Humphries that he was 'really a friend of the girls'. But the women, now knowing Flinn was a company man, had gone straight to Berry when they'd heard of this; they mistrusted Flinn's 'clandestine overtures' and at their request Berry wrote to Flinn to ask him to desist in what the girls considered harassment. Flinn replied that he thought Berry 'impudent' and concluded, 'The other inaccuracies of your letter I will not take the time to answer.'

The women could not avoid Flinn, however, when on 22 April, three days before the trial was to resume, they were summoned to a compulsory examination by the company doc-tors. Flinn, as well as other specialists, including Dr Herman

Schlundt (who was a 'very close personal friend' of Vice President Barker), conducted the tests.

Grace flinched as they pricked her with a needle to take her blood. She was constantly afraid of anything that might result in cuts or bruises, for her skin no longer healed. Some dial-painters had 'paper-thin skin that literally would split open if simply brushed by a fingernail'. A week later, Grace realised she had been right to worry: in the place where the doctors had pricked her, the flesh surrounding the puncture mark was black.

During the examination radioactivity tests were conducted, the equipment deliberately positioned 'so that the table itself was between large portions of the patient's body and the instrument'. Flinn also 'held the instrument two to three feet from the subject, allowing the radiation to dissipate before reaching the device'. Unsurprisingly, the company's verdict was that none of the women was radioactive.

But the girls' case was not finished yet. In three days' time they were back on the stand for the fight of their lives.

29

Katherine Schaub was first up.

'I ascended the steps to the witness stand one by one,' she wrote. 'I felt quite strange to be on the stand; more strange than I had anticipated ... I took the oath.'

As Berry had done with her friends, he eased her into her testimony. She cast her mind back to 1 February 1917, to a cold winter's day when she had excitedly made her way to work for her first day. 'The young lady instructed me,' she recalled, 'told me to put the brush in my mouth.'

Berry took her through her suffering; she revealed she had grown 'very nervous'. The USRC lawyers undoubtedly saw her mental-health issues as a weakness – and that probably explains why they gave Katherine hell.

She had just said that she lip-pointed 'sometimes four or five times [per dial], perhaps more than that', when Markley stood to begin his cross-examination.

'Sometimes more,' he began.

'Yes, sir.'

'Sometimes less.'

'Yes, sir.'

'Sometimes you wouldn't put the brush in your mouth at all, would you?' he exclaimed, spinning round to deliver the line. She must have hesitated. 'You don't *know*?' he said incredulously.

'I am trying to remember,' replied Katherine nervously.

'... Depend on your brush too, wouldn't it? [...] The brushes were supplied there, weren't they?'

'They were supplied, yes, sir.'

'You could get all the brushes you wanted.'

'No.'

'... You would go to [the forewoman] when you wanted a brush, wouldn't you?' he asked, closing in.

'Yes, sir,' Katherine replied, 'but you were not supposed to waste them.'

'Of course you weren't supposed to waste them, but you were supplied amply with them, weren't you?'

The questions came thick and fast. Markley didn't miss a beat and would have his next line of attack prepared even as Katherine stuttered out her answer.

As they had done with Grace, the company lawyers questioned Katherine extensively about her initial dental treatment and whether any connection had been made in the early 1920s between her illness and her job. Perhaps inevitably, under such heated cross-examination the nervous Katherine slipped up. Thinking back to the meeting she and some of the other girls had had in Dr Barry's office, when the condition was considered to be phosphorus poisoning, she revealed, 'There had been some talk about industrial disease ...'

Markley seized on it. 'What do you mean, "There had been some talk"?'

Katherine realised her error. 'I had never connected myself with it in any way,' she said hurriedly, but he wasn't going to let it go that easily. He brought up Irene, who had died in 1923.

'You know Dr Barry told her he thought it might be industrial disease, don't you?'

'Well, he had a slight suspicion that something was wrong,' Katherine conceded weakly.

'He *told* you he had a slight suspicion?' Markley asked.

'He never told me that directly . . . I only know what my folks told me.'

'*When* was it that they told you that?' Markley jumped in, probably hoping for an answer that would kill the case dead.

'Well, I don't know,' retorted Katherine, back on track. 'My cousin was ill so long and *I don't remember*.'

It seemed never to end. She felt worn down by it – so much so that Backes, keeping an eye on his vulnerable witness, interjected at one point to ask, 'Are you tired?'

But Katherine replied firmly. 'No,' she said, 'I try to sit up as straight as I possibly can, because my spine is a little weak.'

She would have been gratified to note that the gathered reporters scribbled down that detail of her suffering as they followed her account.

As at the January hearing, the courtroom was packed with journalists; even more than before, for the women's story was now beginning to reach international shores. The reporters would later write moving descriptions of their testimony, as Katherine, Albina and Quinta all gave evidence. The press called them a 'sadly smiling sorority' and said they 'maintained an attitude of almost cheerful resignation'.

Their composure was in direct contrast to those observing the trial. 'The [women] listened,' a newspaper reported, 'with pensive stoicism, while ordinarily hard-boiled spectators had constant recourse to handkerchiefs to check tears of which they seemed unashamed.'

How could anyone not cry, as Berry took Quinta McDonald through the fate of her friends?

'Were you ever acquainted with Irene Rudolph?' he asked her.

'Yes, sir, while I worked in the radium plant.'

'Hazel Kuser?'

'Yes, sir.'

'Sarah Maillefer?'

'Yes, sir.'

'Marguerite Carlough?'

'Yes, sir.'

'Eleanor Eckert?'

'Yes, sir.'

'... Are all these people dead?'

'Yes, sir.'

It seems Grace may have indicated to Berry that she wanted to be recalled, for she now retook the stand. She had been staring across the courtroom at the gathered USRC executives, and her sharp memory had snagged on one of their faces in particular.

'Miss Fryer,' Berry began, after a quick consultation with Grace, 'you were examined in the summer of 1926 by Dr Frederick Flinn and there was another doctor who you did not know who was present at the examination. Have you seen that assisting doctor since that time?'

'Yes, sir.'

'Is he here in the court today?'

Grace looked across at the executives. 'Yes, sir.'

Berry pointed at the man she had specified. 'Is that the gentleman, Mr Barker?'

'Yes, sir,' said Grace assuredly.

'Do you know he is the vice president of the United States Radium Corporation?'

'I didn't know it then,' she said pointedly.

Barker had been there on the day that Grace had been told by Flinn she was in better health than he was. He had stood by as Flinn had issued the diagnosis there was nothing wrong with her. Barker's presence showed just how involved the company was with Flinn's activities: its own vice president had attended the girls' medical tests.

Elizabeth Hughes, the breath-test specialist Berry had employed, was on the stand next; she testified that it was well known 'that all operatives and all workers should be protected from the radium rays' as 'almost everyone in the field has hand burns'. The newspapers noted of Mrs Hughes, 'She exhibited a thorough knowledge of the subject and convinced Vice Chancellor Backes, at least, that she knew what she was talking about.'

That, of course, was anathema to the company lawyers. They quickly tried to discredit Mrs Hughes, despite her great experience.

'What is your occupation now?' Markley asked her, well knowing the answer.

'Housewife,' she said, for she was currently caring for her young children at home.

And then Markley was off, with question after question to suggest she knew nothing about radium at all. He rather hounded her, undermining not only her qualifications but her skill in handling the breath tests, until he had backed her into a corner and forced her to admit that she 'couldn't define an appreciable amount' of radium.

'All right,' said Markley triumphantly, 'I am perfectly satisfied if you say you do not know.'

But at this Backes, once more, stepped in. 'I want to know what the witness knows,' he exclaimed, 'not merely have you satisfied that she says she doesn't know. I think she said a little more than what your characterisation would purport.'

When the lunch recess fell midway through Elizabeth's testimony, it seemed to come as a relief to both her and Berry. After lunch, Markley returned still in pugnacious mood. The doctor who had conducted Mollie Maggia's autopsy was on the stand, giving his testimony that radium had killed her, and Markley tried to get all evidence regarding Mollie struck out – though he was unsuccessful: 'I will hear it,' said Backes.

'I want to call to Your Honour's attention,' Markley growled, piqued at this decision, 'the fact that this girl was buried on a death certificate for *syphilis.*'

Markley had good reason to fight this hard for the firm. Having shut down the headache of the Orange plant, USRC was now back on track financially; just one single order the firm had recently received, only a few days before, was for $500,000 (almost $7 million). They did not want to lose this case.

The final witness on the stand on 25 April was Dr Humphries, the girls' long-time physician. He was authoritative in describing their unusual conditions. He testified that 'in all these patients' the same condition arose; and not just in them, but in other women he had seen – including Jennie Stocker. Finally Humphries had solved the puzzle of her peculiar knee condition. He now declared, 'I think she died of radium poisoning.'

His testimony was long and something of an endurance test for the five women. For Humphries recounted each of their cases in detail – how they had first come to him with these puzzling pains; how he had 'guessed' at how to treat them; and how now, today, his patients were all crippled. They were not the women he had first seen; though they tried to keep their spirits buoyant, their bodies betrayed them. 'I thought it would never end,' recalled Katherine of his account, 'this excruciating, horrible testimony.' Yet she was brave about it. 'It had to be done,' she went on, 'had to be told – or else how would we be able to fight for the justice that was due us?'

And so the women listened. They listened as, in the public courtroom, Humphries admitted, 'I do not think that anything will cure it.'

The eyes of the many reporters flickered to the women, even as their own filled with tears. Yet the radium girls stoically accepted his pronouncement of certain death.

Like the journalists, however, Backes couldn't seem to bear it. 'You hope to find something every minute?' he said urgently.

'We hope to find something,' Humphries concurred.

'Every minute,' pressed the judge again.

'Yes, sir,' said Humphries simply, but all the judge's urging couldn't magic up a cure. The girls were destined to die.

The only question was whether they would be given justice before they did.

The following day, the trial continued with more expert evidence. Distinguished doctors testified that it had been common knowledge since at least 1912 that radium could do harm. Berry admitted into the court record a host of literature – including articles published by USRC itself – to support the doctors' words.

Though Markley tried to weaken the impact of these documents by citing the curative powers of radium – such as were promoted by USRC client William Bailey in his Radithor tonic – it was apparent there were holes in his arguments. When he quoted a little-known study in an obscure journal and one of the testifying doctors conceded he had never heard of the author, the expert witness added, 'Who is he? What is he connected with?' Markley could reply only defensively: 'I am not here to be questioned.'

The day was going well for Raymond Berry; the doctors were not rattled by their cross-examination in the slightest. One described those using radium as 'fools' and said he thought that radium curatives 'should be abolished'.

'[Aren't they] approved by the Council of Pharmacy?' asked USRC's lawyers indignantly.

'I suppose so,' retorted the esteemed doctor airily, 'but they accept so many things it means nothing to me, sir.'

Andrew McBride and John Roach from the Department of Labor gave evidence regarding their part in proceedings; USRC presidents Clarence B. Lee and Arthur Roeder also took the stand. Roeder confirmed he had been in the dial-painting studio on 'numerous occasions' yet testified: 'I don't recall any instance of an operator putting a brush in her mouth.' He also denied von Sochocky had ever told him the paint was harmful; he said the first he knew of any possible hazard was 'after we heard of some of these early complaints and cases'.

'What was the first case that you heard of?' asked Berry.

'I don't remember the name,' replied Roeder coldly. The dial-painters weren't important enough for him to recall such insignificant details.

And then Berry called on someone very special to testify for the girls: Harrison Martland took the stand. Berry had managed to persuade him to testify. And the Chief Medical Examiner was a superstar; no other word for it. 'His forthright, uncompromising testimony stood out conspicuously,' raved the newspapers; they called him the 'star witness'.

He began by explaining in detail his autopsies of the Carlough sisters, which had confirmed the existence of radium poisoning. It was very difficult testimony for the five women to hear; Quinta, in particular, found it 'excruciating'. 'As she listened to Martland,' one newspaper observed, 'she approached the verge of collapse. Then, by sheer grit, she seemed to regain her composure and sat through the balance of the hearing with only slight traces of emotion.'

Martland was unstoppable. When the company lawyers tried to suggest radium poisoning couldn't exist because 'out of two hundred or more girls, these girls [suing] are the only ones that had this trouble', Martland replied frankly: 'There [are] about thirteen or fourteen other girls that are dead and buried now who, if you will dig them up, will probably show the same things.'

'I ask that be stricken out as an assumption on the part of the doctor without foundation,' said the USRC lawyer hurriedly.

'Let it stand,' replied Backes promptly.

The company tried to say that 'there are no other reported cases' beyond Orange.

'Yes, there are other reported cases,' retorted Martland.

'There is only a stray case, one or two ...' Markley said, wafting a hand dismissively.

But Martland said firmly that the Waterbury cases *did* exist. His testimony was powerful; Backes even referred to the USRC paint itself as 'radium poisoning'; something Markley pounced on indignantly: 'This paint is anything but a radium poisoning!' he exclaimed.

As the day drew towards a close, Berry stood up to redirect Martland. When Markley, predictably, objected, the judge once again overruled him. 'You attempted to weaken [Martland's] opinion,' he told Markley. 'Counsel [Berry] is now trying – *if* you succeeded – to restore it.'

He turned to Berry. 'Proceed.'

Berry could not be happier with how the case was going – and tomorrow he would hammer home the final nail in the company's coffin. Dr von Sochocky was going to take the stand, and Berry couldn't wait to question him on the warning he had given the corporation about the paint being dangerous. That would seal the verdict once and for all – and surely in the girls' favour.

The next morning, towards the end of von Sochocky's testimony, Berry posed the killer question.

'Isn't it true,' he said, his eyes bright as he turned to face the doctor, 'that you said [you hadn't stopped lip-pointing] because the matter was not in your jurisdiction but Mr Roeder's?'

'I object to this, Your Honour,' interrupted Markley at once.

But before the judge could rule, the company's founder answered.

'Absolutely not.'

Markley and Berry both stared at him, open-mouthed. And then Markley confidently retook his seat, crossing his long legs. 'All right,' the corporate attorney said easily, gesticulating for the witness to continue.

'Absolutely not,' repeated von Sochocky.

Berry could not believe it. For not only had Grace and Quinta told him about this, Martland and Hoffman had too: and they had all heard it from the doctor's own mouth. Why was he now backtracking? Perhaps he was concerned about how he would appear; or perhaps something else had happened. 'We should get a line on what [von] Sochocky is doing and where he is,' a USRC memo had noted back in July. Perhaps there had been a conversation behind closed doors that had led to the doctor's change of tune.

Berry quizzed him on his warning to Grace, too. Perhaps here, at least, he could find some traction.

'Well, Mr Berry,' von Sochocky replied, 'I don't want to deny that, but I don't recollect that very distinctly ... There is a possibility I told her that, which would be the perfectly natural thing to do, passing by the plant, seeing the *unusual* thing of a girl putting a brush to her lips; of course I would say ["Do not do that"].'

That account sounded peculiar even to John Backes's ears. 'What reason had you for doing that?' the judge asked.

'Unsanitary conditions,' replied von Sochocky promptly.

'You cautioned this young lady not to put the brushes in her mouth,' Backes said plainly. 'I want to know whether at that time you were apprehensive that the paint with the radium in it might affect her deleteriously.'

But the doctor was unmoved. His choice of pronoun is notable. 'Absolutely not,' he replied to the judge. '[The danger] was unknown to us.'

Berry was bitterly disappointed. Publicly, in court, he denounced von Sochocky as a 'hostile witness'. Grace Fryer, to whom the warning had been given, must have had a few choice adjectives of her own flying through her mind.

Berry gave her a chance to speak again. She was recalled to the stand immediately after von Sochocky's testimony – 'not to discredit [the doctor],' Berry explained, 'but to show actually what he did say.' But Markley objected to her evidence at once, and the judge was forced to sustain it, seemingly against his will. 'Strike out the answer,' Backes commented. 'These rules of evidence have been invented to prevent people from telling the truth.'

There were only a handful of other witnesses, including Katherine Wiley and Dr Flinn, who was there as a paid witness for USRC. And then, at 11.30 a.m. on 27 April 1928, Berry rested his case. Now, for the rest of the day and in subsequent days to follow, the United States Radium Corporation would have an opportunity to put their side of the story and then – *then*, the girls thought hopefully, wondering how they would feel when the time came – the verdict would be given.

Markley stood up, his long body sliding out of his chair effortlessly. 'I was wondering,' he said smoothly to John Backes, 'we may be able to shorten this if we have time for a conference?'

There was a discussion off the record. Afterwards, as the judge's gavel banged, Backes made a pronouncement.

'The hearing is adjourned to September 24.'

September was five months away. *Five months*. To put it bluntly, it was time that the girls, in all likelihood, probably didn't have.

The delay, cried Katherine Schaub, was 'heartless and inhuman'.

But the law had spoken. Nothing further would be done until September.

30

The girls were devastated. Even Grace Fryer, who for so long had stayed incredibly strong, couldn't bear it. She flung her body 'down on the couch in her living room [and] gave herself up to the pent-up tears'.

Her mother tried to calm her, gently touching her daughter's metal-bound back, trying not to bruise her thin skin. 'Grace,' she said, 'this is the first time you have failed to smile.'

But the girls could not believe what had happened. Markley had said 'it would hardly be worthwhile for him to begin his case with only half the day remaining' and thus the case had been postponed until there was sufficient time in the court calendar – the company intended to present approximately thirty expert witnesses. The serial story in the *Orange Daily Courier* that week was 'Girl Alone'; well, all five dial-painters truly felt that way.

But they were *not* alone: they had Raymond Berry. Immediately he fought the decision and, crucially, found two lawyers, Frank Bradner and Hervey Moore, who had a case scheduled for the end of May and were willing to give up their court slot so that the girls' case could be heard instead. Backes agreed at once to the new timing and Berry let the women know the good news.

The United States Radium Corporation, however, was not best pleased at Berry's intervention and said it would be 'impossible' for them to proceed in May; their experts were 'going abroad for several months and will not be back until after the summer'.

Berry was outraged. 'I am sure you must agree,' he wrote to Markley, 'that there is a rather harsh irony in the situation which permits the victims of poisoning to languish and die because certain trained men must disport themselves in Europe.'

Despite the company's intransigence, in Berry's own words he was 'far from finished in this fight'. Aware that USRC's procrastination was arguably cynical – perhaps it wanted the girls to die before a verdict could be given – Berry now drew on the feeble health of his clients to fight their cause, asking four different doctors to sign to sworn statements: 'These girls are all becoming progressively worse. It is very possible that all or some of these five girls may be dead by September 1928.'

It made for horrendous reading for the women. Humphries reported they were 'kept under a constant mental strain'. Yet it was the kind of move that Berry instinctively knew would get results – and he was right. For faced with this kind of injustice, the media were up in arms. Berry's ally Walter Lippmann rose to the occasion magnificently, writing in the *World*: 'We confidently assert that this is one of the most damnable travesties of justice that has *ever* come to our attention.'

His influential editorial provoked immediate support from across the nation. One man wrote to the *News*, 'Open the courts, quash the postponements, give these five women a fighting chance!' Norman Thomas, meanwhile, a socialist politician who was often called 'the conscience of America', declared that the case was a 'vivid example of the ways of an unutterably selfish capitalist system which cares nothing about the lives of its workers, but seeks only to guard its profits'.

'Everywhere,' said Katherine Schaub, almost in disbelief, 'people were asking why justice was being denied these five women, who had but a year to live. What had once been a hopeless case, unheeded and unnoticed, now flashed before the public.'

And the public was transfixed. 'Letters came pouring in from all corners of the earth,' Katherine remembered.

Though most were positive, some swung the other way. 'Radium could not produce the effects ascribed to it,' one radium-company executive wrote bitterly to Quinta. 'It is pathetic that your lawyers and doctors should be so ignorant.' Some quacks were aggressive in their overtures. 'For $1,000 [almost $14,000] I can cure each of you,' a woman proposing a treatment of 'scientific baths' declared. 'If not, I will ask nothing except the $200 [$2,775] I want in advance. This means life or death ... You had better work fast, for when that poison reaches your heart – goodbye girlie.'

Many letters contained suggestions for cures. These ranged from boiled milk and gunpowder to magic words and rhubarb juice. An electric blanket was another suggestion, with its manufacturer envisioning a unique marketing opportunity. 'It is not to make money we wish to cure them,' protested the firm. 'The advertising it would give our method would be amply paid.'

The girls were famous. Undeniably, truly famous. Berry, himself adept at envisioning opportunities, immediately capitalised on it. He broached the subject of courting the press with the girls and they were all for it. And so, as the month of May 1928 dragged on and every day seemed to bring forth another call for justice from the press, Berry ensured that the girls were centre stage. Close friends Quinta and Grace gave a joint photo-shoot and interview; Grace wore a pretty cherry-patterned blouse – with her now-constant bandage on her chin – while Quinta donned a pale dress with a pussy-bow neckline. And the girls, every one of them, talked. They shared the details of their lives: how Quinta had to be carried to her hospital appointments; how

Albina had lost all her children; how Edna's legs were crossed beyond repair. They let their personalities shine through their suffering – and the public adored them.

'Don't write all this stuff in the papers about our bearing up wonderfully,' Quinta said with a cheeky smile. 'I am neither a martyr or a saint.' Grace remarked she was 'still living and hoping'. 'I am facing fate,' she declared, 'with the spirit of a Spartan.'

They weren't always easy interviews. When journalists asked Quinta about Mollie's death, she had to stop for a moment to compose herself. Katherine Schaub said in one interview: 'Don't think I'm crying because I'm downhearted – it's because my hip hurts so. Sometimes it seems as though a knife was boring into my side.'

Yet the tragedy and pain were part of the appeal for the captivated public. Radium poisoning – with its child-killing devastation and disfiguring symptoms – 'seemed to destroy their very womanhood'. The public, shocked and saddened, took the girls to their hearts.

Berry soon realised how much the coverage was helping – because Edward Markley was spitting feathers. 'Personally, I do not like your attitude,' wrote the USRC lawyer huffily to Berry, 'especially the newspaper notoriety which you are giving these cases. The ethical aspect of trying your case in a newspaper is questionable, to say the least. I am quite confident that eventually you will be properly rewarded, either in this world or the next.'

Berry replied only briefly. 'I am surprised,' he wrote innocently, 'that *you* should raise the question of ethics . . .'

Whatever Markley thought of the media, however, the firm he represented knew it had to present its side of the story. Predictably, USRC wheeled out Dr Flinn, who pronounced that his tests showed 'there is no radium' in the women; he was convinced, he said, that their health problems were caused by nerves. This was a common response to women's occupational

illnesses, which were often first attributed to female hysteria. The *World*, for one, was utterly unconvinced by Flinn.
Lippmann wrote that his statement had 'all the appearance of
being timed to support the argument of the [USRC] lawyers'.
He continued: 'It is not part of this newspaper's practice to
attempt to put pressure upon the courts. But this is unmanly,
unjust and cruel.'

Markley was powerless to stop the rising wave of support for
the women. When asked for a comment, all he could say was
that he felt the girls were being 'exploited by a young Newark
lawyer'. Yet the women themselves certainly didn't feel that
way. They were leading the charge to bring their employers to
justice. At last, the world was listening to them – and they were
not shutting up.

'When I die,' Katherine Schaub told the press with heart-
rending pathos, 'I'll only have lilies on my coffin, not roses as I'd
like. If I won my $250,000, mightn't I have lots of roses?

'So many of the girls I know won't own up,' she went on, 'they
say they are alright. They're afraid of losing their boyfriends
and the good times. They know it isn't rheumatism they've
got – God, what fools, pathetic fools! They're afraid of being
ostracised.'

Grace Fryer was also telling truths. 'I couldn't say I'm happy,'
she admitted, 'but at least I'm not utterly discouraged. I intend
to make the most out of what life is left me.' And, when the time
came, she said she wanted to donate her body to science, so that
doctors might be able to find a cure; the other girls would later
follow suit. 'My body means nothing but pain to me,' Grace
revealed, 'and it might mean longer life or relief to the others,
if science had it. It's all I have to give.' She gave a determined
smile. 'Can't you understand why I'm offering it?'

The journalists almost swooned. 'It is not a question of giving
up hope,' one reporter commented after Grace's promise. 'Grace
has hope – not that selfish hope that perhaps you or I might
have, but the hope for contributing betterment to humanity.'

With such a public platform – and public sympathy – the momentum in the case was definitely in the women's favour; and it was at this point that Judge Backes came up with an inspired interpretation of the statute for Berry. He suggested that because the girls' bones contained radium and the radium was still hurting them, they were still being injured; 'therefore, the statute began tolling anew each moment of that injury'. It was brilliant.

Whether it was an argument that would stand up in court was, of course, still to be tested – but Berry found that, in the light of public pressure, the justice system was now willing to support him. No matter the response of the radium company, the trial was scheduled to go ahead. Towards the end of May 1928, Judge Mountain wrote to Berry: 'I will set [the] cases down for trial on Thursday next. Counsel will accordingly prepare to proceed on that morning without fail.'

Nothing was going to stand in the way of justice – of that both Berry and the girls were sure. Carried along on a swell of public favour, it seemed they would soon be home and dry.

Berry was in his office, preparing for the case, when his telephone chimed.

'Mr Berry?' said his secretary Rose. 'Judge Clark is on the line.'

31

Judge William Clark was a hugely respected man. Born with a silver spoon in his mouth – he was the grandson of a senator; the family estate was called Peachcroft – he was thirty-seven years old, with auburn hair, grey eyes and a large nose. He was also Berry's old boss, back when Berry had been a clerk; for Clark had once been a partner in the law firm Lindabury, Depue & Faulks.

'To Judge Clark's office,' read Berry's diary for 23 May 1928, 'and talk re: radium cases with him.' His former boss had a suggestion to put to him.

'Would it not be possible,' Clark enquired lightly, 'to settle the suits out of court . . .?'

Berry wasn't the only party with whom the judge was conferring. On 29 May, Clark met with President Lee and the legal team of USRC; Berry was not asked to attend. When a reporter questioned Berry about the meeting, he commented, 'I know nothing of any such arrangements. I am not even considering a settlement out of court.'

Though he professed to the reporter that he was 'more determined than ever [that the case] will be fought now to

[the] bitter end', privately he was starting to have his doubts. It wasn't that he didn't think he could win; it was whether any verdict would come in time to benefit the girls. Every time he saw them, they seemed weaker than before; Humphries had already told him they were 'not physically or mentally able' to attend the upcoming trial. Even Grace Fryer, who was normally almost effervescent compared with her friends, seemed quieter and less demonstrative. 'I don't dare do much with my hands,' she confided, 'for fear of being scratched. The least scratch will not heal because of the radium.' The girls were becoming like china dolls, wrapped up in the cotton wool of their medical care. Berry wanted them to get justice, but most of all he wanted them to be comfortable in their final days. Perhaps, he thought, he should give Clark's suggestion due consideration, as long as any settlement was fair.

Berry's musings were compounded just a day or so later, when Katherine Schaub collapsed in church. 'Pains like streaks of fire through my whole body!' she cried out. 'I can't go on this way. I wish I wasn't going to live another month.'

It seems Berry made his mind up: it would be inhuman not to try to get the girls a settlement if an offer might be forthcoming. Any legal case could take years to fight and, as Berry well knew from the four sworn statements in his files, the women might not live until September.

On 30 May, Judge Clark was reported as an unofficial mediator. It was a move that provoked considerable comment within the legal profession, for the judge was intervening in a case over which he had no jurisdiction. Clark, however, said he resented criticism. 'Just because I am a federal judge,' he asked rhetorically, 'does that mean that I cannot have a heart?' His motives, he said, were entirely humanitarian.

The following day, USRC held a board meeting to discuss what possible terms it might offer in settlement. Vice President Barker now declared that 'the directors wanted to do what was fair'. He added, however, 'We absolutely deny any liability.'

The company had very good reason to want to settle. Thanks to what it called a 'cleverly designed campaign of publicity' (in which, it said – without any sense of irony – 'the human aspect of live women doomed to die was played up in an appealing manner'), the groundswell of support for the women's case was overwhelming. Settling this famous case out of court would not only make both it and all the negative publicity go away, but it meant the firm could choose *when* to fight its battles in court. Inevitably, there would be future lawsuits from other dial-painters, and the firm no doubt foresaw that they might get an easier ride in a few years' time, when Grace Fryer and her friends were not still plastered all over the papers. A settlement suited the company just fine.

With USRC now happy for the cogs to turn quickly, a meeting between Berry and the firm's lawyers was held the next day, Friday 1 June, at 4 p.m. in Judge Clark's chambers. Two hours later, Clark made a quick statement to the excitable press waiting outside as he ran to catch his evening train: 'There is no definite news but I am confident that the matter will be definitely settled at a conference [on] Monday.'

Everybody seemed happy – everybody but the girls. They were not impressed. RADIUM VICTIMS REJECT CASH OFFERS: WILL PUSH CASES; PARLEYS NOW OFF yelled one headline. The firm had offered them $10,000 ($138,606) each in settlement, but all the girls' medical bills and the costs of litigation were to be deducted from that sum, leaving only a pittance.

'I will not grab at the first thing that comes along,' exclaimed Grace fiercely. 'I will not knuckle down to them now after all I've suffered.' Quinta McDonald simply said, 'I have two small children. I have to see to it that they are provided for after I'm gone.'

No, the women said, *we do not accept*. Grace, as ever, seemed to lead the fight: she declared that she would 'absolutely refuse to accept the company's offer'. Instead, after discussion with the girls, Berry pitched alternative terms to USRC: $15,000 ($208,000) as a cash lump sum for each woman, a pension of $600 ($8,316) a year for life, past *and future* medical expenses, and USRC to cover all court costs. The firm would have the weekend to think it over.

Monday 4 June would prove a hectic day. At 10 a.m., negotiations continued with the world's press camped outside. When, after forty-five minutes, the lawyers exited Clark's chambers, they had to use a rear stairway to escape the massed media.

They were leaving to draw up formal papers. That afternoon, Berry summoned five brave women to his office. They dressed for the occasion: all wore smart cloche hats, while Grace slipped a fox fur around her shoulders. Even Albina made it to this most exceptional meeting; she had barely left her bed in the past month. But better than any outfit, more dazzling than any jewels, were the smiles that wreathed all their faces. For they had done it. Against all the odds, after a phenomenally hard battle – fought while they were in the most fragile health imaginable – they had nonetheless held the company to account.

They spent three hours with Berry and, in that time, the women signed the settlement papers. The company had kept the lump sum in the final agreement at $10,000 – but it agreed *all* their other terms. It was a quite extraordinary achievement.

The media flashbulbs glared as the women posed for a photo to mark the moment. Quinta, Edna, Albina, Katherine and Grace. They stood all in a row: the dream team. The 'smiling sorority' – and, for this one day, not sadly smiling, but beaming, false teeth and all, in pure delight and not a little well-deserved pride.

The formal announcement of the settlement came from Judge Clark himself at 7 p.m. By now, a crowd of perhaps three hundred had gathered; 'all aisles and passageways to the elevators were jammed'. Clark fought his way through the crowds to a good vantage point, from which he could break the news. He cleared his throat and asked for silence, which fell in a soft hush, broken only by the pop of flashbulbs and the papery whisper of pen on pad. Once he had the full attention of the press, the judge announced the exact terms of the deal. 'You can say, if you want to,' he added unctuously, 'that the judge did a good job.'

The settlement specified that the company admitted no guilt. Markley added purposely, '[The firm] was not negligent and the claims of the plaintiffs, even if well-founded, are barred by the statute of limitations. We are of the opinion that [USRC's legal] position is unassailable.' The corporation itself, meanwhile, released a statement proclaiming its motivation in settling was purely 'humanitarian'. The statement ended: '[USRC] hopes that the treatment which will be provided for these women will bring about a cure.'

And therein lay another crucial part of the settlement. The company had insisted that a committee of three doctors be set up to examine the girls regularly: one physician would be appointed by the girls, one by the company, and one mutually agreed. 'If any two [doctors] of this board should arrive at an opinion that the girls are no longer suffering from radium [poisoning],' Berry noted, 'the payments are to cease.'

It was obvious what the company officials planned; they didn't even try to hide it from Berry. 'I fully believe,' Berry wrote, 'that it is the intention of the corporation, if possible, to work out a situation in which they will be able to discontinue payments.'

It all sat extremely uneasily with him, especially because, while he knew his former boss to be 'a very honourable man', he now heard rumours that Clark 'was friendly with certain of the [USRC] directors'. Worse than that, he 'possibly had some indirect business relations with some of the directors of [a company with] a controlling interest in [USRC] who were schoolmates of his' and Berry even learned that Clark 'is, or was, up to a very recent time, a stockholder in USRC'.

'I have,' Berry said with trepidation, 'a great fear in the situation.'

In the Essex County courthouse in Newark, its elaborate murals are dedicated to four things: Mercy, Justice, Peace . . . and Power. In this case, Berry mused, the last seemed cruelly apt.

Clark himself wrote to the women: 'I want to express to you my very great personal sympathy, and my earnest hope that some way will be found of helping your physical condition.' And it was the women, at the end of the day, for whom this settlement was everything. They had come out on top; they had never thought they would live to see the day.

'I am glad to have the money,' commented Albina with a smile, 'because now my husband will not have to worry so much.' Her sister Quinta added, 'The settlement will mean so much, not only to me, but to my two little children and my husband. I want to rest after this ordeal I've been through. I'd like to go with them to some seaside resort.' She pronounced herself 'dissatisfied with the terms', but said: 'I am glad to be free from the worry of the court and am pleased with the thought of receiving the money right away.'

'I think Mr Berry, my lawyer, has done wonderful work,' Edna enthused gratefully. 'I'm glad to get the settlement; we couldn't have waited much longer. It will mean a lot of the things we want, for as long as we can appreciate them.'

Katherine simply said: 'God has heard my prayers.'

It was really only Grace who expressed a more muted response. She said she was 'quite pleased': 'I'd like to get more, but I'm glad to get that. It will help in so many ways; it will alleviate some of the mental anguish.' She added, of their courage in bringing the lawsuit in the first place and of what they had achieved so publicly, 'It is not for myself I care. I am thinking,' she said, 'more of the hundreds of girls to whom this may serve as an example.

'You see, it's got us – so many more of us than anybody knows yet . . .'

32

Ottawa, Illinois
June 1928

The New Jersey settlement made international headlines – and the front page of the *Ottawa Daily Times*. MORE DEATHS RAISE RADIUM PAINT TOLL TO 17! screamed the paper. A STARTLING JUMP IN THE TOLL OF RADIUM-POISON VICTIMS!

The girls in the Radium Dial studio were petrified. It wasn't as though they didn't have anything to worry about; Ella Cruse had *died* last summer and several former workers weren't well: Mary Duffy Robinson; Inez Corcoran Vallat. The Ottawa paper, which the girls pored over with increasing panic, said the first manifestation of radium poisoning was decay of the gums and teeth. Peg Looney, whose tooth extraction from last year still hadn't healed, felt sick to her stomach.

'The girls became wild,' remembered Catherine Wolfe. 'There were meetings at the plant that bordered on riots. The chill of fear was so depressing that we could scarcely work – scarcely talk of our impending fate.'

The studio became a silent, still place: the girls slackening in their work, hands no longer lifting brushes to mouths at breakneck pace. Given they could scarcely work, production declined and Radium Dial took action, calling in experts to run medical tests.

Marie Becker Rossiter watched proceedings with a beady eye. She noted that 'they separated the girls. Some of the girls they took upstairs, away from the other ones. They tested both groups, but separately.' The women didn't know why. Was it to do with those other tests the company had run, back in 1925? But those results had never been shared with the girls, so they did not know.

Divided into their groups, the women apprehensively went to meet the doctors. The physicians checked to see if the girls had radioactive breath, using the tests the Newark doctors had devised; they also took X-rays and blood.

Catherine Wolfe was tested; Peg Looney; Marie Rossiter. Helen Munch, who was about to leave the firm to marry, also blew into the machine. The girls, reassured that the firm was looking after their best interests, returned to their desks and waited for the results that, they hoped, would set their minds at ease.

But the results never came. 'When I asked for a report on the examination,' Catherine recalled, 'I was told that this information could not be given out.'

She and Marie conferred about it. Didn't they have a right to know? Marie, always forthright, determined that they shouldn't take it lying down. Full of fear and indignation, she and Catherine confronted Mr Reed.

Their manager adjusted his glasses somewhat awkwardly and then made an expansive gesture. 'Why, my dear girls,' he said to them paternally, 'if we were to give the medical reports to you girls there would be a riot in the place!' He almost seemed to make a joke of it.

That response clearly didn't settle the girls' nerves, though Catherine later said, 'Neither of us then realised what he meant.'

Mr Reed, seeing their uncertainty, continued, 'Don't worry. There is no such thing as radium poisoning. There is nothing to these stories of radium poisoning!'

'Are the workers in danger?' Marie demanded.

'You don't have anything to worry about,' the superintendent repeated. 'It's safe.'

Nonetheless, the girls continued to devour the newspaper daily, seeing more horror stories that shot bolts of fear right through them.

And then, three days after the announcement of the New Jersey girls' settlement, with tensions in the studio still running high, there was a big piece on page three of the local paper that completely supported their superintendent's statements. The women all pointed it out to each other and read on with ever-lightening shoulders.

It was a full-page ad placed by the Radium Dial Company and here, at last, the girls learned the results of their recent tests. 'We have at frequent intervals had thorough ... medical examinations made by ... technical experts familiar with the conditions and symptoms of the so-called "radium" poisoning,' read the company statement. 'Nothing even approaching such symptoms or conditions has ever been found by these men.'

Thank God. The results were clear. *They were not going to die.* And the company reassured them further: 'If their reports had been unfavourable, or if we at any time had reason to believe that any conditions of the work endangered the health of our employees, we would at once have suspended operations. The health of [our] employees is always foremost in the minds of [company] officials.' The ad continued:

In view of the wide circulation given reports of [radium] poisoning ... it is time to call attention to an important fact that has as yet received only occasional mention in the news ... All the distressing cases of so-called 'radium' poisoning reported from the east have occurred in establishments that have used luminous paint made from mesothorium ... Radium Dial [uses] pure radium only.

That was why Mr Reed had said 'there was no such thing as radium poisoning', the girls now realised. This was why radium was safe – because it wasn't *radium* that had hurt the women out east, it was mesothorium.

Radium Dial, evidencing its claim, cited the work of the 'expert' Dr Frederick Hoffman, who was continuing to promulgate his long-held belief that mesothorium was to blame; a belief Hoffman held even after Dr Martland disagreed with him, von Sochocky changed his mind and Raymond Berry wrote to him, having seen some of his media statements, to say: 'The tests would indicate that there is more radium than mesothorium affecting the [New Jersey] girls.' But Hoffman seems to have ignored all these contradictions to his theory.

Now, in Ottawa, Mr Reed proudly printed bulletin notices with the company's statement, posting them up in the workrooms and deliberately calling the girls' attention to them. 'He said that we should take particular notice of this advertisement,' Catherine recalled.

And he continued to reassure the girls: 'Radium will put rosy cheeks on you!' he told Marie with a grin; then he turned to Marguerite Glacinski and said cheekily, 'Radium will make you girls good-looking!'

The women continued to read the paper – but continued to read only good news. The company re-ran its advert over several days, and the newspaper itself wrote an editorial in support of the community employer, saying the firm had been 'ever-watchful' of its employees' health. The whole town was happy. 'Radium dials' were heralded as one of Ottawa's leading industries; it would have been an awful shame to have lost the business but, thanks to the firm's solicitousness, there was no need for alarm.

In light of all this, the girls returned to work, their panic set at rest. 'They went to work, they did what they were told to do,' said a relative of Marie, 'and that was the end of it. They never questioned it [anymore].'

'The girls,' remembered a local resident of the time, 'were "good Catholic girls" who were raised not to challenge authority.'

And what was there to challenge? The test results were fine and the paint contained no deadly mesothorium. These were simple facts – printed in the paper, pinned up on the notice-board – as certain as the sunrises that bled each morning across the yawning Illinois skies. Back up in the studio, the old routine continued anew. *Lip ... Dip ...*

Only one family, it seems, was unconvinced by the company.

For the day after the advert ran, Ella Cruse's family filed suit against Radium Dial.

Orange, New Jersey
Summer 1928

For those five New Jersey dial-painters who had triumphed over their former firm, life was sweet. From her award, Katherine gave her father, William, $2,000 ($27,700) towards his mortgage: 'I could find, I knew, no greater happiness than that which would be mine by making the folks happy,' she pronounced. 'It made me so happy to see Father relieved of those worries.'

For herself, she declared she would live 'like Cinderella as the princess at the ball ... Today was mine.' The budding author bought a typewriter, as well as splashing out on clothes: silk dresses and lingerie. 'I bought the kind of coat I had always wanted,' she enthused, 'and a tan felt hat to match.'

Edna, who had always loved music, invested in a piano and a radio. Many of the women bought automobiles so they could get around more easily. Yet the girls were also financially astute, investing in building and loan shares.

'Not a cent of [the money] has ever entered this house,' Grace informed a reporter. 'To me, money doesn't mean luxury. It means security. Those $10,000 are safely invested.'

'What for?' asked the journalist.

Grace smiled enigmatically as she answered. 'For the future!'

And the money wasn't the only boon to their spirits, for many of the doctors they consulted now offered hope. Von Sochocky announced that, 'In my opinion, the girls are going to live much longer than they themselves believe.' Even Martland, noting that there had been no deaths for some years of the ilk suffered by Mollie Maggia and Marguerite Carlough, theorised that there were now 'two kinds of dial-painter cases, early ones and late ones. The early ones were marked by severe anaemia and jaw necrosis ... The late cases lacked (or had recovered from) the anaemia and jaw infections.' Martland thought the faster decay of mesothorium accounted for the difference; the girls were under attack ferociously for the first seven years, but once mesothorium moved to its next half-life, the attack diminished sufficiently to spare the girls; almost as though their poisoning was a rising tidal wave, and the women had managed to scramble to safety just as the waters began to recede. Although the radium was still bombarding their bones, radium was notoriously less aggressive than mesothorium. Martland now posited that if the late cases 'survived the early maladies, they had a fair chance of surviving radium poisoning altogether' – although they would always have those moth-eaten bones from the radium rippled right through them. 'I am of the opinion that the girls we are seeing now,' he said, 'while they may be permanently crippled, have a considerable chance of beating the disease.'

That prognosis, bleak as it sounded in some ways, gave the women that most precious of commodities: *time*. 'Someone may find a cure for us, even at the eleventh hour,' Grace said brightly.

Most of the girls went away for the summer. Albina and James set off on 'the dream of a lifetime': a motoring trip to Canada. Louis Hussman took his wife on 'a long, leisurely tour'; Edna wrote to Berry, 'We have a cottage overlooking the lake and enjoy the beautiful scenery.' While Quinta and James McDonald took a few trips to Asbury Park, they didn't go mad;

Quinta was aware that this money was to see her children all right, no matter what happened to her.

However they spent the summer, the girls could rest easy that help was coming to other women similarly afflicted; in the light of the tremendous publicity caused by their case, a national conference into radium poisoning would be held at the end of the year. In addition, Swen Kjaer was now undertaking a much more detailed federal study into radium poisoning. 'There is no question that this is an occupational disease and that there should be a reinvestigation,' commented Kjaer's boss Ethelbert Stewart. He was asked why some firms were still using the old brush application method when others had been invented and replied shrewdly: 'The new methods probably were too slow for the greatest profit to the manufacturers.'

Katherine Schaub, who spent the entire summer away from Newark, experiencing 'real country life', was feeling so much better and declared her summer 'splendid'; 'a vacation like I have never had'. 'I loved to sit on the porch in the sun,' she wrote dreamily, 'and look out over the wide stretches of woodlands and hills.'

While sitting on that porch, she wrote to Berry to thank him for all he had done. 'I myself know,' she wrote, 'that from a humanitarian viewpoint it would be difficult to find another like yourself ... there was nothing too much for you ... and to think that the result now was such a tremendous success overwhelms me indeed.' She also wrote to Martland, as did the other girls, saying simply: 'I am writing to express my sincerest appreciation for your great assistance in bringing it all to a happy ending.'

A happy ending ... if only that could be. Behind the scenes, Berry was most concerned that Katherine's 'happy ending' was as fictional as a fairy tale. 'I think that the matter is not, by any means, over,' he wrote to an associate, 'and that the actual contest has only been deferred.'

Following the settlement, USRC had immediately launched

into damage-limitation mode regarding what it dubbed 'so-called radium poisoning'; it still denied any hazard existed and seemed confident that the medical board appointed to examine the girls would soon give all five a clean bill of health. The firm wasted no time in appointing two doctors who might well give that clearance: one was James Ewing, the radium-medicine specialist who had already spoken out against Martland – one of Berry's doctor friends warned, '[He] must be watched' – and the other, mutually agreed appointment was Lloyd Craver. Both were consultants at a hospital 'closely allied with the use of radium', but Berry found it 'impossible' to keep them out. The girls' appointed physician would be Edward Krumbhaar. Martland wrote: 'The damage is done now and Berry must make the best of it.'

In the fall of 1928, the girls were summoned to a New York hospital for the first committee examination. Since two of the physicians denied the existence of radium poisoning, it is to be wondered what they thought of the suffering women who now came before them. Katherine was 'obviously very lame and bent over'; Grace had 'distinct limitation of motion in her left elbow' and what remained of her jawbone was 'exposed' in her mouth. Quinta was in plaster casts; Edna's legs irrevocably crossed. Yet as the women stripped and underwent invasive medical exams, conducted by these doctors who were strangers to them, it was perhaps Albina's condition that shocked the physicians most. Krumbhaar later said, 'Mrs Larice had marked limitation of motion of both hip joints, so that it was almost impossible for Dr Craver to make a vaginal examination.'

The doctors conducted a breath test; two of them convinced it would clear the company. Yet the results, as Ewing wrote after-wards, 'proved positive, rather to our surprise'. Rather than take this as evidence the girls were telling the truth, he continued, 'The question now arises whether there might be some kind of fraud by the patient ... To make the tests absolutely trustworthy, we think it will be necessary to carry them out at some hotel where the patients can undress.' The girls would have to go through it all again.

In November, the five women attended the Hotel Marseilles for further examination. This time, only Craver from the committee was present; but it was not he who was in command. Instead, Dr Schlundt – the 'intimate friend' of Vice President Barker who had already declared the women to be non-radioactive in the company's breath tests in April – took charge. Barker himself was also present and 'assisted'; there was a further doctor, Dr Failla, in attendance too.

The girls perceived at once that this was not an impartial exam, but what recourse did they have to stop it? It was part of their settlement that they would agree to medical procedures. And so they were forced to strip as directed and went through the tests with the company men watching all they did closely.

The moment they were free from the hotel, however, Grace Fryer telephoned Berry. She was – as in many ways she always had been – the lynchpin of the group, and their leader. Now, she brought their collective protests to Berry.

Their lawyer was outraged. He wrote at once to USRC to say he viewed the hotel setting with 'great suspicion' and believed the presence of Barker and Schlundt 'constitutes a breach of the settlement agreement', for the committee tests were supposed to be non-partisan. As it turned out, however, Dr Failla declared emphatically: 'All five patients are radioactive.'

It was a genuine blow to the company, for every day they seemed to receive another lawsuit; they'd wanted to have these famous dial-painters deemed free of radium as a further defence. Berry himself was representing one of the new cases, acting for Mae Cubberley Canfield, the dial-painter who'd instructed Katherine Schaub. Like the others, Mae's teeth were gone and her gums infected; her jaw also 'felt funny ... like a knock in it' and she was paralysed intermittently on her right side.

Berry won a battle partway through the new war when the

judge in Mae's case ruled that Dr Flinn could not conduct examinations of her for the company; only a physician could do it. It was a small victory, for nothing had come of Berry's complaints to the authorities about Flinn. The lack of action left him free to publish: Flinn next blamed the girls' 'improper diet' for their 'tendency to store radium in their bones'.

No one knew what von Sochocky's diet was, but improper or not, that November he lost his battle against the radium inside him. Martland paid tribute to the doctor: 'Without his valuable aid and suggestions,' he said, 'we would have been greatly handicapped in our investigation.' That was true, for without von Sochocky's help with the creation of the tests, radium poisoning might never have been medically proven. Of course, without von Sochocky's invention of luminous paint in the first place, the girls would have been leading very different lives ...

For the girls, they could not forget what they saw as the doctor's betrayal in the courtroom. Perhaps, then, there was a kind of schadenfreude in his demise. One newspaper described radium paint as a 'veritable Frankenstein in a test tube, which has turned on its creator'; Martland added, 'He died a horrible death.'

It meant he wasn't present at the national radium conference, held in December 1928. All the key players were there: Hamilton, Wiley, Martland, Humphries, Roach, Ethelbert Stewart, Flinn, Schlundt and the radium-company executives.

No one invited the dial-painters.

It was a voluntary conference, organised by the trade, in an attempt to claw back some control. The Surgeon General, who was chairing it, acknowledged that 'anything we draw up here is simply in the form of suggestions, but not any authority in the way of police regulations'. It was, as Wiley's boss later put it, 'a whitewash'.

The issues were debated. Stewart made a passionate speech to the radium industry: 'The luminous watch is purely a fad. Do you want to go ahead with the use of a thing which is so useless;

which has, in spite of everything you can do, an element of serious danger in it? I certainly hope that you are going to agree that it is not worth what it costs.'

But the companies did not agree; one firm said 85 per cent of its business came from luminous dials – it was far too lucrative an industry to abandon. The executives argued that only New Jersey cases had come to light so it wasn't a nationwide problem; with Flinn having silenced the Waterbury girls, Stewart could riposte with only one formally documented case outside USRC, which was evidenced by the Ella Cruse lawsuit in Illinois – yet hers was only a suspected case and not proven. The lack of evidence of an endemic problem meant the girls' supporters were powerless to push through any proposals, even though Wiley's boss called it 'cold-blooded murder in industry'.

The conference didn't confirm that radium poisoning existed or even that radium was dangerous, it simply agreed that further study should continue via two committees – yet there is no record the committees ever met. As the New Jersey girls' stories became yesterday's news, no one was championing the dial-painters' cause anymore. 'The Radium Corp.,' Berry wrote in frustration, 'is playing a game.' And, it seemed, the radium companies were winning.

There were two other delegates worthy of mention at the national radium conference: Joseph Kelly and Rufus Fordyce, of Radium Dial – the executives who had recently signed their names beneath the company statement in the Ottawa press. They appear only to have listened, and not contributed to the debate. They listened as one specialist said, 'My advice to anyone manufacturing watches today would be to cut out the brush because you can paint on in another way.' They listened as the New Jersey girls' deaths and disabilities were debated. They listened, as the industry got away with murder.

And then they went home.

34

Ottawa, Illinois
1929

On 26 February 1929, radium-poisoning investigator Swen Kjaer made his way to the LaSalle County courthouse in the little town of Ottawa. He was surprised by how quiet it was; today, there was a hearing in the Ella Cruse case and given the cacophony caused by the radium lawsuits out east, he had expected more fuss. Yet nobody was around; not an eyelash flickered in the sleepy town.

Inside the courthouse, during the hearing, it was equally undramatic. Here there were no throngs of journalists, no star witnesses, no duelling attorneys. All that happened was that the Cruse family's lawyer, George Weeks, simply stood to request a postponement. With momentum provided by the New Jersey lawsuits, Kjaer was surprised he wasn't pushing it through more quickly.

Afterwards, when Kjaer questioned Weeks, he discovered why he wasn't. The lawyer had needed to ask for postponements several times because he knew nothing about radium poisoning – and could not find any physician in Ottawa who could give him information. The family was claiming $3,750 ($51,977), which wasn't avaricious, but at this rate they would not see a cent. Weeks couldn't find anyone to tell him what radium poisoning was, let alone if Ella had died from it. Her parents were informed that the only way to get proof would be to exhume her body for an autopsy, but it would cost $200 ($2,772); money they simply didn't have. The case was left high and dry.

Kjaer continued on his pilgrimage around town. He called on the doctors and dentists who had promised to alert him should any dial-painters present symptoms of radium poisoning. As before, they all reported no cases.

He also visited the Radium Dial studio. It was still bustling, filled with women painting dials. He met the manager and requested that the company-test data be shared with him. Radium Dial was now conducting regular medical exams of its employees – though the girls had noticed, as before, that they were separated prior to being tested. Catherine Wolfe even remembered, 'Only once [was I] called to report for a physical examination [in 1928], whereas other girls apparently in good health were examined regularly.'

Catherine was not in especially good health; she still had a limp, and just recently she had started suffering from fainting spells. Concerned, she'd asked Mr Reed if she could see the company doctor again, but he'd refused. She told herself she was worrying over nothing. The company had assured her that expert tests showed she was healthy and vowed to close the studio if there was any hazard; yet it was busier by the day. As time passed after the furore in New Jersey, orders swiftly rose again to 1.1 million watches a year. Business was back on track.

However, Kjaer's inspection of Radium Dial troubled him. Two lab workers from Chicago showed changes in their blood, demonstrating that the firm's safety precautions were insufficient. The girls were also still eating in the studio without washing their hands. Kjaer concluded: 'Further steps should be taken to protect the workers.'

He met Joseph Kelly; the president promised him that the firm's 'intention' was 'to assist you in every way possible'. Having now perused the test results, Kjaer wanted to discuss two employees in particular; one was Ella Cruse. Kjaer declared, 'I feel that this case should not be left out [of my study].' He requested further information on both girls.

Yet when Kelly sent him the data, all he enclosed were employment dates; hardly enlightening. Kjaer's time was limited, so he didn't grill the company further; he thought he had enough to go on anyway.

And so, in his report, which was never seen by the Radium Dial girls, he wrote:

> One dial-painter, ML, a twenty-four-year-old female, employed in a studio in Illinois, had been found radioactive in 1925 by electroscopic test. In 1928, another test was made, and she was found still radioactive ... Complete information was not obtainable, and the firm protests against calling the diseased condition radium poisoning, but it seems well indicated by the test.

ML. *Margaret Looney*. She had been told by the firm she had 'a high standard of health'. She had been told that her tests showed nothing to worry about.

She had no idea of what was coming.

Peg Looney smiled up at Chuck Hackensmith. 'Thank you,' she said.

Chuck threw a golden grin over his well-muscled shoulder and picked up the handle of the red metal wagon that Peg was sitting in. 'Here we go ...' he cried with typical verve to his fiancée. *And then the cold marble athlete leapt to life ...*

'Chuck used to put Peg in a little wagon when she got so bad,' remembered Peg's sister Edith, 'and pull her up to where we used to have a picnic. She couldn't walk so he just pulled her, put her in the wagon and away we went ... He was a wonderful fella, Chuck.'

'Chuck felt awful bad about [her illness],' added her sister Jane.

The whole family did. By the summer of 1929, red-haired Peg Looney was not at all well. The teeth extractions that never healed had only been the start of it; she'd developed anaemia and then this pain had settled in her hip so that now she could barely walk – thus the little red wagon that Chuck had commandeered to take her up to the Shack or along to Starved Rock. He was awful kind, but then he loved her fiercely. They were going to be married next June.

Chuck and his red wagon couldn't be there all the time though. When Peg went to the radium studio, she had to walk. Her sister Jean remembered the way that she and all the Looney siblings would look out for her coming home.

'We'd all be sitting out on our porch just watching for her because she looked so bad walking,' said Jean. '[She'd be struggling] all the way home. We'd run to meet her, each one would have an arm to help her.'

When she reached home, borne along by her siblings, Peg could no longer assist her mother with the housework as she once had. She would simply have to lie down and rest. Her mother felt terrible watching her daughter's decline; Peg was wasting away and her family watched in horror as she pulled teeth and parts of her jaw from her mouth.

'My parents took her to a doctor in Chicago,' remembered Jean. The city physician told her she had a honeycombed jaw and that she should change employment.

Perhaps Peg planned to look for a new job, when she felt better. Yet Peg was smart; she knew she *wasn't* getting better. Though the Ottawa doctors seemed clueless – one, who treated her in June 1929, simply put an ice pack on her chest – Peg herself seemed to divine what was happening. 'She knew she had to go,' recalled Peg's mother sadly. 'You could see her slowly dying. There was nothing you could do.'

'Well, Mother,' she used to say. 'My time is nearly up.'

It wasn't just her hip or teeth that caused her agonising pain: it was her legs, her skull, her ribs, her wrists, her ankles ... Though she'd been ill for months, every day she still went to work to paint those dials. To the end, she was a conscientious girl.

Radium Dial – warned by Kjaer that Peg's was a special case in which the government was particularly interested – watched her very closely. They knew she had tested positive for radioactivity in 1925 and 1928; they knew from their own medical tests exactly what was wrong with her. And so, when Peg collapsed at work on 6 August 1929, Mr Reed made arrangements for her to be admitted to the company doctor's hospital.

'We had no say whatsoever about that,' said her sister Edith. 'They wouldn't let us in there.'

'Radium Dial probably paid the bills,' added Darlene, Peg's niece. 'We didn't have money for big medical bills; that was for sure.'

It was so lonely for Peg in that distant hospital, far from her home by the railroad tracks. The girl who had nine brothers and sisters and slept with them all in the one tiny room, three to a bed, was completely on her own. Her siblings weren't permitted to visit. 'I went one time,' recalled Jane, 'and they wouldn't let me into her room.'

Peg had displayed symptoms of diphtheria and was promptly quarantined. In her weakened condition, she also soon contracted pneumonia. Radium Dial, in a show of concern, paid close attention to her progress; to her decline.

At 2.10 a.m. on 14 August 1929, Margaret Looney died. This girl, who was to marry Chuck next year, who loved to read the dictionary, who had once had dreams to be a teacher and was well known for her giggling fits, was no more.

Her family, though isolated from her, were still in the hospital when she died. Peg's brother-in-law Jack White, who was married to her sister Catherine, an imposing man who worked as a car oiler for the railroad, was one of the relatives present. He was the type of man who stood up for the right thing to be done. Which was why, when the company men came in the middle of the night and tried to take her body to bury it, Jack said no.

'No way is she going to be buried that way,' he said firmly to them. 'She's a good Catholic girl and she's going to have a mass and a whole funeral.'

The company men tried to protest. 'They wanted the whole thing done with – just gone,' said Darlene. 'It was like a big cover-up.' But Jack wouldn't allow them to take Peg's body.

Radium Dial lost that particular battle – but did not give up. It seems the firm was concerned that Peg's death would be attributed to radium poisoning, which would scare all the girls at the studio and possibly lead to innumerable lawsuits. The executives needed to take control of the situation. What did the family think, they asked, of having Peg autopsied?

The Looney family were already suspicious, given the Chicago doctor's comments, that it was her work that had killed Peg. They readily agreed, on condition that their own family doctor could be present, because they wanted to find out the truth. Their proviso was all-important: after the firm's midnight machinations, they did not trust them.

The company agreed easily. Yes, yes, they said, no problem. What time?

When the family doctor arrived at the appointed hour, bag in hand, he found the autopsy had been performed an hour before he got there.

He wasn't there to see the multiple fracture lines on Peg's ribs, nor the way 'the flat bones of [her] skull showed numerous "thin" areas and "holes"'. He didn't examine the radium necrosis that was found 'very strongly' in the skull vault, pelvis and at least sixteen other bones. He did not witness the widespread skeletal changes that were evident throughout Peg's battered body.

He was not there to see as the company doctor 'removed by post-mortem resection' the remains of Peg Looney's jaw.

He took her bones. He took the most compelling evidence.

The family was not sent a copy of the report, but Radium Dial received one. It was an incredibly intrusive record for them to have of Peg's last moments. It told them what she was like inside: the weight of her organs, their appearance; whether she was 'normal' or not. When it came to her bone marrow and her teeth, according to the company doctor, *she most certainly was*.

'The teeth are in excellent condition,' read the official autopsy report. 'There is no evidence of any destructive bone changes in the upper or lower jaw.'

Her death certificate was duly signed: diphtheria was the cause of death.

The family may not have been given a copy of the report, but Radium Dial made sure to issue the local paper with a summary of it. And so, in Peg Looney's obituary, the following information was included at the request of the firm:

The young woman's physical condition for a time was puzzling. She was employed at the Radium Dial studio and there were rumours that her condition was due to radium poisoning. In order that there might be no doubt as to the cause of death [there was] an autopsy ... Dr Aaron Arkin ... said there was no doubt that death was caused by diphtheria. There was no visible indication of radium poisoning.

There was a curious final comment, perhaps inserted on a press release by a company executive with a bright idea of how to win support in the community. 'Miss Looney's parents,' read the piece, 'appeared well pleased with the result of the autopsy.'

They were not 'well pleased'. They were devastated by their daughter's death.

'It just killed my mother to lose her,' said Jean. 'She was never the same after she died. My mother was just terrible. We used to walk up to the cemetery all the time, early in the morning, pushing an old push-mower to keep the grass cut up with it; it was a few miles. We'd walk up there all the time.'

As for Chuck, losing his beloved Peg was something he would never get over. He moved on with his life, eventually, and followed the dreams that once they both had shared. He became a professor at a university and published several books; Peg, no doubt, would have loved to have read them. He married, and had children. And he kept in touch with the Looney family for more than forty years. His wife confided in Peg's mother how every year, when it was near the anniversary of Peg's birthday or death, he would become quiet and withdrawn.

'She knew,' Darlene said simply, 'he was thinking about Peg.'

35

Orange, New Jersey
1929

Katherine Schaub rebuttoned her blouse after her medical examination and waited for Dr Craver to speak; he'd said he had something important he wished to discuss. To her astonishment, he proposed that the radium company should stop paying her medical bills; in the settlement they had agreed to cover them for life. He wanted her, instead, to accept a one-off lump sum.

Less than a year after the New Jersey settlement had been reached, the United States Radium Corporation was attempting to renege on its agreement.

The idea of paying a lump sum originated with Vice President Barker, but it had the full support of the company doctors. Dr Ewing deemed the current arrangements 'unsatisfactory' as 'these women are not going to die'. In his lab, Craver now used to Katherine 'the potent argument about the bankruptcy of the company' to induce her to accept – but USRC was not bankrupt; anything but. Such an untruth, Berry later said when Katherine told him anxiously of the doctor's scheme, was 'purely a "painted devil" to compel a settlement'.

Finding the women still alive one year on seems to have been a financial irritation to the firm, for the women, crippled and in pain, were regularly consulting doctors and buying palliative medicines. From USRC's perspective, it was too much; they quibbled every bill. The girls, Ewing warned threateningly, should be 'cautious about assuming that every expense they incur will be paid'.

The board of doctors had been expected to announce that the women were not suffering from radium poisoning, thus freeing the company from its responsibilities. It seemed that Ewing, whom Berry described as having a 'hostile attitude', certainly longed for that diagnosis. But to Ewing's frustration, even though the board kept subjecting the women to test after test, they found each one duplicated the previous results.

Berry wanted the board to issue a formal statement that the girls had radium poisoning: it would be firm evidence that dial-painters as a group were afflicted by it, which Berry and others could then use in the upcoming lawsuits for the girls' friends. But Ewing refused. 'We are quite unwilling to have these findings used in connection with any other case,' he wrote primly.

As for the girls themselves, they were just doing their best to get through it all. They were subjected to a harrowing array of experimental treatments and tests. The physicians tried Epsom salts that made them sick, colonic irrigations and week-long assessments on their spines and excreta. The exams were usually conducted at Ewing and Craver's hospital, which meant the crippled women had to travel to New York. Louis Hussman told Berry that 'it is very difficult for Edna to go so far without injury to her; the last time she went to New York she had to go to her bed as a result'.

Edna's beautiful blonde hair, by now, had become snow-white. All the girls looked far older than they were, with faces that had curiously slack skin around their chins, where their jawbones had been removed. Only Grace seemed better than she had the year before. Although she had now had twenty-five

operations on her jaw, they had failed to break her habit of smiling; she was said to be the happiest of the five by far. When she'd
received the settlement, she'd said with determination, 'People
are now asking me if I am going to stop working: I do not intend
to do anything of the kind. I'm going to keep right on at my job
as long as I can, because I like it.' She still commuted daily, with
the bank being understanding about the time off she needed for
her tests.

Though the tests happened often, the girls never learned
the results. 'The doctors don't seem to tell [me] anything,'
complained Katherine. 'I would like to know if I am getting
any better.' In fact, in many ways, Katherine *was* better, for she
now lived quietly in a rural convalescent home set on a hilltop,
twelve miles out of Newark, which she called 'the jewel of the
east'. She wrote that the setting inspired her to get well so she
could enjoy 'hollyhocks and rambler roses and peonies and sunshine'. The money had helped Albina too; she was described as
being 'the picture of contentment' that summer. Her pleasures
were now her radio, goldfish, the movies and short jaunts in the
country, often with Quinta.

At the present time, however, Quinta had been admitted to
hospital; she was unable to sit up and only family visitors were
allowed. It not only meant that she wasn't available for jaunts
to the country; she was also unable to attend court on behalf of
Mae Canfield, as the other four women did in the summer of
1929. Quinta did, however, ask Berry to represent her.

It was a preliminary hearing. As he worked on Mae's case,
Berry was fast coming to appreciate the sheer canniness of the
radium firm in settling the year before. The second time around,
it was even harder for him to build a case; the Drinkers, Kjaer
and Martland all refused to testify, and there were no champi-
ons in the press hounding the firm into submission.

The five girls were helping Mae by waiving their right to
patient confidentiality; they wanted the committee of doctors
to use their cases to prove that radium poisoning existed. But

not only did Markley object to any reference to the five girls –
both their medical diagnoses and even the fact of last year's
settlement – by saying that they were 'in no way connected to
this case', the company-appointed doctors also declined to give
evidence.

Yet as Katherine had once written, it would be difficult to find
another like Raymond Berry. He summoned Craver and Ewing
to the hearing regardless; they were 'furious'. Even though
Ewing witnessed the women swearing under oath that they
were happy for him to discuss their cases, he refused on the
grounds of patient confidentiality.

Dr Krumbhaar, the girls' ally on the board, was happy to
give evidence. And even though Markley threatened to sue him
if he did, Berry persuaded the doctor to continue. The lawyer's
skill both in handling witnesses and in presenting his case was
growing; he now had all the data and experience to make life
very difficult indeed for the United States Radium Corporation:
he was a most uncomfortable thorn in its side. The executives
had assumed that when they settled the first five cases, Berry
would be off their backs. They now realised they had been very
much mistaken.

Black Tuesday, they called it – 29 October 1929, the day a financial nightmare rocked Wall Street and 'paper fortunes melted
away like frost under a hot sun'.

'Wall Street,' wrote one witness to the crash that day, 'was a
street of vanished hopes, of curiously silent apprehension and
of a sort of paralysed hypnosis.'

More than a hundred blocks north of where America's economy was imploding, Quinta McDonald lay in her room at the
New York Memorial Hospital. Here, too, was silent apprehension and paralysis – but never, Quinta promised herself, *never*
would her hope be gone.

She had been admitted in September 'in a dying condition', but a month on she was still fighting – and how. It was

incredible for her friends and family to witness. 'She was a Spartan,' said her sister-in-law Ethel, who was caring for the McDonald children while their mother was in hospital. 'She always said "pretty good" when I asked her how she was. Never did she think she was going to die.'

'Her one thought was to live for the children,' commented Quinta's husband, James. 'The thought of them lent her courage to fight for her life.'

The McDonalds were now reconciled, yet the past year had been turbulent. Although James had been awarded $400 ($5,544) in the 1928 settlement, that sum was dwarfed by his wife's new wealth – and it seems the difference had rankled. Then unemployed, James had spent his money in speakeasies over the summer, while Quinta had invested hers in a trust fund for the children. One night in September 1928, his resentment had come to a head. When Quinta refused his demands for money, James had viciously struck his crippled wife and threatened to gas her to death, turning on every gas jet in the house as she lay helpless in her plaster casts. He was arrested. Quinta, however, did not press charges; it was not the first time she'd been hit. She did begin divorce proceedings, with Berry's help, but it seems James later won her round and they were eventually dropped. 'My husband tries to be brave,' she'd said of him once. 'But it's harder on men than women.'

Now, in the fall of 1929, it was Quinta who had to be brave. 'For the past three weeks,' Ethel said in early November, 'she could not move. She had to be fed with a spoon.' But in a turn-around that amazed doctors, Quinta now began winning her desperate fight.

She might have been inspired in her recovery by Grace and Albina, who were both doing well. When Grace visited Quinta in hospital one evening, she granted the reporters waiting outside a brief interview, revealing proudly that she no longer always wore her back brace. 'Doctors told me I had great resistance to disease and that's why I got along so well,' she told them, and then added jokingly: 'I had resistance enough to get up and vote for Hoover when I was supposed to be sick in bed!'

Quinta, too, hoped to be up again before too long – or at least well enough to go home. She improved rapidly; so much so that James got the house ready for her return, and the family celebrated Thanksgiving and their daughter Helen's tenth birthday with the cheering thought of her homecoming uppermost in their minds.

'Each time we [saw her] during the last several weeks,' Grace enthused, 'she has been stronger. And today she was her old self again. It's been a long time since she has been so well.' Quinta asked Grace to buy Christmas gifts for the children on her behalf; she was determined to make it a holiday season they would never forget.

By 6 December, Quinta was almost perky. James visited her on that Friday evening and they chatted about Christmas; they hoped she would be home to enjoy the festivities with the family. Midway through their conversation, she suddenly sighed.

'I'm tired,' she remarked.

James was not surprised. He bent to give her a kiss, careful not to touch her leg. She had a swelling of some size at the top of her thigh, and it gave her a lot of pain. They both glanced at the clock on the ward; it was not quite the closing hour for visitors.

'Would you mind leaving a little early?' she asked.

He did as she requested, departing with no sense of foreboding.

That swelling on Quinta's leg ... Had Martland seen it, he might have recognised it. For it was a sarcoma – the kind of bone tumour that had killed Ella Eckert on a cold December day almost two years ago.

Just before 2 p.m. on 7 December 1929, Quinta McDonald sank into a coma. The hospital telephoned James and he left home immediately, driving as fast as he could; he was stopped twice for breaking the speed limit, but the police let him go after they learned his mission.

His efforts were all in vain. When James arrived at the Memorial Hospital, 'tears streaming down his face', he was a few minutes too late. Quinta McDonald was dead. He oscillated between rage and depression before settling on simple grief.

'I am heartbroken,' he later said. He added quietly, 'I am glad she has found peace.'

Her friends were devastated. They had become a tight-knit unit: the five of them against the company; against the world. Quinta was the first of them to fall. Albina collapsed when she heard the news; Katherine Schaub was greatly shaken too. Katherine chose not to attend the funeral, but returned to her country home 'to find forgetfulness and to continue [her] studies'. She was taking an English correspondence course at Columbia University; she planned to write a book about her experiences. 'For a time,' she said, 'I succeeded in losing myself entirely in my writing.'

For the girls remaining in Orange, there was no such forgetfulness. In a way, they *wanted* to remember: to remember Quinta. On Tuesday 10 December, Edna, Albina and Grace arrived at St Venantius Church for her funeral. The variance in their fortunes was clear for the waiting reporters to see: Grace 'walked briskly and unaided', while Edna 'seemed to be the most affected by the disease'. For Albina, this was the second sister she had lost to radium poisoning and even to attend was a struggle. Yet she was determined to pay her respects. There was a long flight of stairs leading to the church door, but Albina fought her way up every single step, even though she was 'apparently near collapse'. This was more important than her comfort. This was for Quinta.

It was a brief service. Helen and Robert, the McDonald children, 'kept close to their father, both too young to realise their loss, yet sensing'. In the weeks to come, they would indeed have a Christmas they would never forget.

Immediately after the Mass, the family and close friends progressed to Rosedale Cemetery, where Quinta would join her sister Mollie in rest. It was a simple burial, without fuss; as she would have wanted.

There was one other thing she had asked for. She wanted her death to be of service to her friends; 'she could thus,' said Ethel sadly, 'leave a parting gift to other victims.' And so Martland conducted an autopsy and discovered that Quinta had died from the same rare sarcoma that had killed Ella Eckert. Quinta's may not have been on her shoulder, but it was the same thing; it was just that the radium had chosen a different target in her bones. Martland gave a statement about this new threat. 'The bones of the victims,' he revealed, 'had actually died before they did.'

One might have thought that on learning of Quinta's death – this woman the company doctors had professed was not going to die – the United States Radium Corporation might, at last, have softened. But you would be wrong. Berry did manage to win a settlement for Mae Canfield in the New Year of $8,000 ($113,541) but the company had a straitjacket clause attached. The only way they would pay his client any money, they said, was if Berry himself was incorporated into the deal. He was far too knowledgeable about their activities – and becoming far too skilled in court – to be left off a leash.

And so Raymond Berry, legal champion, the pioneering attorney who had been the only lawyer to answer Grace's call for help, found himself forced into signing his name to the following statement: 'I agree not to be connected with, directly or indirectly, any other cases against the United States Radium Corporation, nor to render assistance to any persons in any actions against said Company, nor to furnish data or information to any such persons in matters against said Company.'

Berry was gone. He had been a serious fighter against the firm; an irksome thorn in their side. But now, with surgical precision, they had plucked him out and banished him.

They were two settlements down, but the United States Radium Corporation was winning the war.

36

Ottawa, Illinois
1930

Catherine Wolfe gave a huge sigh and rubbed her hands in tiredness over her face and along her short dark hair. She watched idly as the clouds of radium dust flew up around her from the layer on her desk, disturbed by her discontented sigh. Then she returned reluctantly to weighing out the material for the girls. Catherine wasn't a full-time dial-painter anymore; the change in her duties had been directed by the studio bosses.

They'd been good to her really, she thought; Mr Reed had been so understanding. He'd called her in to see him one day last year and said that, on account of her poor health, she was to take a six-week vacation. Radium Dial knew that she was sickly and, as they had done with Margaret Looney, they were keeping a close eye on her.

Yet the vacation hadn't helped. And so her job, eventually, had been changed. Her work now, in addition to weighing, was to scrape out the compound from the girls' dishes, often simply using her fingernails. As was to be expected, her bare hands became 'luminous bright', and as it was her habit to run them through her hair her whole head glowed fiercely. If anything, she often thought, when she peered at herself in a mirror in a dark bathroom, the new job got her even more covered in radium than the old one had.

It wasn't as fun as painting had been, but then again Radium Dial was different: most of Catherine's clique had left by now, only she, Marie Rossiter and Marguerite Glacinski remained. Catherine tried to see her new role as a promotion: the radium was very valuable, so to be the worker chosen to distribute and salvage it was an accomplishment. After eight years' employment, she was one of the most trusted workers.

Even so, she knew some girls gossiped about the reason for the change in her duties. 'I believe,' said one of her colleagues, 'she was transferred from dial-painting because she was a poor worker.'

Not so poor that she didn't still paint some dials, Catherine thought defensively. Every week there seemed to be some emergency order that required an extra pair of hands. Then Catherine would slip her brush between her lips, dip it in the powder and paint; the girls all still did it that way at Radium Dial, for their instructions were never changed.

There was a sudden movement in the studio; Catherine looked up to see the girls going for medical tests. Catherine stood up to join them, but Mr Reed intercepted her easily. 'I was excluded from the examination,' Catherine remembered. 'Mr Reed told me not to go.'

She had asked Mr Reed personally several times now for examination by the company doctors, but was always refused. She'd gone to a local physician and he'd told her that her limp was caused by rheumatism. Catherine felt she was too young for that; she was only twenty-seven. 'I knew I was suffering from some disease, but what it was I did not know,' she said in frustration.

At least, she thought as she sat back down again with a heavy sigh, none of it had put off Tom Donohue. She gave a little smile at the thought of him. One day soon, she knew, they would marry. She allowed herself to daydream a little. Maybe they would have a family – though you couldn't assume what blessings God would give you. Marie Rossiter had lost two babies already; she'd just discovered she was pregnant for the third time, and Catherine prayed fervently that this baby would survive.

Charlotte Purcell and her husband Al, too, had had a terrible time. They'd had a son, Donald (nicknamed Buddy), in August last year, but he'd come two months early and weighed only two-and-a-half pounds. The doctors had kept him in an incubator for six weeks; at last, the little trooper had pulled through.

As Catherine sat lonely at her desk, the other girls gone to be examined, she felt disappointed at Mr Reed's rejection of her plea. Maybe she should do what Inez Vallat had done, she thought lightly. Inez had gone to the Mayo Clinic in Minnesota for examination, on account of her bad headaches and locking hips. Though she was only twenty-three, Inez was unable to work at all now; she'd lost twenty pounds over the past year and looked as skinny as anything when Catherine had seen her in church. More worryingly, her teeth had started coming loose and she now had an infected mouth; Inez had to hold a bandage constantly to her seeping jaw.

It was a bit like what Peg Looney had suffered; though she had died of diphtheria. Poor old Peg; she still missed her dreadfully. Catherine possibly didn't know it, but Peg's family had consulted a lawyer to bring suit against Radium Dial, just as Ella Cruse's parents had done. (The Cruse case, incidentally, was still no further forward.)

'The family,' Peg's sister said, with some understatement, 'felt the death certificate was in error.'

Their lawyer was a man named O'Meara. A single hearing was held in 1930, but nothing ever came of it; perhaps O'Meara ran into the same trouble as George Weeks. 'He couldn't do nothing for us,' remembered Jane Looney.

Jean added: 'My dad said at the end, "You can't beat them. There's no sense in trying. It's not worth going through this kind of mess."'

'Just forget it,' he'd say bitterly. 'We won't go any further.'

There was nothing they could do.

There was nothing the doctors could do, either, for Mary Vicini Tonielli. She had quit Radium Dial when she became sick; she had sciatica, she thought, but when she prodded gingerly at her back, she realised she had some kind of lump on her spine.

'The doctor said it was a sarcoma,' Mary's brother Alphonse later recalled.

Mary had had an operation on it in the fall of 1929. But sixteen weeks on, she was no better. In fact, Alphonse revealed, 'She suffered like a dog for four months. There was never any more peace for her.'

On 22 February 1930, Mary Tonielli died; she was twenty-one. Her husband of less than two years, Joseph, buried her in the Ottawa Avenue Cemetery.

'We thought it was radium poisoning,' said Alphonse bleakly. 'But her husband and the old people didn't investigate. They felt so bad about her death.'

37

Orange, New Jersey
1930

Katherine Schaub placed her cane gingerly on the low step in front of her; she could now walk only with the aid of a cane or crutches. She had been forced to return to Newark: having spent large sums trying to regain her health, she was now entirely dependent on her $600 ($8,515) annuity, but it didn't provide enough money for a rural residence. She hated being back in the city, where she felt her health declined.

She started up the low step, but slipped and came down hard upon her knee. It would have been painful for anyone, but Katherine was a radium girl: her bones were as fragile as china. She felt the bone fracture, but when Dr Humphries examined her X-rays, he had worse news to tell her than the broken bone.

Katherine Schaub had a sarcoma of the knee.

She was admitted to hospital for ten long weeks, while they treated it with X-rays. It seemed to reduce the swelling, but Katherine was utterly demoralised. Encased in plaster for months, she was eventually told the bone 'didn't knit the way it should' and that, from now on, she would have to wear a metal brace. 'A lump came into my throat,' Katherine recalled, 'as the doctor fastened on my leg the strange contrivance. I cried a little bit, but my faith consoled me.'

Despite the consolation of her faith, however, she found her prognosis profoundly depressing. That old cine-reel from years gone by started up again in her mind, now with an ever-growing cast of ghost girls. Where Katherine had once gained relief from being in the sunshine, now, she said, she was 'having difficulty with the light and sun up here on the roof'. 'My head,' she stammered, 'had me full of fears – couldn't tell if it was mental or real ... I couldn't stand the light in my eyes; was a wreck by 4 p.m.' Perhaps all this was why she began to develop what she called 'this craving of alcohol of mine'.

The committee of doctors was, as ever, on hand to help, but Katherine now refused the treatments Ewing and Craver suggested. 'They say you do not know a person,' she wrote assertively, 'until you have lived with them. I have lived with radium ten years now and I think I ought to know a little bit about it. So far as [the suggested] treatment, I think it's all bosh.' She would not kowtow to their demands.

Ewing and Craver were mad about it – and not just Katherine's stubbornness, but increasing boldness from all four remaining women. 'Relations are far from satisfactory,' Krumbhaar wrote. 'It is difficult to get them to come to see us and they will not accept our treatment.'

Yet in standing up for themselves the women were playing a dangerous game; the committee had control of the purse strings for their medical care. It wasn't long before Grace was told she could no longer call on Dr McCaffrey; the board also raised concerns about Dr Humphries, writing: 'It might be that, even though [Humphries] has the women's confidence, it would be better, all things considered, for someone else to take care of them.'

The company was 'kicking' about every bill, yet the firm itself was in good financial health. Despite the Wall Street Crash, the use of luminous dials had not diminished and the firm was also still supplying radium for the Radithor tonic and other medicines; the craze for these had continued, after a momentary dip when the girls' stories first hit the headlines.

The year 1930 flowed fluidly into 1931. At New Year Katherine was still in hospital, though her tumour was shrinking thanks to Humphries's attentions; it currently measured 45 centimetres. Come February, she was still unable to do much in the way of walking, but it seemed she had beaten the worst of it.

The spring of 1931 found Grace Fryer in good spirits too; partly because she had made a new friend at her hospital appointments. By chance, the famous aviator Charles Lindbergh was working on the floor above, and once in a while he would visit her. 'My impression,' said Grace's brother Art, who drove her to her appointments, 'was that these occasional visits made her feel many times better, even if it was only for a short while. Seeing Grace's spirits higher was perhaps one of the greatest feelings I've ever had.'

Grace was still determined to be as positive as she could. It was true that she'd had to revert to wearing her brace, but she did not let it slow her down. 'I work and I play and I "dance" a bit,' she said. 'I go motoring. I even swim – but I can stay in the water only two minutes at a time. I can't leave the brace off my back any longer than that.'

In the hospital in Orange, however, there were no such diversions for the latest patient who was now wheeled through its doors. Irene Corby La Porte, who had worked with Grace during the war, now followed her friends to Dr Humphries's office.

It was the summer of 1930 that she'd noticed something wrong. She and her husband Vincent, longing for a family – Irene had suffered three miscarriages by that time – had made love while they were staying in a cottage in Shark River Hills. But it didn't feel right to Irene, inside. There was a swelling in her vagina, which interfered with intercourse.

Vincent took her to Dr Humphries, who diagnosed a sarcoma, then about the size of a walnut. Despite the doctor's efforts, her decline was swift. 'Her whole leg and side began to swell rapidly and paralysed her,' recalled her sister. 'She was getting worse every minute.'

Irene was admitted to hospital, but by March 1931 the doctors said there wasn't very much they could do for her except try to relieve the pain. By then, the area around the top of her thigh had become four times as large, as the sarcoma grew unstoppably inside her. Doctors found that 'a vaginal examination could hardly be made on account of the tumour blocking the entrance to the genitals'; Irene had great difficulty urinating and the pain was 'terrific'.

In April, they called in Dr Martland. 'I found a bedridden patient extremely emaciated and filled with a huge sarcoma,' he remembered. His diagnosis was immediate and absolute.

'He told me definitely,' remembered Vincent La Porte, choking up, 'that she did [have radium poisoning] and she had about six weeks to live.'

They did not tell Irene, wanting to spare her, though she was smart enough to know. 'She was always saying, "I know I am dying from radium poisoning,"' remembered one of her physicians. 'I convinced her she wasn't; that she was going to get better. It is tact of a physician not to reveal a fatal prognosis.'

Martland wasted no time enlightening the world about the evolution of radium's MO. He had seen enough cases now to know that these latent sarcomas – which could leave a victim healthy for years after her exposure to radium, before coming horribly to life and taking over her body – were the new phase of this terrifying poisoning. He added: 'When I first described this disease, there was a strong tendency among some of those interested in the production and therapeutic use of radium to place the entire blame on mesothorium ... In the cases autopsied recently, the mesothorium has disappeared while the radium persists.' He could reach only one conclusion: 'I am now of the opinion that the normal radioactivity of the human body should not be increased; [to do so] is dangerous.' It had to be, for each week another dial-painter presented another sarcoma, each in a new location – her spine, her leg, her knee, her hip, her eye . . .

Irene's family couldn't believe how fast she was fading from them. But she still had grit in her. On 4 May 1931, as she lay dying in hospital, she filed a claim for damages against USRC; she was willing to settle.

But the company was just about done with settlements. Now they had seen off Berry, they were not too concerned about the adversaries to follow.

Only a month later, after fighting such a hard, hard battle that she was destined never to win, Irene died on 16 June 1931. At the time of her death, Martland said her tumour had become 'a huge growth'. So much so, he went on, that 'you couldn't take the whole mass out together without taking the woman apart. The whole mass was larger than two footballs.' That was how Irene La Porte died.

Her husband Vincent was filled with such a rage that he knew not what to do with himself. It was red hot at first, searing him with pain and grief, but as time went on it cooled into an icy, diamond-hard desire for vengeance. And Vincent La Porte would fight on for his wife. He would fight on through the courts – through 1931, and '32, and '33 and even beyond.

Irene La Porte's case against USRC would be the one that finally led to a judgment for all the Orange girls. Vincent didn't know it when he started, but that fight was going to take years yet. The company was in no rush.

Then again – neither was he.

Martland had one final statement to give on the sarcomas; on the insidious time bombs that he now knew were lurking inside any and all dial-painters who had ever once lifted a brush to their lips.

'I believe,' he said, 'before we get through, the number will be appalling.'

38

Ottawa, Illinois
August 1931

Catherine Wolfe paused for a moment on her way into work, stopping at the corner of East Superior Street to catch her breath. It was normally a seven-minute stroll to the studio from her house, but these days it took her much, much longer. As she limped down Columbus Street, the sight of the white church lifted her spirits; it was like a second home to her. It was where she had been christened and baptised; where she took communion; where she would marry one day ...

She had lots of blessings, she thought to cheer herself as she made her way along, counting them off as though they were the beads of her rosary. There was her health: for Catherine, despite the limp, was in fairly good health otherwise. There was Tom Donohue: the couple were due to be married in January 1932. There were her friends' blessings: Marie had had a healthy little boy, Bill; and Charlotte Purcell a girl, Patricia, who was not born early. And there was her job. Six million Americans were currently unemployed: Catherine earned $15 ($233) a week and she was grateful for every cent.

She had made it, finally, to Radium Dial. There was only Marguerite Glacinski from the old gang to say hi to now. As

Catherine made her awkward way over to her desk, she felt the other girls' eyes on her. Her limp, she sensed, was 'causing talk', but Mr Reed never criticised the quality of her work, so she tried not to let the gossip bother her.

She had just begun weighing the material when the girls nearest the window sent the message round that Mr Kelly and Mr Fordyce had come on a visit: the president and vice president of the firm, all the way from Chicago. The girls straightened their blouses and Catherine ran a nervous hand through her dark hair before she pushed herself up from her desk and limped across the studio to the stockroom.

She was partway there when Mr Reed and the executives came into the studio. Mr Reed was pointing out various aspects of the work, but Catherine had this funny feeling that the visiting officials were looking only at her. She got what she needed and made her slow way back to her desk. Mr Reed and the other men were still standing there, having an inaudible conference under their breath. She felt inexplicably anxious, and turned to face the windows, lit by the August sun.

The sunlight was blocked by a shadow.

'Mr Reed?' asked Catherine, looking up from her work.

He wanted her to come to the office; she made her tortuously slow way there. Mr Kelly and Mr Fordyce were also in the office. She fiddled with her hair again.

'I'm sorry, Catherine,' said Mr Reed suddenly.

Catherine looked at him in confusion.

'I'm sorry, but we have to let you go.'

Catherine felt her mouth drop open, suddenly dry. *Why?* she wondered. Was it her work? Had she done something wrong?

Mr Reed must have seen the questions in her eyes.

'Your work is satisfactory,' he admitted, 'it's your being here in a *limping* condition.'

She looked from one officer of the company to the next.

'Your limping condition is causing talk,' Mr Reed went on.

'Everyone is talking about you limping. It's not giving a very good impression to the company.'

Catherine hung her head, though whether with shame or anger or hurt, she was not quite sure.

'We feel ...' Mr Reed broke off for a moment, to make eye contact with his bosses, who bestowed on him an agreeing nod of endorsement: they were all in this together. 'We feel it is our duty to let you go.'

Catherine felt stunned. Shocked, wounded. 'I was told to go,' she remembered later. 'I was told to go.'

She stepped out of the office, left the radium men behind. She picked up her purse and limped back down the stairs to the first floor. All around her was familiarity – for nine years, six days a week, she had spent her life in this studio. The walls of the old high school seemed to ring for a second with the laughter of the girls she had known there: of Charlotte and Marie; Inez and Pearl; of Mary; of Ella; of Peg.

No one was laughing now.

Catherine Wolfe, fired for being sick, swung open the glass door at the entrance of the studio. It was six steps down to the sidewalk, and on every one she felt her hip ache. Nine years she had given them. It had meant nothing.

No one watched her go. The men who had fired her got on with their day, Mr Reed no doubt enlivened by the presence of Messrs Kelly and Fordyce; he was a company man, and the opportunity to rub shoulders with the bosses was not to be missed. The girls were too busy painting to put down their brushes. Catherine knew, as she reached the final step, what they would all be doing inside. *Lip ... Dip ... Paint.*

No one watched her go. But Radium Dial had underestimated Catherine Wolfe.

The firm had just made a very big mistake.

39

Orange, New Jersey
February 1933

Katherine Schaub bit down hard on her lip to keep from crying out, her eyes squeezing shut with the pain.

'All done,' said the nurse reassuringly, having changed the dressing on Katherine's knee.

Katherine opened her eyes warily, not wanting to look down at her leg. All through the past year, doctors had been keeping tabs on her tumour: it was 45 centimetres, they told her; 47 and a half; 49. Its earlier reduction had been reversed. In the past week or so, the bone tumour had broken through her paper-thin skin; now, the lower end of her femur was sticking out of the wound.

She tried to focus her mind on happier things. Before she'd been admitted to hospital, she had spent some time at a private sanatorium, Mountain View Rest, for her nerves, and that had been quite wonderful. She had finished writing her memoir; had even had an excerpt of it published in a social reformers' magazine. She, Katherine Schaub, was a published author: it was what she had always longed to be. 'I have been granted,' she wrote with peaceful pleasure, '[a] priceless gift – I have found happiness.'

If only she could have stayed in the mountains; she felt so

much brighter there. Yet as her health worsened and she'd had to take regular taxis into Orange to see Dr Humphries, the board of doctors had baulked at paying the bills. In fact, they'd had enough of the women's expenses altogether.

The previous February, 1932, Katherine, Grace, Edna and Albina had all received a no-nonsense letter from Dr Ewing: 'We wish to inform you that no bills will be approved by the Commission for any services which have not been specifically approved by Dr Craver. The Commission feels that they must scrutinise expenses more carefully.' The board now refused to cover medicines 'we do not feel are useful', routine doctors' visits and home nurses; the latter was a service the women increasingly relied on to help clean and dress themselves. The board was acting, it said, 'to prevent this "exploitation" of the radium corporation'.

There had been fallout from the committee's decision. For Katherine, it made her even more determined not to submit to their experiments: 'I have suffered my share ... I don't think that I should be at the mercy of these New York doctors.' The physicians moaned heartily about her behind her back: '[She is] one of the most difficult patients to handle,' complained one. 'I am really at a loss what to do with this highly hysterical woman.'

Katherine's suspicion of medical men appears to have made her nervous of accepting any therapeutic advice. Dr Humphries recommended a leg amputation, but she refused. 'I have made no headway with her,' Humphries wrote, 'and doubt very much being able to do so.' Katherine could be as stubborn as a mule when she wanted to be; it was perhaps partly why she had been one of the five girls who'd won a settlement from USRC in the first place.

Ewing's letter had mentioned 'the very depressed state of business' as a reason for the withdrawal of expenses. Inevitably, as the economy crumbled, sales of radium watches declined along with everything else. But it wasn't only that which was

Katherine Schaub.

Grace Fryer.

Edna Bolz Hussman.

Hazel Vincent Kuser.

Albina Maggia Larice.

Quinta Maggia McDonald.

Helen Quinlan
and her boyfriend.

Irene Rudolph.

Marguerite Carlough.

The dial-painters at a company social, including Ella Eckert (second left), Mollie Maggia (third from right) and Sarah Maillefer (second from right).

The dial-painting studio in Orange, New Jersey in the early 1920s.

Sabin von Sochocky (centre), the radium firm's founder, at a company picnic.

Top left: Arthur Roeder.
Top right: Dr Frederick Flinn.
Middle left: Katherine Wiley.
Middle right: Dr Harrison Martland.
Bottom left: Raymond H. Berry.

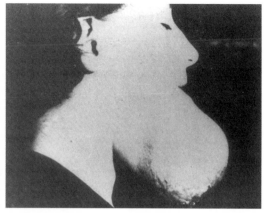

A dial-painter with a radium-induced sarcoma of the chin
(front and side views).

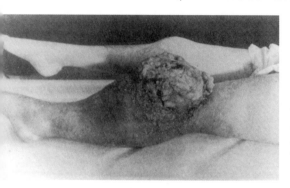

A dial-painter with a radium-induced
cancer of the knee.

Mollie Maggia's lower jawbone,
riddled with holes and crumpled
from the radium.

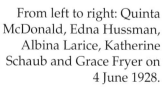

From left to right: Quinta
McDonald, Edna Hussman,
Albina Larice, Katherine
Schaub and Grace Fryer on
4 June 1928.

Peg Looney and Chuck
Hackensmith, with two
of her little sisters, Edith (far
left) and Theresa (right).

Marie Becker Rossiter.

Mary Ellen 'Ella' Cruse.

A section of the Radium Dial company photograph.
First row: Mr Reed (far left, seated on ground, wearing white flat
cap). *Second row:* Catherine Wolfe (second from left in black dress),
Miss Lottie Murray (fourth from left), Marguerite Glacinski (tenth
from left). *Third row:* Margaret Looney (first from left), Marie Becker
Rossiter (eighth from left), Mary Duffy Robinson (second from right).

Pearl Payne 'during a sick spell', *circa* 1933.

Charlotte Purcell, 1937.

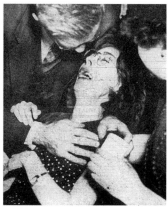

Five Ottawa women in Chicago, 21 July 1937. From left to right: Marie Rossiter, Frances Glacinski O'Connell, Marguerite Glacinski, Catherine Wolfe Donohue, Pearl Payne, Grossman's secretary Carol Reiser, Leonard Grossman.

Catherine collapses on 10 February 1938; Tom Donohue and Pearl rush to her aid.

From left to right: Pearl Payne, Frances O'Connell, Marguerite Glacinski, Helen Munch and Marie Rossiter listen to the evidence at the LaSalle County courthouse, 10 February 1938.

Charlotte Purcell demonstrates lip-pointing for Grossman, 11 February 1938.

The bedside hearing at the Donohue home. Charlotte sits behind Grossman, next to Pearl.

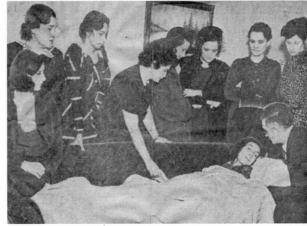

The Donohue family: Tom, Catherine, Tommy and Mary Jane.

United: Catherine surrounded by her friends and Tom.

now sucking the dollars from the firm's bank account. It was the case of Eben Byers.

It had been all over the papers last March. Byers was a world-renowned industrialist and playboy; a wealthy man who raced horses and lived in a 'magnificent home': he was high-profile and important. After receiving an injury back in 1927, his doctor had prescribed Radithor; Byers was so impressed with it he consumed several thousand bottles.

When his story made the news, the headlines read: THE RADIUM WATER WORKED FINE UNTIL HIS JAW CAME OFF. Byers had died of radium poisoning on 30 March 1932, but before he died he gave evidence to the Federal Trade Commission (FTC) that Radithor had killed him.

The authorities reacted with much more alacrity than they had in the cases of the dial-painters. In December 1931, the FTC issued a cease-and-desist order against Radithor; the US Food and Drug Administration would go on to declare radium medicines illegal. Finally, the American Medical Association removed the internal use of radium from its list of 'New and Nonofficial Remedies', where it had remained even after the discovery of the dial-painters' deaths. It seemed wealthy consumers were much more worthy of protection than working-class girls; after all, dial-painting was still going on, even in 1933.

Katherine had read the stories about Byers with sadness for the victim, but also an overwhelming sense of vindication. Radium *was* a poison. The girls knew it, intimately, but until the Byers case, public opinion had swung the other way. Indeed, with four of the famous radium girls still alive – almost five years on from their case – there had been much muttering that their lawsuit had been nothing more than a fraudulent scheme to get money from the company.

For that company, the Byers case was a disaster. USRC supplied the radium for many of the products that had now been banned. The whole radium industry collapsed. It may or may not have been connected, but in August 1932, having failed to find a buyer for the old Orange plant, the firm had it razed to the ground. The dial-painters' studio was the last building to come down.

The women had mixed feelings, seeing it gone. It was a bittersweet triumph of sorts; except, for them, erasing the studio and all it had done was not as simple as covering the site with anonymous asphalt. Lying in her hospital room in February 1933, Katherine Schaub forced herself to look at what the radium had done to her. Her leg was a mess. Finally, after much consideration, she decided to have it amputated.

It was a decision for her future. 'It is my ambition to continue writing,' she said. She could do that, she thought, with or without a leg.

But Humphries had bad news for her. 'There is no question of an amputation,' he now said. Katherine and her leg had worsened of late and both were now in far too serious a state for such a major operation to be performed. Subsequently, Katherine took another slide downhill. On 18 February 1933, at 9 p.m., she died at the age of thirty.

Two days before her funeral, perhaps distracted by grief, her beloved father, William, fell down a flight of stairs at his Newark home. He was rushed to hospital, but just one week after Katherine died she was joined on the other side by her dad. His funeral was held in the same church as Katherine's, and they were both buried in the Holy Sepulchre Cemetery. They were together, in the end, at the close of Katherine's long journey; of her 'adventure' as she herself put it.

Katherine Schaub was just fourteen when she'd started work with the radium company on that long-ago February day. She had dreamed of writing and of fulfilling her potential – and she did publish her work and she did fulfil that potential; it was just that her destiny was not quite the one she'd dreamed of as a girl. In taking on the company, she became a celebrated example of standing up for your rights.

40

I t could be worse, Grace Fryer thought, it really could be a lot worse.

Just lately, in this July of 1933, Grace had become bedridden, unable to get about at all. But, as she kept telling herself, it really could be a lot worse. 'I feel better when I'm at home,' she said brightly. 'I guess it's because I like home better than anything else.'

Grace's friend Edna shared her feelings about that. 'Home always makes me feel better,' Edna commented. 'I have my good and bad days, but I can endure them when I'm home.'

Edna was doing very well, all things considered. Despite her crossed legs, she still managed to move about with the aid of a cane, calling on friends and even hosting bridge parties. She'd taken up crochet; it was something she could do for hours on end without leaving her chair. Although her spine was now affected by the radium, she kept her spirits up and even believed she would live 'quite a few years yet'. Her optimistic attitude had a lot to do with Louis: 'He helps me so very, very much,' she said quietly.

Edna declared she never thought about her illness being fatal.

'What good would that do?' she exclaimed. She was leaving things to fate.

Albina Larice, meanwhile, found herself feeling surprised. She'd been expected to die before all the others; but here she was, six years on, still living. Katherine Schaub had died and so had Albina's little sister Quinta – but she was still here. It was a peculiar, confusing thing.

As with Edna, Albina's spine had become affected and she now wore a steel corset, but she was able to hobble about with her mouse-size steps as long as she had a cane. Though she was only thirty-seven, Albina's hair, like Edna's before her, had turned entirely white. She was known to be less cheerful than Edna, but then she had lost far more. Three children. Two sisters. It was a terrible, tragic tally.

But, thanks to her husband James's careful attentions, she was much happier these days than she had once been, when she'd taken only to her bed and thought of how she might end the half-life she'd been given. 'I know they say,' Albina said shyly, 'there is no hope of being cured – but I'm trying to hope there is.'

By September 1933, Grace was clinging to the same hope; but it seemed to fade with every passing day. Though her mother kept her home as long as she could, Grace was eventually admitted to hospital under the care of Dr Humphries.

He was worried, he said, about the growing sarcoma in her leg.

'I am not going to live much longer,' Grace had once said. 'No one has ever been known to recover from this trouble. So, of course, I will not either. But why worry?'

'It was not death of which Grace was afraid,' said her mother. 'It was the dread of the suffering – the eternal suffering – the years of torment. She was brave, until the last.'

The last came on 27 October 1933. She died at 8 a.m., a somewhat typically helpful time, ready for the doctors who were starting their day. It meant Dr Martland was able to attend her autopsy, to conduct his final careful consideration of this most special of patients. Grace's death certificate stated she was killed by 'radium sarcoma, industrial poisoning'. It was a fact, black and white: it was the radium industry that killed her. It was the company.

Grace was buried in Restland Memorial Park; her grave marked with a stone that had a gap beneath her name. When her mother passed away, fourteen years later, her name was added to her daughter's, so they could both rest in peace.

Her death was reported by the local papers. The family supplied a photograph of Grace to accompany the news; a picture from before the poisoning had set in. She looked forever young: her lips smooth and shining; her eyes piercing, as though she could see into souls. She wore a conservative set of pearls, a lace-shouldered blouse. She was beautiful and bright and unbroken, and it was how she would always be remembered by those who'd loved her.

'The family all just seemed so sad,' her nephew Art remembered. He'd been born after Grace had died, the son of her little brother Art, who used to take her to hospital appointments. 'My father really did not talk about it. But I think he was affected his whole life by this. This was his big sister, who was a beautiful girl.'

And Grace Fryer was not just beautiful. She was brilliant. She was smart. She was determined and forthright and strong and special.

Her little brother did once speak of her, when his grandson had asked him to. 'I will never forget her,' he'd simply said. 'Never.'

And Grace Fryer was never forgotten. She is still remembered now – *you* are still remembering her now. As a dial-painter, she glowed gloriously from the radium powder; but as a woman, she shines through history with an even brighter glory: stronger than the bones that broke inside her body; more powerful than the radium that killed her or the company that shamelessly lied through its teeth; living longer than she ever did on earth, because she now lives on in the hearts and memories of those who know her only from her story.

Grace Fryer: the girl who fought on when all hope seemed gone; the woman who stood up for what was right, even as her world fell apart. Grace Fryer, who inspired so many to stand up for themselves.

She was buried in Restland Memorial Park. But even as she was laid to rest, her story was not over. For her spirit lived on, 800 miles away, in the women who came after her. When Grace died, no radium company had been found guilty for killing their workers. No firm had been found at fault. Now, as Grace slept peacefully, others would take up her torch. Others would follow in her footsteps. Others would battle on as they continued the fight. For recompense. For recognition.

For justice.

PART THREE

Justice

41

Ottawa, Illinois
1933

The executives of the Radium Dial Company had had confirmed knowledge of radium poisoning since at least 1925, less than three years after their studio first opened in Ottawa. That was the year Marguerite Carlough first filed suit in New Jersey and Martland devised his tests. The executives had read Kjaer's studies, attended the radium conference and seen the Eben Byers story: they *knew* radium was dangerous.

When their employees had found out about the New Jersey cases in 1928, the company had lied. There was a full-page advertisement in the paper: the girls are safe, their medical exams prove it; the paint is safe, for it is 'pure radium only'. When Peg Looney died, the company lied. There was 'no visible indication of radium poisoning'; but only because her jawbone was no longer visible to anyone, having been cut from her after death.

With these assurances plastered across the papers, the company had been supported by the town. After all, the executives had promised they would close the studio if there was *any* danger. No wonder the town was behind them, when they took such good care of their employees and were willing to put people before profits. It must be really, *really* safe to work there, everyone thought.

For eight years on from Marguerite's lawsuit, Radium Dial was still trading daily in the little town of Ottawa.

Oh no, the local doctor said, it was definitely *not* radium poisoning that Catherine Wolfe Donohue had. She limped out of the consulting room, still no wiser as to the cause of her illness, and made her slow way home to East Superior Street. She was not alone; she pushed a stroller in which lay her baby son, Tommy, born in April 1933, just over a year after she'd married Tom Donohue. 'God has sure blessed me,' Catherine wrote, 'with a grand husband and [a] lovely child.'

She and Tom had been married on 23 January 1932 in St Columba. It was a modest wedding with only twenty-two guests; Catherine's uncle and aunt had both passed away by then and Tom's family did not approve of the union. As their niece Mary remembered it, 'None of Tom's family wanted him to marry her because they saw her as not being in good health.' But Tom Donohue adored Catherine Wolfe and it was a love match; he married her no matter what his relatives said.

The Donohues seem to have come around to the idea by the time the vows were exchanged: Tom's brother Matthew was the best man, while his twin sister Marie also attended. The local paper deemed it 'one of the prettiest weddings of the mid-winter season'. As Catherine had limped down the aisle to marry Tom, dressed in a green crêpe gown and with a bouquet of tea roses clutched in her hands, she'd thought, despite her hobbling steps, that she had never felt so well; a feeling topped when they'd then been blessed with Tommy. If it wasn't for her declining health, she would be on top of the world.

Today's appointment had been the third doctor she'd consulted, but he was as uninformative as the rest. 'They were just guessing,' commented one dial-painter's relative about the town's physicians. 'They had no idea [what the trouble was] – especially any doctors in Ottawa.'

It was certainly true that the local physicians, perhaps as a result of the isolation of the tiny town, were not the most knowledgeable. Some of their cluelessness was apparently due to ignorance, despite the fact that by this time Dr Martland had published many articles on radium poisoning. As an example, one Ottawa doctor – a former physician of Peg Looney, as it happened – had recently stated: 'It has never been brought to my attention that the use of luminous paint could in any way be responsible for the production of sarcoma.'

Whether it had been brought to their attention or not, Ottawa doctors were now seeing peculiar conditions in former Radium Dial girls. Sadie Pray had had a big black lump on her forehead; she'd died back in December 1931 – of pneumonia, her death certificate said; Ruth Thompson had supposedly passed from tuberculosis. The doctors thought it a coincidence that the girls had all worked at Radium Dial, but nothing more than that; they had all died from different things and their symptoms varied so much that there was no possible connection.

Catherine pushed the stroller home dejectedly and let herself in the front door: 520 East Superior Street, which had been left to her by her uncle when he'd died in 1931, was a two-storey detached white clapboard house, with a pointed roof and a covered porch. It was on a quiet residential street. 'It wasn't a big home,' remembered Catherine's nephew James. It had a galley kitchen and a modest dining room, where Tom would read books of an evening; this room was furnished by a blue couch and a round oak table. It was a perfect family house. 'We were so happy just staying home with Tommy,' remembered Tom with a fond smile.

As Catherine set Tommy down on a rug and watched him play, her mind went over her appointment. Mindful of the dial-painters' deaths out east, she had asked the doctor today if radium poisoning could be what she had; but he'd said clearly that he didn't think so. He – like all the others – had 'repeatedly advised her he was not informed as to radium poisoning thus to diagnose her case'. Perhaps the doctors were influenced by what they'd read in the paper: no Ottawa girls could possibly have radium poisoning, for the paint Radium Dial used was not hazardous.

Every time she went to church, Catherine was aware of the Radium Dial studio just across the road. It was a much quieter place these days; the economic downturn had the little town of Ottawa firmly in its grip – tenaciously so, in fact, as Illinois was such a big farming state. Many dial-painters had been laid off. Those who remained no longer lip-pointed; perhaps because of the Eben Byers case. Some were using their fingers instead; this doubled the amount of paint each woman handled. But given the financial hardships, the workers would paint any way they could: those lucky enough to have a job were fiercely loyal to the firm. There was a feeling that the whole town needed to support such an employer; there were very few of them about in these straitened times.

Though most of the original girls had been laid off or quit, their friendships had not faded. Catherine's close neighbours included Marie Rossiter and Charlotte Purcell; they often spent time together and, when they met, they talked. They talked of Catherine's tender jaw; of Charlotte's achy elbow; of Marie's sore legs. Marie and Charlotte had also gone to various doctors. And as the women discussed what the different physicians had said, they realised they'd all had the same response. And it wasn't just them: Mary Robinson's mother said the doctors 'scoffed' when she mentioned radium poisoning as a possible cause of her daughter's disease.

As had happened in Orange, mysterious illnesses were plaguing the girls of Ottawa – but here there was no Dr Martland making pioneering medical discoveries; not even a Dr Barry who was familiar with phossy jaw. The conditions the girls were experiencing were completely novel in this town.

Although ... there *had* been that visit from the national investigator Swen Kjaer. He had visited the local dentists and doctors – and he had visited not once, but twice. He had told them what he was looking for; described the tell-tale signs of radium poisoning. Yet the doctors do not seem to have joined the dots, nor notified the Bureau of Labor Statistics of these curious cases, as they had once promised to do.

An oversight? Or was it as some of the women were now starting to fear: that 'none of the local physicians will admit it'. One dial-painter's relative thought so: 'They didn't want anything to happen to the company,' he said.

'They were all bought off,' claimed another.

'It was confusing,' remembered Catherine's niece Mary. 'I only remember that no one seemed to know what was wrong. But we knew *something* was wrong; really wrong.'

42

Charlotte Purcell heaved the bags of groceries into her arms and set off home. She was already thinking of how many meals she could eke from the food she'd bought. Times were tough, and everybody was tightening their belts.

The newspapers were full of more bad news in that February of 1934: the country was experiencing the worst-ever drought in its history. For Charlotte and Al, who now had three kids to feed, it was a precarious situation. Charlotte paused on her way home to rest, rubbing cautiously at her left arm. It had begun to bother her last year, but now there was a continuous achy pain. 'The local doctors told her to use hot towels,' recalled her husband, Al.

Hot towels, however, had had zero effect. Charlotte concentrated on her fingertips, running them gently over her arm. Yes, she thought, it was definitely bigger. She peered closely at the little swelling nestled in the crook of her elbow. It was just a little bump, but it seemed to her it was growing larger. She would show it to Al later, she thought, see what he had to say.

Suddenly, Charlotte cried out in pain. The bag in her left arm dropped fast to the floor, spewing groceries onto the sidewalk. She had felt a 'sharp, knife-like pain which went through the elbow'. She bit her lip, rubbing again at where the pain was, and then bent to clear up her shopping. This was happening more and more frequently; when holding something it would drop out of her hands. It was the last thing she needed. The kids were four, three and a year-and-a-half old. She needed to get well.

Maybe prayer would help. That Sunday, she slid into her pew at St Columba with her usual piousness and bent her head to pray. There was a bit of a commotion further ahead, and Charlotte glanced up to see Catherine struggling; at that time her friend's legs had been stiffening so she had trouble kneeling in church. Catherine could barely bend her legs on the solid wooden plank in the pews. Tom had his arms around her, trying to help her; he looked alarmed at his wife's condition.

In fact, Tom found himself 'in a frenzy of anxiety'. Catherine was still just about able to kneel and get around, but some days it was a close-run thing. She kept saying that they didn't have the money to get better medical care, but Tom now decided something had to be done. Catherine owned their house outright, after all. They could always mortgage it; that would free up some cash for doctors' bills.

Tom helped his wife slowly back to her feet. She was panting from the effort, blowing out little pained breaths as she tried to force her limbs to straighten. Yes, this had gone on too long. If the doctors in Ottawa wouldn't help, Tom was determined to find somebody who would.

He went to Chicago, the nearest city. It was 85 miles away, but Tom travelled 85 miles there and 85 miles back – and he brought a doctor back with him: Charles Loffler. A 'reputable medical man' and blood specialist, Loffler was kindly-looking with sticky-out ears. He first saw Catherine at the Ottawa office where she worked on 10 March 1934. Despite his experience, he was initially flummoxed by her symptoms but adamant he would learn the cause. He took a blood specimen and, on testing it in Chicago, noticed 'a toxic quality in her blood'.

The following Saturday, he returned to Ottawa – and found Catherine had declined significantly in the intervening week. She became so ill that just at the time her doctors' bills soared – Loffler's invoice, by the end of it, would be for some $605 ($10,701) – she was forced to leave work. Loffler did what he could to alleviate her anaemia and increasing pain, while he continued to hunt for a diagnosis.

The lump in Charlotte Purcell's elbow, meanwhile, had now swollen to the size of a golf ball. She had 'terrible pain' all down her arm; it was worse at night, when she'd lie awake, scared and confused. She and Al also went to Chicago, like their neighbour Tom Donohue, but found 'fifteen Chicago specialists were puzzled by her case'.

Catherine told her friend about Dr Loffler, so the next time he was in Ottawa, Charlotte also went to him for treatment – and it seems she persuaded many of her former colleagues to do the same. 'She got them together,' remarked a relative. 'She was kind of pushy about it.' The girls had been a clique at work and those who were left alive had not forgotten the bonds of their sisterhood. In the end, Loffler hosted several informal clinics for the women at a local hotel.

Helen Munch attended; no longer married because, she said, her husband had divorced her because of her illness. She confessed her legs felt 'hollow ... as if air was rushing through'. Though she was a woman who 'wanted to be going all the time', she said miserably, 'Now I have to be quiet, still. I *never* wanted to be quiet.'

Olive West Witt, a dark-haired motherly woman, was distraught. 'I'll tell you how I feel,' she said. 'I'm just thirty-six, but I live like an old woman of seventy-five.' Inez Vallat also hobbled to the hotel; since last February one side of her face had drained constantly with pus, while her hips were now so locked that she was almost at the point that 'she could move neither backward nor forward'. Marie Rossiter told the doctor how she 'would love to dance, but I can't because of my ankles and the bones of my legs'. Charlotte persuaded the Glacinski sisters, Frances and Marguerite, to come too. 'Charlotte never felt sorry for herself,' said a relative. 'She would just take over and take care of [everyone].'

Though Loffler travelled to Ottawa every weekend throughout March and into April 1934, he was still not ready to present a diagnosis. Come 10 April, Charlotte could wait no longer. The growing mass in her arm was excruciating. 'We finally took her to Chicago to Dr Marshall Davison,' her husband, Al, remembered.

It was there, at the Cook County Hospital, that Dr Davison presented Charlotte with a choice. In order to live, he told her, there was only one option. He would have to amputate her arm.

Charlotte was twenty-eight years old; she had three children under five. Yet what choice did she have? She chose life.

They cut off her arm at the shoulder. 'There was no way,' a relative later said, 'that they could use a prosthetic arm or a hook because they had nothing to attach it to.' It was gone. Her limb, which had always been there, scratching her nose, carrying shopping, holding a watch dial, was gone. The doctors were mystified by the arm itself. With all the grim fascination of medics, after the operation they kept it in formaldehyde because it was so odd.

For the Purcells, there was simply a strange relief. 'Dr Davison says we're lucky to have her still with us,' Al Purcell said quietly.

But his wife had been left 'helpless'. Before the operation, she had slid from her left hand, for the last time, her wedding ring. Now, she wore it on her right hand, and asked Al to safety-pin her left sleeve to cover up the missing limb. 'My husband,' she later said, 'is my hands.'

Charlotte and Al only hoped that such a huge sacrifice would be enough. But already it didn't help with one thing: 'She still feels,' Al remarked, 'the terrible pain of the hand and arm they removed.' The ghost girl had phantom pains from the limb that was no longer there.

'There is some possibility,' Al added, 'of recurrence on the right side. We're not sure yet.'

Only time would tell.

43

It arrived by letter at 520 East Superior Street. A slim, unre-markable envelope addressed to Mr Thomas Donohue. It looked innocent enough, but the news it held was anything but.

Having run his tests, including X-rays on her jaw, Dr Loffler could now confirm it. Catherine Donohue was suffering from radium poisoning.

'Tom was devastated,' remembered his niece Mary. 'Just absolutely devastated. I don't know how the man functioned.'

'After that,' Tom himself said, 'I took care of [Tommy] when [Catherine] could not do so.'

Catherine herself never spoke publicly about how she felt. She probably prayed, as did many of her fellow sufferers. 'I firmly believe,' wrote one of her friends, 'that prayers is all that brought me through.'

Yet just days after Catherine and Tom had received that letter from Chicago, Catherine's disease took even the solace of prayer away from her. On Wednesday 25 April 1934, she hobbled the short distance to St Columba – but found herself unable to kneel in church. Her hips had become so locked that she could no longer bend her legs to pray; for Catherine, so devout, it was profoundly distressing. At about the same time, Charlotte Purcell came home from hospital for 'the first time with her arm off'. The doctors had confirmed radium was to blame for all of this – and Tom Donohue felt somebody should tell Radium Dial.

Ottawa was a small place. Mr and Mrs Reed, the firm's superintendent and instructress, didn't attend St Columba, but they were forever walking past it as they went to work.

'I saw him on the street,' remembered Tom of running into Reed. 'I told him the women were in a bad way, and that the doctors were finding it was from the material in the paints they were using.'

But Mr Reed refused to admit any responsibility. He refused even after he saw Charlotte and her husband walking past the studio, where they met him coming down the steps. Al was 'very angry' at what had happened, but Mr Reed brushed off everything they said.

Dr Loffler tried to communicate with the firm too. Going above Mr Reed's head, he telephoned Vice President Fordyce. 'I told him from the cases I had seen, I thought it would be wise to investigate all the [other] cases.'

Loffler's phone call was not unexpected to Rufus Fordyce. After all, the firm had in its possession the results of the radioactivity tests of all the women at Radium Dial, taken back in 1928. The results that showed that, of the sixty-seven girls tested that day, thirty-four were suspiciously or positively radioactive. Thirty-four women: *more than half the workforce*.

The company had said in its press statement at the time: 'Nothing even approaching symptoms [of radium poisoning] has ever been found.' That declaration was not some miscalculation, caused by a misunderstanding of the data. The data was clear: most of the employees were radioactive; a telltale sign of radium poisoning. But though the women's breath betrayed the truth, the company had deliberately and unashamedly lied.

The company still had the women's names on its secret list of results, each numbered according to how radioactive she was. Ranked at number one for positivity: Margaret Looney, Mary Tonielli ... Marie Rossiter. 'Very suspicious' were the results of Catherine Wolfe and Helen Munch.

For almost six years Radium Dial had known the women were radioactive. Yet 'the knowledge of the discoveries had been carefully concealed by the firm, who feared disruption of their business if the facts became known ... the victims had not been informed of their condition, nor the cause, through fear of panic among the workers'.

It all meant that when Loffler's call came in, Fordyce was ready. He refused to do anything.

Catherine, Charlotte and all the other girls, however, were determined to make the company pay. In many ways, they had no choice: Catherine had already expended large sums in a vain effort to be cured of the disease and she and Tom were stone-broke.

It was Loffler who helped the women take the next step, connecting them with an acquaintance of his: the stenographer of a Chicago lawyer, Jay Cook. Cook was formerly with the Illinois Industrial Commission, which oversaw all industrial-compensation cases, and he agreed to represent them 'virtually on charity'.

Though the women never met him, he nevertheless gave advice from the big city. Like so many New Jersey lawyers before him, he saw at once that the women's case was complex and that an early settlement might be to their advantage. The girls told him there were rumours that their former colleague Mary Robinson might have been given some compensation, after she'd had her arm amputated at the start of the year. 'The Dial people gave her some money,' Mary's mother confirmed. 'They sent it to her husband, Francis. Not very much, probably not over $100 [$1,768] altogether.'

It may not have been much, but it was an open door through

which the other women hoped to find some financial aid. There was another reason to approach the company: the statute of limitations. Under Illinois law, at the earliest instance of diagnosis, the women had to give Radium Dial notice of their condition; such notification should then lead the company to act lawfully in providing medical care and compensation, since the women had been injured at work.

It was Charlotte and Catherine, as they had done from the start, who led the way. They only hoped that the company would now be fair. With the help of Jay Cook, and working with their husbands, the women came up with a plan. Catherine wrote a letter on behalf of them all on 1 May 1934 and then Al Purcell telephoned the studio so that Catherine could read it down the line to the manager. Immediately afterwards, Tom took the hard copy and ran it down the street to the mailbox. The company had been given their notification. Now, the women only had to wait.

They waited ... and waited ... and waited. By 8 May, there had been no reply: not one word.

On the advice of Cook, the women now took matters into their own hands – and headed back to Radium Dial to confront their old manager Mr Reed.

It was a journey Catherine had undertaken so many times before. Turn right out of the house, walk straight to Columbus Street, turn left and walk one block to Radium Dial. But it had never been a journey like this before. She felt nervous, but knew she had to stand up for herself – and for all the other girls; they had agreed that she and Charlotte would be 'spokesmen for the other women'.

Charlotte walked slowly by her side, keeping pace with Catherine's limp. It felt so strange walking, Charlotte thought; she had never realised before how much you used your arms when you walked. Now, there was nothing but air by her side.

Charlotte was a woman who didn't dwell on herself. 'She

never felt sorry for herself, ever,' said a relative. Though she had said after her amputation, 'I can't do housework,' already she was finding ways to cope: she had managed to open and close her baby's diaper pins with her mouth; washing up the frying pan, she had discovered, could be achieved if she set the handle under her chin. It was Al, of course, who picked up the rest of the slack.

But Al wasn't here now. It was just the two of them: Catherine and Charlotte. The women walked along, so different from how they'd been when they'd first entered the studio. Catherine hobbled up the six front steps and tried to straighten up as much as she could. They made their way inside and found Mr Reed.

'I have received a letter from my doctor, who has been treating me for weeks,' Catherine said formally to him. She had a 'cultured voice' and her words were sure. 'He has come to the definite conclusion that my blood shows radioactive substance.' She gestured at Charlotte to include her: 'We have radium poisoning.'

There it was: fact. Hard to say out loud, but fact. She paused to see if there was a reaction, but there was nothing from the man who had been her manager for nine years.

'Having consulted legal advice,' Catherine went on, in spite of Mr Reed's silence, '[my lawyers] have advised me to ask the company for compensation and medical care. We have legal advice that we are entitled to compensation.'

Mr Reed looked over his former employees. Catherine had barely been able to get into the studio; Charlotte no longer had an arm.

'I don't think,' he said slowly, 'there is anything wrong with you.'

The women were gobsmacked.

'There is nothing to it at all,' he said again.

'He refused,' Catherine remembered angrily, 'to consider our request for compensation.'

She notified him about the condition of the other women, too, but he didn't back down. He didn't back down even when, two days later, Mary Robinson died.

Her death was important. 'Mary's was the first case definitely called radium poisoning,' her mother Susie recalled. '[Her doctors] sent a sliver of bone to a New York laboratory. They sent back word it was radium poisoning. The Ottawa doctors couldn't deny it then.'

But Susie had reckoned without the stubbornness of the Ottawa physicians. Just because these hoity-toity New York and Chicago folk were saying Ottawa girls had radium poisoning, it didn't make it true – not in their eyes. The Ottawa doctors remained sceptical and 'steadfastedly refused to admit that radium poisoning was the cause of the women's illnesses and deaths'. When Mary's death certificate was signed, the attending doctor answered 'no' to the question: 'Was disease in any way related to occupation of deceased?'

But although the local doctors were not convinced, the women absolutely were. In light of the company's refusal to aid them, in the summer of 1934 a large group of dial-painters – including Catherine, Charlotte, Marie and Inez Vallat – filed suit for $50,000 ($884,391) apiece. Jay Cook thought they stood a good chance of winning: Illinois law was progressive and a pioneering act passed in 1911 had long commanded companies to protect their employees.

But not everyone was pleased with the possibility of bringing the firm to its knees. The town 'bitterly resented these women's charges as giving a "black eye" to the community'. Ottawa was a close-knit and folksy town, but the girls soon realised that when it turned against you, it turned *hard*. 'They weren't treated too nice,' commented a relative of Marie with understatement.

After all, Radium Dial had long been a valued employer. With the country in the middle of its worst-ever economic depression – what some were now calling the Great Depression – communities were even more protective of the firms that could give them work and wages. The women found they were disbelieved, ignored and even shunned when they spoke out about their ailments and the cause.

Day after day, former colleagues and friends lined up to dismiss them. 'Margaret Looney was one of the girls that appeared to [me] to look as if she had one foot in the grave when she was hired!' exclaimed one Radium Dial worker bluntly. 'The [girls] that people thought died from radium and looked so terrible looked terrible when they were hired.'

'Some of them shun us as if we had the plague,' remarked Catherine's friend Olive Witt. Catherine lived just a few paces from Division Street in Ottawa and it was painfully apt, given the way the women had split the town – and the disapproval went right to the top, with 'business interests, politicians and the clergy' all against the women bringing suit.

In her little house on East Superior, however, Catherine ignored everything that was going on in the outside world. Her world, now, reduced down: down to the four walls of the clapboard house, down to the room she was standing in, down to the dress hanging on her body ... down to her body itself. She stood quite still, as though listening. Then she felt it again.

She recognised that feeling. She knew what that was.

Catherine Wolfe Donohue was pregnant.

C atherine's treatment from Dr Loffler stopped immediately. There would be no more injections for her severe anaemia and no more pain-killing sedatives; they might hurt the baby. There was absolutely no question of a termination. Catherine and Tom were devout Catholics and would never have considered it. This child was a blessing from God.

Catherine continued to consult Loffler, however; he was the only physician she could trust. He was very expensive, though. The mounting doctors' bills became too much for her husband, though Tom tried not to let that show.

As people in the local area surrounding Ottawa came to learn of the dial-painters' lawsuit, censure of their actions heightened. Yet for every citizen disapproving of the news, there were women for whom the gossip brought the most enormous sense of relief, for it provided a solution to a long-unanswered question.

'It came to my attention,' wrote Pearl Payne, 'that girls who had formerly worked at the Radium Dial were dying prematurely and of mysterious causes. I began to put two and two together ... I then came to the conclusion that *I* had radium poisoning.'

Pearl had dial-painted for only eight months, in the early 1920s. She didn't live in Ottawa but LaSalle, some thirteen miles down the road; a fair distance if you didn't have a car, which most people did not in the 1930s. Pearl had left Radium Dial to nurse her mother, and then focused on having a large family with her husband Hobart. She'd been thrilled when they'd had their first child, Pearl Charlotte, in 1928.

But to Pearl's despair it had all gone wrong the following year. She began staggering when she walked and was sick throughout 1929. In 1930, she underwent an abdominal operation to remove a tumour; afterwards, her head swelled to twice its normal size – and it did not go down. 'There were big black knots behind her ears,' recalled her husband. A specialist was summoned. He cut Pearl's ears inside and out 'for drainage'; the cuts had to be opened up every few days. Though it eventually reduced the swelling, said Pearl, 'one side of my face was paralysed'. In time, this paralysis left her – but then another problem began.

Pearl started bleeding continuously, down below. Another tumour was removed and a 'curettement' of the womb performed, which meant a scraping out of tissue. Yet it didn't help. The next time she bled, she bled for eighty-seven days straight.

'During this time,' she remembered, 'the doctor was perplexed and said I must have had a miscarriage.' He persisted in this argument as Pearl, again and again, bled and endured yet another curettement. 'I knew this was not so,' Pearl cried in frustration at the doctor's diagnosis, 'because nothing had been done to cause me to be pregnant.' Instead, the problem seemed to be the tumours growing inside her – growing where her children should have been.

Her condition was serious. She endured 'five years of continual doctoring, six operations and nine trips in all to the hospital'. At one stage, she had been moved to write to Hobart from her deathbed, believing the end was near. 'Dearest Sweetheart,' she wrote:

*I love you, and am laying here thinking of you and wishing I
was in your dear arms. I am afraid I was very impatient with
you for some time and I am heartily sorry. Please forgive me, as
I have been very nervous and ill for a long time. Beneath it all I
have loved you very deeply and dearly.*

*Pray for me daily that I may get well perfectly. If not, do not
grieve, as we must bow our heads to the Lord's will ... Be good
to our baby girl, teach her to love and remember me, and above
all to be a good, virtuous girl.*

Tell her I loved her dearly.

The emotional pressure was unbearable. Pearl never knew if
today would be her last; in time, her sickness affected both
her body and her mind. 'I am unable to enjoy life as a normal
woman should,' she wrote dully.

The doctors told her she 'belonged to a class of women of
which the medical profession does not know the reason for their
illness'. She was treated for malaria, anaemia and other conditions. The doctors' guesses were especially frustrating for Pearl
because she had been trained as a nurse: she knew none of the
theories was right, but she was at a loss as to what could be the
true cause.

By the April of 1933, Pearl had become desperate. 'I notified
my doctor [of more bleeding],' she remembered, 'and he advised
the removal of the uterus. I refused and lay in bed for several
days debating what to do.' A hysterectomy: it would mean the
end of her dreams for more children. No, she thought, no, not
yet. She needed more time; more hope.

She called in other doctors, had other treatment, hoping for a
different outcome. But it was all to no avail. '[In] July 1933,' she
wrote numbly, 'I was completely sterilised.'

Pearl was heartbroken. 'I was attacked by severe heart
and sinking spells,' she recalled. As she read of the radium-
poisoning cases in Ottawa, she grasped that her devastating
condition might prove fatal – but at least she had an explanation.

'I believed,' she wrote of her case, 'that radium had attached to the tissue of certain organs, causing them to be destroyed through tumorous growth.'

She decided to get in touch with her old friend Catherine Donohue. The two women, who were very similar in nature, now became extremely close. Not long after, Pearl joined the fight for justice. The lawsuit was gaining momentum; the women were gaining friends.

In Chicago, however, Joseph Kelly, president of Radium Dial, was finding the opposite. By October 1934, perhaps in light of the lawsuits, he had clean run out of friends in his company. An executive called William Ganley wrested control of Radium Dial and Kelly and his associates were voted out. 'There were very hard feelings,' recalled one company officer, 'because of the corporate shenanigans that went on.'

But, Kelly had decided, he was not finished with Ottawa. Every current dial-painter at Radium Dial now received a letter. Mr Turner – a manager at the plant under Mr Reed – invited them all to a restaurant, where they were fed while he talked to them. He had an announcement – a new dial-painting business was going to be opening in town – and he had a question: how about you highly skilled girls join us at Luminous Processes?

It seems the women weren't told it would be run by Joseph Kelly and Rufus Fordyce, who'd had charge of Radium Dial during the radium-poisoning scandal. They *were* told something extraordinary though. Mr Turner 'informed them that earlier dial-painters had died because they put brushes in their mouths, and since brush-licking was no longer permitted, exposure to radium would not be harmful'. It was an admission of guilt, but the original dial-painters never got to hear of it.

The new studio opened just a few blocks over from Radium Dial in a two-storey red-brick warehouse. Thanks to the clandestine meeting in the restaurant, most of the dial-painters moved across, thinking the new operation safe. They applied paint using hand-held sponges and wooden spatulas, using fingers to smooth it over, and wore thin cotton smocks to give them some protection from the dust.

Not every worker went, however. Mr Reed stayed on as superintendent of the old firm. Loyal to the end, he and Mrs Reed stuck with the company that had made them. They faced 'a fiercely competitive situation', for Radium Dial now competed directly with Joseph Kelly's new business in the same small town.

Just down the road from this battle of big business, however, Catherine Donohue cared not a jot for the corporate infighting going on that fall. All that mattered to her was the tiny little girl she was cradling in her arms. She and Tom named their daughter Mary Jane, after Tom's mother. 'We always called her Mary Jane,' remarked her cousin. 'Never Mary. Just Mary Jane.'

Catherine Donohue vowed that she would make her daughter proud.

45

As 1935 began, Jay Cook was busy working on the women's lawsuit. He filed two separate claims for them: one in the normal law courts and a second with the Illinois Industrial Commission (IIC). The lead case, Cook determined, would be that of Inez Vallat. 'She was a living corpse,' Catherine said of her former desk-mate, 'hobbling around like an old woman.'

But, as the case got going, almost immediately the women ran into trouble. Radium Dial was being represented by a team of top lawyers who found several legal loopholes, through which they twisted the case. There was that old chestnut: the statute of limitations. Inez had filed suit years after she left Radium Dial and her disability did not occur while she was employed. There was the fact that radium was a poison; injuries caused by poison were not covered by the Occupational Diseases Act. And there was the law itself: Radium Dial charged that its antiquated wording was 'vague, indefinite and did not furnish an intelligible standard of conduct'.

'When Attorney Cook filed his test case,' the *Chicago Daily Times* later wrote, 'Radium Dial did not even bother to deny the women's charges. In effect, the company's reply was, "Even if it is true, what of it?"'

On 17 April 1935, a ruling was given. 'The court ruled the legislature failed to establish any standards by which compliance with the law could be measured,' reported the *Ottawa Daily Times*. The women had lost, on a legal technicality. They could not believe it – yet they fought on. Cook, at his own expense, took the battle all the way to the Supreme Court. Yet it was all to no avail: the law was declared invalid.

The *Chicago Daily Times* called it 'an almost unbelievable miscarriage of justice'. But there was nothing the women could do: they'd had their day in court, and the law itself had been found wanting. 'There was never a trial of this case on its merits,' lamented the newspaper.

Cook, reluctantly, had to drop their case, even though the girls still had a claim filed with the IIC and legislators now vowed to rewrite the law in light of the women's case. 'I hated to have to do it, but I simply couldn't afford to keep on,' Cook later said. 'If I had the money, I'd fight their case free. It's one of those things that should be fought through to a finish. I hope they get another lawyer.'

But finding another lawyer was easier said than done. There were forty-one attorneys listed in the Ottawa town directory, but not one would help them. Just as the native physicians had done, the legal profession was shutting down what was seen as a scandalous attack on a loyal local business.

As if to rub it in, on the day the Ottawa paper reported the loss of the women's case, it featured an article on Clarence Darrow, one of the nation's leading lawyers. *That* was the kind of person the girls needed, but they had no money to secure legal aid.

The Donohues now found the mortgage on their house had crept to $1,500 ($25,000). 'There are medicines that ease the pain,' Catherine said; and on these she and Tom were spending hundreds of dollars. They found themselves playing make-believe, trying to ignore what was happening to their family of four. 'We never talk about it,' confessed Tom. 'We just go along as if we were all going to be together forever. That's the only way.'

'We're so happy together,' Catherine said with a too-big smile. 'As long as we're together, it doesn't seem so bad. We just pretend I'm the way I was when Tom married me.'

They hadn't given up looking for a cure. Catherine found herself trying various Chicago hospitals and dentists, pushing herself to get to the appointments, even though she frequently 'fainted during the course of the examinations' due to the pain. 'She was looking for help,' a commentator said, 'any way she could get.' But no one could stop the disintegration of Catherine's mouth, which grew more serious by the day.

The women struggled on: downcast about the devastating loss of their court case; in denial about the looming demise that seemed inevitable. And then, just at the end of the year, word of another legal judgment reached them. It didn't necessarily affect their case, but it was of considerable interest nonetheless.

On 17 December 1935, the ruling finally came back on the Irene La Porte case in New Jersey, which her husband Vincent had been fighting for more than four years. *This* was the case the United States Radium Corporation had chosen to see through to judgment: the one on which they had placed their bets. The company, by now, did not deny the cause of death – it simply cited the statute of limitations as the reason why it should not have to pay. '[Once Irene's] employment [was] terminated,' the USRC lawyers stated, 'all duty we had to that girl as our employee ceased. There is no relation afterwards; she is a perfect stranger.'

Several dial-painters testified at the trial; many had their own lawsuits pending. All pinned their hopes on Irene winning her case, for if she did the verdict would apply to them too. Everyone gathered to hear the judgment.

The judge began:

Naturally, there is no question as to where the sympathies of any human being would lie in a case of this sort. It is tempting, in the light of the knowledge of today, to create the thought that the [company] must have been negligent in some way. But it should be noted that this case *must* be decided on the facts as they existed in the light of the knowledge of 1917. A court has no power to adjust the law to meet the needs of a time when no such case as this could be foreseen.

He concluded bluntly: 'The [case] must be dismissed.'

USRC had chosen well. Seven years on from Grace Fryer's case, there was now no censure in the media; no censure even from the judge. The company had the answer it was looking for: not guilty.

Justice had been denied to Irene La Porte – but not just to her. For all those New Jersey dial-painters with lawsuits still pending; for all the families battling on for loved ones who had died; for all those New Jersey women who had not yet found a worrying lump on their hip or their leg or their arm, but who would in future, justice was denied.

It had been, the USRC executives reflected, a very good day indeed.

46

You fight and you fall and you get up and fight some more. But there will always come a day when you cannot fight another minute more.

On 25 February 1936, Inez Vallat died; she was twenty-nine. After eight years of agony, she finally succumbed to a 'haemorrhage from sarcoma of the neck', bleeding out as medics desperately tried to staunch the flow of blood. 'Mr Vallat,' dial-painter Frances O'Connell recalled, 'would not talk at all about his wife because she had died a very horrible death and he did not want to think about it or talk about it.'

The Ottawa doctors completed her death certificate. *Was death in any way related to occupation of deceased?*

No.

Inez's death, coming on top of the lawsuit defeat, left the Ottawa women reeling. Many of the original clique were too ill to attend her funeral, as much as they wanted to say goodbye. Catherine Donohue, these days, was 'rapidly becoming too weak to move about her home very much' and seldom even left the house.

There was some coverage of Inez's death in the Chicago papers.

The press called the girls, rather dismally, 'The Suicide Club'. A senator commented that he would try to interest the Industrial Commission in their case, but added: 'Unfortunately, any proposed legislation cannot be made retroactive. It is pitiful indeed.'

The girls couldn't even get excited when the governor signed the new Illinois Occupational Diseases Act, which now included a provision for industrial poisoning. The new bill was the direct result of the women's case and would protect thousands of workers – but it would not become law until October 1936. Given how quickly the women were dying, they hadn't much hope they would be alive to see the day.

The same month the new law was signed, the girls had an approach from a journalist that lifted their spirits somewhat. Mary Doty, a leading reporter with the *Chicago Daily Times*, now gave them a voice. She turned the public spotlight back on their suffering in articles that ran for three days in March 1936. 'We'll always be grateful to the *Times*,' Pearl Payne would later say, 'for helping us when everything was so black.'

The *Times* was 'Chicago's Picture Newspaper', a populist publication. Doty knew just how to write for her readership: 'They shoot to kill when it comes to cattle thieves in Illinois, and fish and fowl are safeguarded by stringent game laws – but womenfolk come cheap.' She decried the fact that dial-painters had been 'dying off for thirteen years in Ottawa without any official comment or investigation'. And she painted a picture of the women's conditions that would haunt her readers: 'Some [girls] creep along, unable to move beyond a snail's pace; another with an empty coat sleeve or a mutilated nose, withered hands, a shrunken jaw.'

The girls posed for photographs, many with their children. Mary Jane Donohue looked absolutely tiny – Doty called her a 'wizened little baby'. At a year old, Mary Jane weighed only 10 pounds and had 'match-thin arms and legs'. 'Her parents,' wrote Doty, 'hope against hope her mother's illness will not leave its permanent mark on her.'

Catherine herself said to the press, 'I am in constant pain. I cannot walk a block, but somehow I must carry on.' When the journalist asked about her friend Inez, 'it brought tears'.

Marie Rossiter spoke of her son, Bill. 'I'm frightened to death, but I want to live as long as I can for the sake of my little boy,' she told the press. Though Marie now had five bad teeth, 'the [Chicago] dentists say they won't touch them because of the radium poisoning eating into the bones of my jaw'.

Charlotte Purcell was pictured with her daughter Patricia. She was gradually coping with having only one arm; 'Having three babies, she adapted,' said a relative. In time, she would relearn how to make beds, peel potatoes and even hang out washing, the clothespins stuffed into her mouth. As she told reporters, she was haunted by the thought that the sacrifice of her arm had not been enough; the radium ran right through her, and she didn't know where it might strike next.

The final piece in Doty's series focused optimistically on Catherine Donohue: 'She waits hopefully for another call to the city for an operation.'

Privately, Tom had to whisper to Doty: '[It] will never come.'

The women found the publicity got them motivated again. Charlotte's son Donald remembered: 'Mom used to get dressed up and take her friends and they'd go up to Chicago to see these lawyers.' A few months later, Charlotte, Catherine and Marie engaged a new attorney, Jerome Rosenthal, for their case before the IIC. They also decided to approach the government for help: their target was Frances Perkins, the Secretary of Labor; the first woman ever to serve in a presidential cabinet. It was Tom who contacted her, having 'telephone conversations and personal correspondence' with the Secretary. Whatever this quiet man said clearly made an impact, for no less than three federal departments began investigating.

The case was snowballing, and Tom now dug deep for the most important act of all. His wife had told him about the company tests and he judged – since it was clear Radium Dial had lied about the results – that getting hold of the original data would provide powerful evidence in court. On 20 May 1936, he decided to ask Mr Reed outright for the results. He felt they should have been given to the women anyway; or at least to him as Catherine's husband. He was only asking for what was rightfully theirs. 'This day,' Tom said, 'I wanted to find out the name of them doctors, who was supposed to examine them women that was working there, that didn't give them a report.'

Reed might have seen him coming. At any rate, the two met each other not in the studio, but on the streets of Ottawa.

Tom started out calmly enough. 'Why wasn't the report given to me?' he asked.

Reed, taken aback by Tom's direct question, did as he had always done and tried to ignore the situation. He brushed by him.

'I only have another question to ask you!' shouted Tom at the superintendent's retreating back – then he ran to catch up with him. 'I only want to help the women!'

Mr Reed had had enough. Perhaps there was a guilt eating away at him which led to what happened next. 'He started to swing at me,' Tom remembered with some astonishment.

Tom, though small, had an 'Irish temper'. 'I don't think anybody in our family,' one of his relatives later said, 'would go out of their way to cause a confrontation, but they wouldn't let one go if it came to them. I'm sure he'd have been angry. I'm surprised he stayed as level-headed as he did.' With Reed – the man who had overseen his wife's slow murder and then fired her when the poison's effects started to show – now hitting him, Tom dropped all pretence of civilised conversation. 'I swang at him,' he remembered with some satisfaction. He said Reed 'got excited'.

The two men brawled in the street, hitting out at one another in a 'fisticuff encounter'. Tom found himself landing blows for Catherine, for Inez, for Charlotte's lost arm, for Ella, for Mary, for Peg. Reed floundered under the attack and the police were called. Even though Mr Reed had started it, the respected superintendent of Radium Dial had Tom Donohue arrested. He brought charges of assault and battery and disorderly conduct.

Tom was now in the hands of State Attorney Elmer Mohn, facing two criminal charges.

A ssault, battery, disorderly conduct ... and insanity. The 'controlling interests' in the affair now even tried to bring a charge of insanity against Tom. In Hobart Payne's opinion, it was because he had 'vigorously opposed the operation of [the Radium Dial] plant'; he considered Tom had been 'persecuted'.

Tom's relatives thought such a move 'typical of a company with its back against the wall'. 'They know that they're gonna go down,' said his niece Mary. 'They'll do anything. They'll try anything.' Fortunately for Tom, the police case against him was not reported to progress further than a handful of initial hearings; perhaps because there was no foundation to the trumped-up charges.

Like all cowards with their backs against the wall, the company now chose to turn and run. In December 1936, Radium Dial abruptly closed its doors and upped sticks – to where, nobody knew. Nobody left behind did, at least. The Reeds followed the company out in the New Year, packing up their house on Post Street. No longer would the Donohues and Purcells run into the girls' old boss when they made their way about town.

Radium Dial had been 'run out of business' by Joseph Kelly's new firm, Luminous Processes. After more than fourteen years of the radium company operating in the old high school, the rooms fell silent. No chatter from the girls, no laughter from the darkroom: just empty rooms, haunted by the memories of all that had gone on.

With Radium Dial gone, Joseph Kelly had a monopoly on radium dials in the little town of Ottawa. It may have been the Great Depression, but things were turning out rather well for the company president. The same, however, could not be said for the husbands of the former dial-painters. They had just about managed to cling on to work through the Depression so far, but in 1937 their luck ran out. Workers were laid off from the Libbey-Owens glass factory and Tom Donohue and Al Purcell were among them.

For the Purcells, who had three children to feed, it became almost impossible to cope. 'They struggled really hugely, financially,' said a relative. Charlotte ended up feeding the children mustard sandwiches. 'You took whatever you could get,' remembered Tom's niece Mary of that period. 'It was a very tough time.'

Charlotte and her sisters agreed a solution: move to Chicago. But even in the city it was challenging. Charlotte's son Donald recalled: 'We used to go to a bakery and ask for day-old [bread]. We heated the apartment with a coal stove, and we used to walk around the [train] tracks in Chicago and pick up coal.'

It was hard – but it was harder still back in rural Illinois. Pearl Payne said there was no 'steady work; just periodic streaks of work'. Tom Donohue was not lucky enough to land even those streaks. With the house already mortgaged to the hilt, he was running out of ideas. 'Tom was nearly bankrupt,' remembered a brother-in-law. 'Catherine was full of radium and dying by inches. She suffered agonies, and [he spent everything] buying medicines to try to relieve Catherine.' The family now had debts of some $2,500 ($41,148).

There was nothing else for it. 'They were on relief for a while,' confided their niece Mary. '[They felt] very ashamed. Not wanting people to know about it.'

Yet they weren't the only ones needing help: lines of desperate people queued outside the soup kitchens in Ottawa. Everyone was living hand to mouth. The Donohues had almost no thought of a lawsuit anymore – this was a battle for survival. By the spring of 1937 their lawyer, Rosenthal, had dropped the case anyway. The women were due to have a hearing before the Illinois Industrial Commission later that year but, as things stood, they had no attorney to represent them.

Time passed. On 28 March 1937, Catherine Donohue and her family marked Easter Sunday, one of the most important dates in the Catholic calendar. Someone gave a gift of a 'timid-looking bunny rabbit' to Mary Jane and Tommy, who were then aged two and almost four. Tommy liked to paint, just as his mother and father had once done; he had a watercolour set that he played with often.

Catherine took her communion gratefully from the visiting priest – she received it at home now, being unable to get to church – and prayed. Easter Sunday was all about Christ being reborn: salvation, hope, the repairing of a broken body.

It was all the more horrible, then, that this was the moment Catherine's body fell further apart. 'Part of her jawbone,' Hobart Payne wrote, 'broke through the flesh and [came] out into her mouth.' Her tongue stumbled over it: unfamiliar object. Catherine picked it out with tears in her eyes. It was her jawbone. *Her jawbone.*

'It was so horrible,' remembered her niece Mary. '[It] just dropped out. I mean it was just . . . You thought, Oh dear God. Can't even eat! Just so sad.'

Tom Donohue was forced to watch his wife literally disintegrate before his eyes. It was horrifying – yet, on this supposed celebration of renewal, Tom found himself renewing one thing at least: his desire for justice. And he knew just who Catherine needed to help her now.

Her friends.

*

'[Catherine's] husband called me,' remembered Marie Rossiter, 'and wanted to know if I could call some of the girls [who] might want to help hire a lawyer.'

Tom Donohue had chosen the friend he had reached out to wisely. Marie Rossiter 'would always take the bull by the horn' – she was a fighter. 'If she [thought] maybe she could help [a] person,' said a relative, 'she would get involved. She was a protector.' And not only a protector, but a hugely popular girl.

'I said [to Tom], "Well, I knew enough of them [girls],"' recalled Marie. 'So I did [help]. I called some by phone . . .'

Charlotte Purcell took up the story. Though she was now living in Chicago, she was still very much involved: 'All through [it] she [was] with us,' recalled Marie. 'She was faithful all through it.' Charlotte continued, 'And they said no.'

No, they wouldn't help. For there were dial-painters who didn't want to face up to what was happening. While there were countless people in the town who denied radium poisoning existed, the reasons for that denial could vary. 'They pull back in fright,' said Olive Witt, 'asking if it's catching.'

Marie was frustrated by the townspeople's attitude. 'She used to say,' recalled a relative, '"Nobody wants to listen to us!" And I think that hurt.' Nonetheless, she kept trying with the dial-painters. 'We got a few,' Marie said in the end. 'We continued with the few we had and we kept plugging right along with [Catherine].'

That band of girls now shot for the moon, targeting the best lawyer they had ever heard of. The women felt the approach would be best coming from the men who backed them, so Hobart Payne and Tom Donohue wrote to the most famous American lawyer of the era, the one who 'always took the impossible cases'.

They wrote to Clarence Darrow.

'Dear Sir,' Hobart's letter read. 'It is as a last resort that I turn to you for assistance or advice . . . These cases are to come before the Industrial Commission for a final hearing [soon] and there is no attorney to represent these girls. Would it be possible for you to take up this case?'

But Darrow was turning eighty in 1937 and not in good health. Though he said he was sympathetic to the women, he was unable to help – he did, however, promise to refer the case to another lawyer.

Next, remembering their experience from the year before with Mary Doty, the women turned to the media to generate publicity of their plight.

RADIUM DEATH ON RAMPAGE! cried the front page of the *Chicago Daily Times* on 7 July 1937. WALKING GHOSTS JILTED BY JUSTICE! Charlotte Purcell, with her single arm, was the cover girl for the piece; she told the paper she 'lives in daily fear of [the] end that is inevitable'. Charlotte, Marie and Catherine were only three of the girls involved; others were the Glacinski sisters, Pearl Payne, Olive Witt, Helen Munch (who now lived in Chicago) and a handful more.

As the girls had requested, the paper reported that they had no lawyer for their upcoming hearing before the IIC, which was scheduled for 23 July: sixteen days away. The hearing was 'their last stand – their last hope of collecting damages'. 'Without a lawyer,' the paper wrote, 'the women fear legal trickery. Indeed, so hopeless is their outlook that many of them may stay away.'

Catherine Donohue spoke up. 'That's what the company's lawyers would like, I suppose,' she said archly, 'for all of us to stay away.'

'The Radium Dial Company,' the piece went on, 'has closed its plant in Ottawa [and] "skipped out from under", leaving only a $10,000 [$164,595] bond posted with the Industrial Commission.' That $10,000, in the light of Radium Dial's vanishing act, was the sole pot of money available to the girls for compensation and medical care.

Though Joseph Kelly had set up an identical business now doing a roaring trade, Jay Cook, the women's former lawyer, explained: 'This is a "new" corporation. Under the law, the "new" company isn't liable for any of the acts of the "old" concern.' It was Radium Dial, not Joseph Kelly, that was being sued. 'All they've really got to levy on is the $10,000,' said Cook. 'That is, of course, unless they were able to locate other assets of the "old" company somewhere ...'

The following day, the girls' ally in the media struck again. OTTAWA RADIUM COMPANY NOW IN NEW YORK! chorused the *Times* in triumph. 'The Radium Dial Company,' the article read, 'was found here by the *Times* today doing business on New York's lower east side.' They were hiring young women to paint dials ...

Having been located, Radium Dial's new president, William Ganley, came out fighting. 'These women's claims are invalid and illegitimate,' he stated defiantly. 'A lot of those women were with us only a few months; practically all of them have been out of our employ for many years.'

And then, dismissing the firm's secret test results, dismissing Peg Looney's company-led autopsy where the doctor had been instructed to destroy the evidence of the real cause of her demise, he declared, 'I can't recall a single actual victim of this so-called "radium" poisoning in our Ottawa plant.'

Radium Dial was not going down without a fight. They had won this case before in the courts and were supremely confident they would win again.

The president's attitude underlined to the women just how much they needed an attorney. Yet as the clock counted down to the all-important hearing, no lawyer came forward. Letters and press appeals and word-of-mouth had so far had zero effect. Despite their crippling illnesses, the girls decided they would have to take matters into their own hands.

It was time for the Suicide Club to take a trip to the Big City.

48

Chicago: a land of steel and stone and glass, where a forest of skyscrapers stretched above the ant-like actions of its citizens below. As the five women made their way through the thrusting city streets, everywhere Chicago's urban architecture dominated the view. Here there were no yawning horizons, such as those with which they were familiar, where the sun would hang in the sky like a citrus fruit above the endless fields. Here there were no fields – just opportunities, ripe for the plucking.

It was two days before the hearing: Wednesday 21 July. The women headed to N LaSalle Street, right in the heart of the theatre district. They had dressed smartly, many in tailored jackets – all wore hats with beribboned bands – and given the heat of the July day they were glad when they reached the address they sought: number 134. So this was the Metropolitan Building.

Even with their necks stretched back fully, they could not see its roof; the building was twenty-two storeys high. And it wasn't just any office building; as they hesitated outside the lobby, their eyes drank in the details: the gold panelling on the walls; the

'M' emblazoned on the ground; the building's own name picked out in solid gold letters above the door. It was all very different from the place they'd started out in this morning; that was for sure.

Catherine Donohue had just about managed the journey, for this was one meeting she would never have missed. The remaining girls had 'formed an organisation to band together in the prosecution of their cases' and despite her fast-fading health Catherine was its chairman. It was essential that she led this quest to secure a lawyer who could represent them.

She had chosen to wear a smart black dress with white polka dots; it was the nicest one she had. She had slipped it on that morning with nervousness – and some concern. That lump on her hip, she'd thought anxiously, as the fabric slid over her increasingly thin body, it was definitely a little bit bigger than before.

With her in Chicago were Marie Rossiter, Pearl Payne and the two Glacinski sisters, Frances and Marguerite. These five were representing all the litigating dial-painters, including the estate of Inez Vallat, whose claim for damages had been added to that of the living girls. Hats straightened, dresses smoothed, the women walked fearlessly into the lobby and rode the art-deco elevator up to the office they required.

That office was lined with bookcases full of heavy legal tomes; on the walls hung framed qualification certificates. Dominating the room was an enormous desk, made of a lustrous reddish wood and topped with glass. Yet all these furnishings faded out of consideration when the women made eye contact with the man standing behind the desk. He wore a three-piece tweed suit and glasses atop his large nose; his dark hair was neatly styled with a side parting. The man was somewhat full-figured, and he had kind eyes.

'Ladies,' he announced, stretching a welcoming hand across the desk to greet them, 'I'm Leonard Grossman.'

*

The women may have been referred to him (or vice versa) by Clarence Darrow. Like Darrow, Grossman was a larger-than-life and flamboyant attorney, whose concern was for people on the bottom rung. He was born in Atlanta in 1891, making him forty-six when the five women knocked on his door; his birthday was on Independence Day.

That quirk of birth in fact characterised his personality and passions in many ways. He had been an early supporter of the Suffragettes; an article on their major march on Washington was headlined 200 WOMEN AND 1 BACHELOR – and that bachelor was Leonard Grossman. He was the type of person who always managed to get into the picture when a newspaperman happened to be near; he had worked as a stringer for various papers when he first left law school and his nose for a story never quite left him. He was a brilliant orator. Grossman had been involved in politics in the past, yet it was workmen's compensation cases that really inspired him. 'He had a passion for the working person and for people who were in trouble,' said his son, Len. 'He never went for the big buck.'

Sometimes, he didn't even go for the small buck; 'he took shoes as fees,' remembered his son. 'He did that way too often.' This may explain why in July 1937 – despite the seeming glamour of his office – Grossman was 'kind of on the outs; scraping it together'. But it didn't matter to him: money wasn't what drove Grossman; his principles were his fuel.

This was the man – his passions and priorities – into whose office the five women of Ottawa walked. It was, perhaps, the most perfect meeting of minds.

'We were at our wits' end when he came to our rescue,' remembered Catherine Donohue. 'He had no thought of money. He just wants to help us girls, to help humanity.' Grossman declared to his new clients: 'My heart is for you; I am happy to be in this fight for you.'

At last, the women had a legal champion. And they had found him not a moment too soon; in two days' time, Grossman and the girls had a hearing to attend before the Illinois Industrial Commission.

Step by step, inch by inch, Catherine and the other girls made their slow way to the yellow-stoned LaSalle County courthouse on Friday 23 July. It was just four blocks south of St Columba, so they did not have far to go. When they arrived, they were gratified to see that their story was being covered by the press.

It was a boost that Catherine in particular needed. In the short time since she'd been in Grossman's office in Chicago, another piece of her jawbone had come out into her mouth. Not knowing what to do with it, she had put it in a small paper pillbox.

Despite her trials, Catherine that day seemed to take inspiration from Grossman, finding fuel in her principles – in standing up for what was right. She 'presided over the reporters' as she and the other girls spoke to the press. As the women entered the courtroom and saw Grossman there, ready to do battle for them, they knew that, this time, they stood a fighting chance.

Some of the women sat at the counsel table with Grossman as he prepared to open their case. Representing Radium Dial were the same legal firms who had fought – and won – the Inez Vallat case two years earlier: the leading lawyer was Arthur Magid, a young-looking man with thick dark hair and glasses; another was Walter Bachrach.

Grossman's first job was to ask for a postponement to give him time to 'familiarise himself with the case and to trace, if possible, the "old" company's assets'. Magid readily agreed: the firm was in no rush for the trial to begin, for the longer the legal process could be stretched out, the weaker the women would be.

There was not much more to that first hearing – although Bachrach did now reveal what the company's defence would be. He said he would 'contend the paint was not poisonous and that none of the women actually were suffering from radium poisoning'.

Not poisonous. Even from the little Grossman knew of the case, he realised that this position was a complete volte-face on the arguments the exact same lawyers had used in the Vallat suit. Then, the company had said radium *was* a poison – because poison was not covered by the law and the court had to find against the girls. Now that the law had been rewritten to *include* poisons, the company was trying the opposite tack.

It was just the sort of slippery, unjust finagling that Grossman had fought against before. Inspired, he rose to the occasion. And even though this was a modest hearing, Grossman now revealed how apt it was that his office was located in Chicago's theatre district. For he was a showman, 'a silver-tongued orator', and as he stood centre stage in the courthouse, he exhibited his skills. Seeing him shine, many of the girls were seen 'weeping' that an accomplished attorney was finally on their side.

'We should have laws,' Grossman started, in his sombre, melodious voice, 'that will do away with things that rack, ruin and destroy bodies.'

He turned and scanned his eyes across the crippled women sitting at the desk. He gestured at them with feeling. 'We do not need to have martyrs such as we have sitting around this table,' he said, 'and the many dead who worked with these girls.'

He paused dramatically, and then went on. 'It is a heavy cross to Calvary,' he declared, 'but we *will* bear it. And, with the help of God, we *will* fight to a finish.'

49

Work began on the case immediately. That same day, straight after the hearing, Grossman and the women met for a conference so he could gather more information. Then he packed up his big brown leather briefcase, swivelled in his spats and headed on back to Chicago.

Assisting him in his preparations were his loyal secretary Carol Reiser and his German wife, Trudel. Much of the historical radium literature was in German, so Trudel spent hours translating documents as Grossman got up to speed on the intricacies of the case. He regularly turned in eighteen-hour days and his team worked hard to keep up.

Since Al Purcell now lived in Chicago, he nipped over to Grossman's office to see if the women needed to do anything. 'For God's sake,' Grossman had declared, 'get a doctor's statement!'

They followed his directions, but securing their medical records turned out to be difficult. 'I have written my doctors,' reported Catherine later that year, 'and no reply came back.' Pearl Payne also found that the hospitals where she'd been treated refused to release her records. She ended up begging her doctors: 'Please help me get these records. This case is up for final hearing.'

The women were not the only ones requesting records. That fall, Grossman served notice on Radium Dial to 'produce [the results of] all physical examinations of employees'. The company had concealed the true test results: Grossman wanted to know how much the firm had known, and when.

The women were delighted by his diligence. 'At a great sacrifice,' Pearl Payne wrote to commend him, 'you have continued daily to lay other engagements aside to formulate the great mass of information necessary to properly present these cases.'

Grossman decided that the lead litigant would be Catherine Donohue; followed by Charlotte Purcell, whom Grossman described as 'my next best case'. Catherine didn't necessarily have the most evidence behind her, nor was she the most compelling personality to take the stand. It wasn't even that she had the most fire in her belly for the fight. It was simply believed that she was the woman who would be next to die. 'She hasn't long to live,' Pearl said quietly of the decision. 'We want her to have her day in court.'

Although Catherine was no more an extrovert than her husband, she nevertheless seemed accepting of the responsibility. 'The strength for the women in my family,' said one of her relatives, 'has always been to do the right thing and stand up for what you believe in. [Catherine] saw a huge wrong and [she wasn't] going to be quiet about it.'

While Grossman beavered away in Chicago, it seemed a long and lonely fall to Catherine Donohue. Her condition continued to deteriorate, more and more rapidly. 'My hip is very bad, Pearl,' Catherine admitted to her friend. 'It is all I can do to get around at all.' That hard lump on her hip was growing undeniably bigger. She took X-ray treatments for it but later said, 'Well, I took thirty of them and it sure failed to give me any relief.' Her physicians seemed unable to stop her decline, but Catherine refused to give up hope. There'd been some coverage a while back about a treatment that might eliminate radium in victims' bones – she just needed to hang in there, and a cure would come.

With Catherine unable to manage the stairs anymore due to her misshapen hip, Tom brought her wrought-iron bed downstairs to the front room; he slept on a couch at the foot of it. He made it as comfortable as he could for Catherine; there was a makeshift lamp at the head of the bed, as well as a radio, and he hung a very large wooden crucifix on the wall above the bed. It had Jesus on it, so He could look over and look after Catherine as she slept. Her crutches were set against the wall, ready for when she was assisted to the bathroom; a 'well-worn pair of slippers' rested by their feet. The 'timid-looking bunny rabbit' given to the children last Easter kept her company on the bedside table.

The room had two windows in the front and a window to the west. 'It had good light,' recalled her niece Mary, 'but they kept the shades drawn; I suppose that was what she wanted.' It made for a rather dim setting – but then Catherine had a light of her own.

'Even now,' she said numbly, 'my body gives off a faint luminous glow when surrounded by darkness.'

'You could see every bone in her body,' remembered her nephew James. 'She was just lying on the bed.'

Marie Rossiter had once remarked of the girls' games in the darkroom at work: 'You don't see nothing, no body, all you see is the radium.' Her words now seemed strangely prophetic.

'People are afraid to talk to me now,' Catherine confessed. 'Sometimes it makes me terribly lonesome – they act as though I'm already a corpse. It's hard to have people around and still to be alone.'

Even when the family came to visit – the Donohues had always hosted meals after church on Sunday, when they'd serve eggs and bacon and Catherine would pour tea from her white china teapot printed with pink rosebuds – James remembered that they talked in the other room so Catherine could rest. Someone else now poured the tea.

As the year drew towards its end, Catherine's isolation became even more intense. She now spent 'nearly all of her days and nights lying down, venturing outside only with help, generally that of her husband'. 'He used to carry her around in his arms,' recalled James.

In such a condition, there was no way she could mother her children as she wanted or needed to. Although the Donohues had no money, a housekeeper was arranged; this live-in nanny, Eleanor Taylor, now became a surrogate mother to Tommy and Mary Jane. Catherine tried to direct her children's care from bed.

'I think it made her feel so sad that she couldn't take care of her baby girl,' commented her niece Mary. 'She had been able to somewhat take care of the boy, and so he got to really have a mother's love. It was just a very sad situation, it really was.'

It wasn't even simply Catherine's health that now kept her from the kids. Mary Jane was still very small, and her mother worried desperately that the glow she gave off in the dark was harming her baby. 'They were almost afraid,' remembered Mary, 'to have Mary Jane interact with her mother. They really didn't quite understand the radium sickness [and what it might do]. That was the sad part.'

'I suffer so much pain,' Catherine wrote to Pearl, and she may not simply have meant her aching hip and jaw, 'that at times I feel my life was pretty hard to bear.'

Stuck on her own in bed all day, Catherine was incredibly lonely. Charlotte now lived in Chicago; Pearl lived miles away in LaSalle. Though the girls wrote to each other, it wasn't the same. Catherine exclaimed in a letter to Pearl that December: 'I have so much to say, one cannot give it all on paper.' Her loneliness leapt off the page: 'It has indeed been a long time since I have heard or seen any of you girls that it seems like writing to a stranger. I only wished we lived nearer one another.' Still, at least she could be honest with them: 'As to my health,' she wrote bluntly, 'I am still a cripple.'

Her isolation meant she had no idea what was happening in the court case. 'We have not heard from Grossman ourselves and I can't understand,' Catherine wrote to Pearl. 'Tom is not working now or I would call him long-distance and find out if he is coming down. Seems funny he has not written, doesn't it?'

But Grossman had been too busy to write. 'This is the first of the Radium Dial cases,' he later said, 'and I can leave no stone unturned in reaching for all the light and truth and all the facts of record.' He did, however, drop the girls a festive card 'with every good wish for a Happy Holiday Season'.

And Catherine took his advice to make that Christmas a happy one. Though Tom was still unemployed, she wrote in upbeat words to Pearl: 'It makes it bad around Christmas, but one mustn't complain.' When Father Griffin visited to give her Holy Communion, Catherine sent a little prayer up to God to give thanks for all her blessings. She and Tom and Tommy and Mary Jane might be poor, and Catherine might be sick, but they were together at Christmas, and that was something for which she was simply very, very grateful.

The New Year, 1938, was all about preparing for the trial. The court date was set for 10 February, six days after Catherine's thirty-fifth birthday. Grossman was as busy as ever, and now spending more time in Ottawa as he prepared the women for their testimonies. Since it was wintertime and Illinois weather could be fierce, on occasion he had to pull out all the stops to make it there. 'They went back and forth,' recalled his son Len. 'I know one time the roads were bad so he rented a private plane and somebody flew him down there [in] a two-seater or four-seater plane.' It was a typically flamboyant Grossman gesture.

The day after Catherine's birthday, she and Tom struggled to make what was now an extremely laborious journey to Chicago for examinations by three physicians: Dr Loffler, Dr Dalitsch (a specialist dentist) and Dr Weiner; the latter took X-rays of her radium-filled bones. This trio of doctors had agreed to testify in court, and they would base their testimony on the exams.

They were shocked as Catherine staggered into their offices that Saturday morning. She was, Sidney Weiner recalled, 'a woman appearing much older than her given age and walking with the assistance of two people; markedly emaciated; with an ash colour [face]'. She had no fat on her body at all. Unable to eat – for it was too painful to do so – the weight simply fell off her frame and left her skeletal beneath her loose dresses. Catherine knew she had lost weight. But even she was shocked when she stepped on the doctor's scales; she weighed 71 pounds (5 stone).

From his dental examination, Dalitsch found the 'destruction' of Catherine's mouth went 'right through the body of the lower jawbones'. These fractures had led to 'displacement of the fragments' – which was why Catherine kept having to pick out pieces of her jawbone from her mouth. There was also, Dalitsch noted, 'considerable discharge of pus and foul odour'.

Loffler, meanwhile, ran tests on her blood, finding 'an alarming loss of blood powers'. He discovered that she had a white blood cell count of only a few hundred, whereas normal levels are about 8,000. She is, he thought to himself, 'near death from exhaustion caused by the lack of these [cells]'.

Yet it was her X-rays that troubled the doctors most. The hard tumour on her hipbone, which had so been concerning Catherine over the past few months, was now 'about the size of a grapefruit'.

The doctors didn't share their findings with the Donohues. Catherine was a sick woman; she needed to get home to bed. Just as Irene La Porte's doctor had felt, the physicians did not believe it was right to share their prognosis with Catherine, for fear of accelerating her decline. Far better that she stayed hopeful and positive: that, the doctors believed, would help her fight this disease far more than knowing the facts.

   Catherine and Tom made the difficult journey back to East
Superior Street. Tom carried his wife into the front room and
laid her gently down on the bed. She needed to rest. For in five
days' time, she would have her day in court. Catherine Wolfe
Donohue was holding Radium Dial to account for what it had
done to her and her friends – and she was determined, no
matter what, to make a difference.

50

Thursday 10 February 1938 dawned as a cool and cloudy day. In the front room of East Superior Street, Tom Donohue helped his wife to dress. He helped her slide on her knee-high nude stockings; lace her flat black shoes. Catherine had picked out her best outfit: once again the black dress with white polka dots slipped over her head and she slowly fastened its black belt around her emaciated waist. The dress hung so much more loosely than it had done in July when she'd first met Grossman, but she wasn't going to think about that today.

As a final addition, around her left wrist she looped a silver-banded watch that Tom had given her before their marriage; it was not luminous. With her spectacles on, a black hat pulled onto her head and a dark fur coat wrapped round her shoulders, she was ready.

Her husband, too, took care with his clothes. Tom usually wore the garments of a labourer: dungarees and rough work-wear. Today he donned a dark three-piece suit with a sober striped tie; his thick hair and moustache were neatly combed and he also wore glasses. Having added a light-coloured trilby, he was set to carry Catherine to court.

But he could not do it alone. Clarence Witt, the husband of Olive, helped him. Catherine was seated on a blond wood chair as they lifted her; her skin bruised so easily and her bones were now so fragile that it was difficult for Tom to carry her in his arms, next to his chest: the chair was a safer choice. They carried her all the way to the courtroom and then went up to the fourth floor, where Grossman greeted them, coming to assist.

As they helped her into one of the courtroom's black chairs, Catherine gazed around at the nondescript room. As it was a hearing before the Industrial Commission, it looked more like a meeting room than a court; it was, in fact, the office of the county auditor. It had a diamond-patterned tiled floor and was dominated by a large wooden table with sturdy legs; chairs were set around it for the key players and then ranged in semi-circles beyond that for spectators.

Catherine's friends were already there, including Pearl Payne and Marie Rossiter, yet the women weren't the only ones present. Just as the New Jersey girls' case had done a decade before, the women's plight had captured the imagination of the nation: reporters and photographers from across the country thronged the room.

Although the media had turned out for the trial, it seemed the Radium Dial executives had not. Neither had all its legal team, for only Arthur Magid was present, seated next to the arbitrator (judge) at the big table. There was no Walter Bachrach, no Mr Reed, no President Ganley; no one but Magid to represent the firm. Perhaps they thought it was beneath their attention, or perhaps some other reason kept them from the court.

Catherine looked closely at the judge: this was the man who would decide her fate. George B. Marvel was sixty-seven years old: a round-faced gentleman with white hair and spectacles, which he wore positioned towards the end of his small nose. He had been a lawyer and bank president prior to joining the Industrial Commission; Catherine wondered what he would make of her case.

As she took in her surroundings, waiting for the trial to begin at 9 a.m., the press took in the sight of her. 'Mrs Donohue,' the *Chicago Herald-Examiner* later wrote, 'could hardly stand alone. Her arms were no larger than a child's and her face was drawn and pinched. Her dark eyes burned feverishly behind rimless glasses.' The *Times*, somewhat unkindly, called her a 'toothpick woman'.

Catherine sat at the main table, with Tom seated just behind her. She carefully pulled off her big fur coat and placed it neatly on her lap, but she kept her hat on; she seemed to be cold all the time these days, frozen by the lack of fat on her body and by her failing heart. Feeling the pus starting to ooze again in her mouth, she pulled out a patterned handkerchief and kept it by her. She seemed almost constantly to have to hold it to her mouth.

Grossman checked with her to see if she was ready and she nodded briskly. The lawyer was dressed in his usual three-piece tweed suit, his eyes bright with anticipation of the job ahead. For more than half a year he had worked tirelessly on the women's case: he knew both he and Catherine were well prepared.

'We do not belong,' Grossman stated in his opening to the case, 'to that resigned class of victims who stretch forth unsuspecting throats to the sharpened sword of even so distinguished an adversary as the law firms of record for respondent in this case ... Under the intrepid Illinois Industrial Commission, larger and larger grows the brightening rainbow of our hopes for the right against the wrong, and the weak against the strong.

'... Human lives,' he continued, bringing his introduction round to the woman at the centre of the case, 'were saved among our country's army of defence, because Catherine Donohue painted luminous dials on instruments for our forces. To make life safe, she and her co-workers [are] among the living dead. They have sacrificed their own lives. Truly an unsung heroine of our country, our state and country owe her a debt.'

Now, it was that unsung heroine's turn to speak. Sat at the central table, with Grossman by her side and Magid and Marvel opposite her, Catherine was the first to give evidence. Though she wanted desperately to come across as strong, her voice, projecting through her battered mouth, betrayed her. The papers commented on her 'weak and muffled voice', which was 'faltering' and 'barely audible even to [her friends] who sat in a circle behind her chair'.

But speak she did; describing her work, the way the powder covered the girls all over and made them glow, the practice of lip-pointing. 'That's the way this terrible poison got into our systems,' she cried. 'We never even knew it was harmful.'

Grossman gave her an encouraging smile; she was doing brilliantly. While Catherine took a quick drink of water, her lawyer now introduced into evidence the deceitful full-page advert that Radium Dial had printed in the local paper.

'Objection,' said Magid, rising, but George Marvel allowed it to stand.

'After those New Jersey people died from radium poisoning in 1928,' Catherine continued, 'we began to get alarmed. But shortly after that Mr Reed called our attention to [this] advertisement. He said we did not have to worry.'

Marvel nodded slowly, taking notes and reviewing every word of the controversial notice. Catherine kept on with her testimony, looking over her shoulder at her friends, who sat in a row listening intently to her speak. 'After Miss Marie Rossiter and I had been examined the first time,' she recalled, turning back to face the judge, 'we wanted to know why we didn't get our reports. Mr Reed said to us, "My dear girls, if we ever give a medical report to you, there will be a riot in this place." Neither of us then realised what he meant.'

But they did now. As Catherine described the encounter in court, Marie 'paled at her words'.

'Oh!' she cried aloud, the implication of her manager's words sinking in.

'That is the Mr Reed,' Catherine added pointedly to the judge, 'who is still with the company in New York.'

The papers had found him there, overseeing the dial-painting girls. He had 'assumed responsibility for the operation'; which could well have been a promotion, since the New York plant was far more prestigious than the one in Ottawa. The company, it seemed, rewarded loyalty from its employees.

There was a disturbance then, as the chief security examiner of the commission came rushing into the room, bringing documents that Grossman had subpoenaed. The lawyer quickly flicked through the files. He could see at once that the girls' test results from 1925 and 1928 were not included. There were, however, some letters of especial interest.

Kelly, the president of Radium Dial, had written to the Illinois Industrial Commission in 1928:

We have not been successful in obtaining compensation insurance since the cancellation of our policy [on] August 18 1928. In view of the publicity given the so-called radium poisoning cases of the United States Radium Corporation of New York, the [insurance company] decided they did not care to carry the insurance any longer and incur the risk of our having such cases at our plant in Ottawa, Illinois.

Kelly had applied to ten different insurance firms. All had turned him down.

'You can readily see,' Kelly continued, 'that it makes it rather an unfortunate situation for *us*. Can you advise us how WE may obtain protection? Does the State of Illinois have any compensation insurance?'

Kelly's only thought was how he could protect the financial assets of his company; he didn't seem to consider that perhaps the insurance companies were refusing to cover him because what he was doing was too dangerous to support. In response, the commission told him: 'The only thing you can do is to carry your own risk.'

Kelly had decided it was worth a punt. That was why there were no insurance-company lawyers at this trial: because Radium Dial *had* no insurers. On 30 October 1930, the IIC gave notice to Radium Dial that it had not complied with the Workmen's Compensation Act, which required insurance; in response, Radium Dial 'was forced to post securities and offer guarantees with the Industrial Commission that it was carrying its own risk'. And this is when Radium Dial paid to the commission the $10,000 that Catherine and her friends were now trying to share between them. This was how that meagre pot of money came to be.

And there was no more money. Grossman had had no luck in tracing Radium Dial assets for the girls to claim from; now the firm had fled to New York, it seems the Illinois Industrial Commission had no power to reach across state lines to commandeer any of the company's funds. It was financially disappointing, but in many ways this case was not about the money. It would make a difference, sure – Tom and Catherine in particular would be saved from destitution if they won – but it was more important to the women by far that what had happened to them was recognised. The girls had been shunned, told they were liars and cheats and frauds; they had seen the company literally get away with murder. The truth was what they were fighting for.

To the almost continuous objections of Arthur Magid, which were all overruled, Catherine now told of her and Charlotte's visit to Mr Reed after they'd been diagnosed. 'Mr Reed said he didn't think anything was wrong with us,' Catherine whispered, as angrily as her weakened voice would allow. 'He refused to consider our request for compensation.'

Marvel nodded, transfixed by Catherine. 'Her emaciated body [was] shaking' but she didn't let it stop her.

'After two years,' she said, remembering back to 1924, 'I began to feel pains in my left ankle, which spread up to my hip. Fainting spells also occurred. At night the pain became unbearable.'

She told of how her pains had spread, all across her body: her ankles, her hips, her knees, her teeth; how she had become a bedridden invalid, unable to eat, unable to care for her own children. And then, as her fingers twisted a scapular medal – a Catholic talisman – she told of no longer being able to kneel to pray. With immense pathos she described her suffering – and not just her own. Catherine told the court how her two children were also affected.

Shortly before her testimony ended, Catherine reached for her purse and withdrew a small jewellery box, which she held discreetly on her lap. She and Grossman had discussed this beforehand, so he asked her about the exhibit she had brought.

Catherine bent her head to the box and lifted it up with her thin hands. The court leaned in, wanting to know what was inside. Slowly, very slowly, she opened it. And then, from within it, she withdrew two fragments of bone.

'These are pieces of my jawbone,' she said simply. 'They were removed from my jaw.'

C atherine's friends, watching her hold up the pieces of herself, 'shuddered' in the courtroom.

Her bones were admitted into evidence, along with several of her teeth. After such staggering testimony, Grossman now allowed her to rest. She sat quietly in her chair, dabbing her handkerchief to her mouth and watching as Dr Walter Dalitsch came to the table to give evidence on her behalf.

He was a clean-featured man, with a strong forehead, thick lips and dark hair; he gave evidence authoritatively. Grossman took him through his dental treatment of Catherine and then they proceeded to a more general discussion of radium poisoning. When Magid objected to Dalitsch's assertion that many dial-painters 'became sick and died with diagnosis different to the truth', Marvel overruled him. The judge added emphatically: 'The doctor is skilled and testified as an expert.' It seemed the arbitrator was on Dalitsch's side.

Dalitsch gave his expert opinion on the cause of Catherine's disease. 'The condition,' he said plainly, 'is a poison from radioactive substances.'

With the killer statement in the bag, Grossman began more quick-fire questioning.

'In your opinion,' he asked, 'is Catherine Donohue today able to do manual labour?'

The dentist looked across the table at Catherine, who was huddled in her chair, listening to him speak. 'No,' he said sadly, 'she is not.'

'Is she able to earn a livelihood?'

'No,' said Dalitsch, refocusing on Grossman.

'Have you an opinion as to whether this condition is permanent or temporary?'

'Permanent,' he answered swiftly. Catherine dropped her head: *this is forever.*

'Have you an opinion,' Grossman asked now, 'if this is fatal?'

Dalitsch hesitated and 'glanced meaningfully' towards Catherine, who was only metres from him. Grossman's question hung in the air, suspended in time. Five days ago, after the examinations in Chicago, Catherine's three doctors had indeed determined that her condition had reached its 'permanent, incurable and terminal stage'. Yet the physicians, who in all kindness sought to spare her, had not told Catherine Donohue.

'In her presence?' Dalitsch now asked, uncertainly.

But he had said enough. He had said enough in the way he had paused. Catherine 'sobbed, slipped down in her chair and covered her face' with her hands. At first, silent tears ran down her cheeks, but then, as though the full weight of what he hadn't said hit her, she 'screamed in hysteria'. She screamed aloud, as she thought of leaving Tom and her children; as she thought of leaving this life; as she thought of what was coming in her future. She hadn't known; she had had hope. She had had *faith.* Catherine had truly believed she was not going to die – but Dalitsch's face said otherwise; she could see it in his eyes. So she screamed, and the broken voice which had struggled to speak was now made powerful in her fear and distress. Tom 'broke down and sobbed' at the sound of his wife's cries.

The scream was a watershed; after it, Catherine could not keep herself upright. She collapsed and 'would have fallen had not a physician nearby caught her'. Dr Weiner had leapt to his feet to hold her up, and as he did so Tom seemed released from his paralysis. He rushed to Catherine's side as she lay slumped in her chair. While Weiner felt for her pulse, Tom's concern was only for Catherine. He cradled her head with his hand, touched her shoulder to try to bring her back to herself; back to him. Catherine was sobbing hard, her mouth wide open, showing the destruction inside: the gaps where her teeth should have been. But she didn't care who saw; all she could see was Dalitsch's face in her mind. *Fatal. This is fatal.* It was the first time she'd been told.

Pearl, seeing her friend so distraught, was just a moment behind Tom. The two of them bent over Catherine, Pearl proffering a cup of water that was not acknowledged. Tom had his arms around Catherine, trying to get through to her as she cried. His labourer's hands supported her, one on her back, another pressed to her front, trying to show her he was there.

The press photographers wasted no time in capturing the moment. Tom was suddenly aware of them, suddenly aware that he wanted to get his wife away from all this. Leaving Pearl to care for Catherine – she gently stroked her friend's dark hair – Tom summoned Grossman and Weiner and together the three men lifted Catherine's chair and carried her from the courtroom, Pearl clearing the way through the people.

'The woman's sobs,' a paper commented bleakly, 'could be heard from the corridor.'

As the judge called an immediate recess, Catherine was carried to the county clerk's office and laid out on a desk. Pearl spread Catherine's fur coat beneath her so she had something soft and pleasant against her skin; books of birth records propped up her head as a makeshift pillow. Tom gently eased his wife's glasses from her face and stood by her side; he had both hands on her: one holding her hand which wore his watch and the other softly caressing her hair to calm her. Pearl held Catherine's other hand, trying to reassure her friend. They both murmured soothingly to the woman they both loved.

Catherine, by now, was too weak for tears, but as she felt her husband standing near her, she did have one thing to say. In a 'feebly wavering' voice, she clutched his hand and murmured: 'Don't leave me, Tom.'

He wasn't going anywhere.

Catherine was unable to return to the hearing. 'She is in total collapse,' said an attending doctor. 'She will not, cannot, live very much longer.'

Tom was not there to hear his words; he had carried Catherine home to East Superior Street. Yet when the papers printed the pictures of Tom and Catherine the next day, they pulled no punches. Above a photograph of the stricken couple was the headline: DEATH IS THE THIRD PERSON HERE.

The hearing resumed at 1.30 p.m. in Catherine's absence. Having settled his wife at home, Tom had made his way back to the courthouse, wanting to represent Catherine at this hearing which was so very important to her. If she was not well enough to be there, then he would stand for her instead.

The hearing picked up where it had left off, as Tom sat numbly in a chair at the back of the room.

'Is her condition fatal?' asked Grossman of Dalitsch.

The doctor cleared his throat. 'It is fatal in her case,' he acknowledged.

'In your judgement,' asked the lawyer, 'what might be Catherine Donohue's reasonable [life] expectancy?'

'I don't think we can state definitely,' Dalitsch began to bluster, perhaps conscious of her husband sitting in court, 'depending on the care she had and so on – treatment ...'

Grossman fixed him with a look. This was a courtroom, not a clinic, and it did not help Catherine's case to beat around the bush. Dalitsch straightened up under Grossman's glare.

'I would say ... months,' he stated bluntly.

Tom felt tears at his eyes again. *Months.*

'There is no cure in its advanced stages?' queried Grossman.

'No,' said Dalitsch. 'There is none.'

As the afternoon wore on, the other doctors were questioned. With each new testimony, the words became a litany of loss that Catherine's husband was forced to hear.

'She is beyond a doubt in the terminal stages of the disease,' testified Dr Weiner.

'She has but a short time to live,' concurred Loffler. 'There is absolutely no hope.'

No hope. No cure. No Catherine.

Tom listened to it all with tears streaming down his cheeks. He endured it all. By the end of the afternoon, he was near collapse and had to be led from the courtroom.

The company lawyer, for his part, made no showy gestures. He limited his cross-examinations of the doctors to what Radium Dial considered the critical issue: was radium a poison? It seemed irrelevant to Magid that Loffler declared, 'There is a definite causal relationship between her employment and the condition in which I found her.' Magid instead declaimed that, 'Radioactive substances may be *abrasive*, but not poisonous.'

'The company's position,' the slick attorney explained, 'is that [the women] cannot recover compensation under the new section of the law because that relates only to diseases incurred from *poisons* as a result of occupation.' With the firm determining radium was *not* a poison, they held themselves 'not liable'.

Magid called radium poisoning a 'phrase' that was 'merely a convenient method of describing the effect of radioactive substances upon the human body'. It was a position he maintained even when Loffler said angrily, 'The radioactive compounds had a poisonous effect upon [Catherine's] system and their effect was not merely what is ordinarily termed abrasive, but [comes] under the medical definition of poisons!'

It was not lost on Grossman that this was the *same* lawyer who only a few years before had maintained radium *was* a poison in the Inez Vallat case. Grossman dubbed Magid's attempt to twist the truth 'brilliant sophistry and attempted magic' by a 'past master in the wizardry of language and poison'.

The women's lawyer added: 'For evidence in support of respondent's theory that radium is not a poison, the record is as silent as the Sphinx of Egypt.' No testimony was presented to back up the company's claim.

In contrast, the women had plenty to say and with Catherine having recovered a little from her collapse, she was determined to continue giving evidence. Yet her physicians pronounced her too ill to leave her bed and said she was 'in a state of complete collapse that might prove immediately fatal to her, were she forced to continue as a witness'.

But Catherine was adamant. At this point, Grossman suggested that the hearing be resumed tomorrow at her bedside. If she couldn't come to the courtroom, then Grossman would bring it to her. George Marvel, after considering the request, agreed.

It fell to Grossman to inform the press. As he announced that a bedside hearing would be held the next day, he added a final comment that he knew would provoke the press into copious column inches.

'That is,' he said darkly as he surveyed the gathered media, 'in case she is alive . . .'

52

When Friday 11 February dawned, Catherine Donohue *was* still alive. The weather outside on East Superior Street was 'unsettled', but Catherine, despite her weakened condition, felt certain of what she had to do.

'It's too late for me,' she said bravely, 'but maybe it will help some of the others. If I win this fight, my children will be safe and my friends who worked with me and contracted the same disease will win too.'

It had been agreed by Radium Dial that Catherine's would stand as a test case. If the court found for her, then all the other victims would find justice as well. It made it all the more important to her that she did not fall at this final hurdle; she had to fight on, come what may.

Tom supported her decision to testify, but he was worried sick. 'All this is too late for us,' he echoed, 'but Catherine wants to do all she can to help the others. Even if the excitement—'

His voice broke off abruptly. He had heard what the doctors had said: that continuing as a witness could prove fatal. But Catherine was determined; and who was he to stand in her way? 'We've had so little time together,' he simply said, quietly. They had been married just six years.

Tommy and Mary Jane, then aged four and three, were at home. They played upstairs as the mass of visitors was directed to the dining room, where Catherine lay on the blue sofa, pillows propping her up and a white blanket covering her to her chin. One after another the guests crowded into the room, some thirty people in all – lawyers, witnesses, reporters and friends.

Catherine barely had the strength to open her eyes to welcome them. She made a 'pathetic spectacle'; her friends greeted her with evident concern. They usually visited this place socially, but today was a very different occasion. The women sat on chairs lined up by the sofa: Charlotte Purcell, who had come down from Chicago, was closest to Catherine; she sat next to Pearl. Charlotte had declined rapidly of late, having lost a tooth only the week before. She sat huddled in a thick grey coat, its left sleeve hanging empty by her side.

The lawyers drew up chairs at the round oak table and spread their papers across it: Grossman, Magid and Marvel, with Grossman's secretary, Carol, taking notes. Conscious of his children upstairs, Tom hovered halfway between the dining room and the rest of the house, leaning disconsolately against the door jamb.

The scene thus set, the hearing now began. 'Weak but determined, Catherine Donohue was ready to resume her story.'

As Grossman questioned his client, he knelt by her side so that she could hear him better. She answered him 'through closed eyes'. Only occasionally would she open them, and even then she didn't really seem to see.

'Show us,' Grossman encouraged, 'how you were taught to point [the brush], as you described in testimony yesterday.' He held out a child's paintbrush towards her, taken from Tommy's watercolour set.

As Catherine reached out a skeletal hand from beneath her

blanket to take the brush, Arthur Magid rose from where he was seated at the table. 'Objection,' he said. 'We object to the use of the brush, as there is no proof it is the same type as that used in the plant.'

Marvel turned to Grossman. 'Is there one you could get?' he asked.

'Yes,' replied Grossman somewhat tartly. 'They are being used now at the Luminous Processes plant, which is using all the equipment of the Radium Dial Company, and employs some of the company's girls. There's even an official there who was at the Radium Dial Company.'

'It was decided,' wrote a reporter after witnessing this exchange, 'that the brush could be used for the demonstration.'

Catherine took the delicate paintbrush proffered by her lawyer. She paused for a moment, feeling its barely-there weight in her hand, the way her fingers curled familiarly around it.

'Here's how it's done,' she croaked, after a beat. Her voice sounded tired. 'We dipped it in the radium compound mixture.' Catherine dabbed the brush into an imaginary crucible and then, very slowly, bent her stiff arm back and raised the brush to her lips. 'Then shaped it,' she said with some emotion, 'like this.'

She slipped the brush between her lips and twirled it. *Lip ... Dip ... Paint*. When she was finished, she held it up with a shaking hand: the bristles now tapered to a perfect point. Seeing it, 'a shudder ran through her trembling frame'.

Her friends and former colleagues watched her with their faces 'drawn with emotional intensity'. The women were visibly affected by her demonstration and fought back tears.

'I did this,' Catherine said dully, 'thousands and thousands of times ... That was the way we were *told* to do it.'

Tom watched his wife from the doorway; watched as she demonstrated how she had been killed. Though he had thought himself wrung out of tears, he wept quietly, unashamedly, as Catherine showed off the simple movement that had left her little more than a living corpse.

Grossman cut through the chilling atmosphere in the room with a question. 'Did any official of Radium Dial ever tell you that the US Government had condemned the use of camel-hair brushes in painting with radium compounds?'

Catherine looked shocked to hear it. 'No,' she replied. The girls sitting behind her exchanged looks of anger.

'Objection,' piped up Magid, almost speaking over Catherine.

'Sustained,' responded Marvel.

Grossman wasn't thrown off-track; he had another query. 'Was there any notice posted arising from the dangers of radium dial-painting with the hair brushes?' he asked.

'No, sir,' Catherine replied surely, 'there was none. We even ate our lunches on the work tables near the luminous paint. Our superintendent, Reed, told us it was all right to eat there, but not to let the food spot the dials. All they told us' – she was panting now with the effort of speaking – 'was to be careful not to get any grease spots on the dials.'

Grossman touched her gently on the shoulder. She was exhausted, he could tell. He carefully took her through the remaining key points, including the debacle of the glass pens and the way she had been fired for limping, and then he let her rest.

He called Charlotte Purcell to take the oath.

'Objection,' called out Magid immediately. He didn't want the other girls to give evidence, citing the fact that this was Catherine's case alone.

'This is a test case, Your Honour,' Grossman cut in smoothly, appealing to Marvel. 'I don't know that I will have these girls with me at any future time.' His eyes scanned the row of young women sitting alongside Catherine's makeshift bed. 'Not *all* of them,' he added pointedly.

Marvel nodded. He allowed the women to be questioned, though the girls 'were not permitted to testify directly about their own conditions'.

As Charlotte stood up to give evidence, Pearl helped her to

slip her grey coat from her shoulders. Underneath she was wearing a green blouse with a fussy white collar; its sleeve 'hung
limply, revealing her amputated arm'. She came to the table
to take the oath. Then she too twirled the brush in her mouth,
showing her missing teeth as she did so. She testified calmly, as
the anxious eyes of her friends followed her evidence. One girl's
eyes filled with tears as Charlotte talked.

'Were you employed,' Grossman asked her, 'at the Radium
Dial Company when Catherine Donohue was employed there,
in the same room?'

'Yes, sir,' said Charlotte. Her stronger speech was in direct
contrast to the strained whisper that had been all Catherine
could manage.

'Did you have your left arm then?'

Charlotte swallowed hard. 'Yes, sir.'

'How long were you employed there?' he asked.

'Thirteen months,' she said, almost spitting out the words.

He asked her about the confrontation she and Catherine had
had with Mr Reed. 'Did you have your arm then?'

'No, sir,' she replied bluntly.

'. . . What did Mr Reed say?'

'Mr Reed,' said Charlotte, her eyes burning angrily, 'said he
didn't think there was any such thing as radium poisoning.'
She testified 'the loss of her arm was due to the poisonous compound used'.

One by one, Grossman called the girls to give evidence, which
they did sitting alongside the lawyers at the Donohues' dining
table. Marie Rossiter clenched and unclenched her fingers as she
reported what had gone on.

'Mr Reed said radium would put rosy cheeks on us,' she
remembered in disgust, 'that it was *good* for us.'

Grossman asked each of them in turn if the painting demon-
strations had been an accurate re-enactment of the technique
they'd been taught. Like a row of doppelgängers, each nodded
her head.

All the women testified on Catherine's behalf: Pearl Payne, the Glacinski sisters, Olive Witt and Helen Munch too. As each woman stood up, literally and figuratively, for her friend, she was mirrored by Arthur Magid, who continually objected to their evidence. Tom Donohue spoke only briefly, to confirm the calamitous amount of debt that he and Catherine were in due to her medical bills.

Throughout it all, Catherine herself lay dumbly on the sofa, sometimes dozing to the lullaby of her friends' voices as they lilted around her. At last, it came to an end. Across the two days, fourteen witnesses had given evidence for Catherine. Now, Grossman rested his case and everyone turned expectantly to Arthur Magid.

But the company lawyer presented no evidence and called not a single witness. The firm was standing solely on its legal defence that radium was not poisonous.

With no further evidence to hear, shortly after 1 p.m. Marvel formally closed the hearing. He would, he said, give his verdict in a month or so; before then, both sides would have opportunity to submit complex written legal briefs, which set out their arguments in full.

There was just one final element to the proceedings; an opportunity that the gathered reporters would never have let pass them by. Before the throng of people departed from the Donohue home, the media requested a photo call. George Marvel and Arthur Magid both moved behind the sofa as Grossman knelt beside Catherine; a cigar was already between his fingers now the case was at rest. As the men came into her line of vision, Catherine stretched out a thin hand towards George Marvel. He took it, clutching her fingertips gently, shocked at her emaciated bones, at how delicate her hand was. Catherine, later, deemed him 'so sympathetic'.

It wasn't just the attorneys who were required in shot. Catherine's friends, too, surrounded her once the lawyers moved away. Charlotte perched on the arm of the sofa at her feet, while the others stood behind. Pearl Payne was in the centre, holding her hand. All the women were looking at Catherine – but Catherine was looking at Tom. He had come forward now that the hearing was over, and seated himself by her side. As the camera clicked, the husband and wife had eyes only for each other.

'Suddenly,' a reporter later wrote of seeing Tom and Catherine together, 'I forgot her crumbled teeth, the shattered jaws ... I forgot the tragic remnants that radium poisoning left of a once-handsome woman ... I saw briefly [instead] the soul that holds her husband's love – [a] love grown blind to the fragile shell of a woman that is all other people see.'

There was just one other photograph. Hearing the meeting was over, Tommy and Mary Jane came running into the room. Tom lifted them up, one in each arm, and sat them on the back of the sofa so that Catherine could see her children. And now, for the first time all morning, she came alive again, reaching out to hold Tommy's hand with an animated expression on her face as she chatted to her boy and girl. Mary Jane had sweet bobbed hair with a ribbon in it and wore a coloured dress; Tommy was in a long white shirt. Both seemed somewhat overwhelmed at all the guests and the photographers, and shortly afterwards Tom ushered everybody out.

Grossman and the other girls went directly to a downtown hotel, where they conferred at length before Grossman left for Chicago. The women knew that whatever happened next, it would affect them all. Even that day at the hearing, Magid had confirmed once again that, whatever the judge's decision, the firm would abide by it in handling the other dial-painters' claims.

With the hullabaloo of the court all gone, Tom shut the door of 520 East Superior Street. Somehow, the house seemed even quieter than it had before the hearing.

Now, all he and Catherine could do was wait.

S PRING IS IN THE AIR chorused the *Chicago Daily Times* the weekend after the trial. The newspapers were full of Valentine's adverts for romantic gifts, bridge parties and dances, but there was only one date that the dial-painters of Ottawa were keeping; and that was with Catherine Donohue.

They found her in good spirits when they called on her. When a reporter tagging along asked Catherine 'what slim thread binds her to life', she replied, 'It's the fighting Irish,' with a fond glance at Tom. 'I *will* live,' she said determinedly. Doctors had said that she would 'never leave her bed alive', but she was not done fighting yet.

Together, the women prayed for a cure, yet 'there was no horror among them of death itself'. 'Each declared,' said the *Chicago Herald-Examiner*, 'that if the fates decree, she will face the next world with the realisation that her sacrifice may have saved others.'

For the women, somewhat to their surprise, had become poster girls for workers' rights. Already, they had effected a significant change in the law that protected thousands of vulnerable employees and removed a loophole by which corporations could shirk their responsibilities. Inspired by what they had achieved, that same day Pearl Payne wrote to Grossman with an idea:

> *Sensing your humanitarian zeal for helping those on the*
> *lower rungs of life's ladder, it has occurred to me as well as the*
> *other participants in the Radium Dial suit that you forge the*
> *beginning of a society, whereby those, of which there must be*
> *thousands, could band together, secure legal aid and in general*
> *use our organised presence to simplify, promote and improve*
> *the laws relative to those who are maimed due to occupational*
> *hazards.*

Grossman thought it brilliant. And so, on Saturday 26 February 1938, the society had its first meeting. The founding members were Pearl Payne, Marie Rossiter, Charlotte Purcell and Catherine Donohue. Three of the four went to Chicago to meet Grossman; Catherine, far too ill to travel, was represented by Tom. They called themselves, with an instinct for a media hook that possibly came from Grossman, The Society of the Living Dead.

'The purpose of the society,' announced Grossman to the gathered press, 'is to obtain better protection by legislation and otherwise for persons endangered by occupational diseases.'

The meeting coincided with the filing of Grossman's first legal brief to Marvel, which was probably deliberate ('He loved the press,' said his son). As the camera bulbs flashed, Grossman gave the girls their own copy of the pale green brief, signing Pearl's with the slogan 'In Humanity's Cause'. The dense document was some 80,000 words in length, and it saw Grossman in full flow.

'The circumstances call,' he wrote, 'for the sharpest pen I can unsheathe. I ask simply that [the law's] protecting folds ever serve as a shield to protect, and not a sword to destroy the human right of Catherine Donohue to compensation. Give

Catherine only what is justly and rightly due her under the law of God and Man, and you will give her the award we ask!'

The brief was filed in the late afternoon, just in time to catch the evening papers, and the press were all over it; coverage of the case jostled for space on the front page with stories about the Nazis in Germany. And if it was a trial by media, the girls would have won hands down – the papers called Radium Dial 'criminally careless'.

The press asked Tom Donohue if there was any hope of a cure. He replied that Frances Perkins, the Secretary of Labor, had 'sent medical authorities to investigate'. There'd been some hope that a calcium treatment might prolong Catherine's life, but her illness was too advanced for her to survive the process.

The federal investigations Perkins had ordered into the women's poisoning appear to have come to nothing. The government, struck by a double-dip recession within the Great Depression, had other priorities. One politician admitted they were 'floundering' about the economy: 'We have pulled all the rabbits out of the hat, and there are no more rabbits,' he said. It was cold comfort to Tom, who was still without a job.

Although the calcium treatment was not possible, Catherine still refused to surrender. 'I am hoping for a miracle,' she said. 'I pray for it. I want a claim to life, and to stave off the end for the sake of my husband and children.' Catherine's own mom had died when she was six; she knew what it was like to grow up without a mother and she was determined her own children would not suffer the same fate.

But for all Catherine's courageous talk, as the weeks passed and they all waited for the verdict, her health deteriorated rapidly. 'Once that [stage of the] illness began,' remembered her niece Mary, 'it was just like a spiral down, down, down ... It wasn't a gradual thing. It was *fast.*'

It left Catherine utterly unable even to direct her children's care via the housekeeper. 'She was so ill,' said Mary, 'I truly do not remember her interacting with the children. She wasn't *able*. You can't imagine ... It had just sapped all the energy and *everything* out of her.'

All Catherine could do was lie weakly on her bed in the front room with the shades drawn. Her days were punctuated by the taking of her medicine, and by the frequent rattle of the train tracks behind the house: the sound of carriages bearing people away on journeys that Catherine Donohue could never now undertake. The house had 'a smell of urine'. Her entire world was that front room. She lay under a blanket, the tumour on her hip a malignant mountain rising beneath, with every bone in her body aching. She was in so much pain.

'I just remember her moaning, moaning,' remembered Mary quietly. 'You knew that she was in pain, but she didn't have the energy to scream. Moaning was about the best she could do. I think she just didn't have the energy to cry or cry out. She'd just moan.

'I can't describe,' she went on, 'how *sad* that house was. You felt the sadness when you came in there.'

As Catherine's illness worsened, some of her relatives considered her condition too horrific for her young nieces and nephews to see. 'She was falling apart from the radium,' recalled her niece Agnes. 'They didn't want us to see her; they said she looked so terrible.' So although Agnes's parents would visit Catherine once a week, she always had to wait outside.

One relative who became a frequent visitor was Tom's big sister Margaret. She was a stocky fifty-one-year-old woman who was 'the boss of the family'. 'She was the only woman that I knew of that could drive,' recalled her nephew James. 'She had an automobile called the Whippet.' Another relative remarked: 'She'd go take care of [Catherine] and take care of the kids. She did what a good sister-in-law would do.'

Father Griffin was another regular visitor, and Catherine welcomed the nuns from the convent too, who brought her a relic of the True Cross. 'It is like having God in the house with me,' she exclaimed in elation.

She also took solace from an unexpected source: the public. With her story emblazoned across the newspapers, readers were horrified by it in a way that some of the women's neighbours had never been. Catherine received hundreds of 'lovely letters' that came from coast to coast. People sent her trinkets and ideas for cures; money for flowers to brighten her sickroom; some simply wrote in the hope that 'my letter will cheer you a little'. 'You have my sympathy and my greatest desire for a complete victory,' read one. 'And I know millions of people think the same.'

Her friends also lifted her spirits. Marie would spend an evening with her, sitting by the side of the wrought-iron bed; Olive 'brought me a chicken all cooked up lovely', Catherine wrote to Pearl with pleasure. 'She, like you dear, are truly pals, and may God bless you both.'

By March, Catherine had cheered considerably. 'I'm sitting up for a few minutes today,' she wrote proudly to Pearl, 'and oh, how good it feels after so long in bed!'

Leonard Grossman had not seen his bed for a very, very long time, or so it seemed. Throughout February and March there was a convoluted exchange of briefs as he and Magid duelled with pen-swords in further book-length submissions to Marvel. 'He worked around the clock for a week,' said Grossman's son. 'He had three or four secretaries in.' This crack team of assistants took dictation as Grossman paced his office or sat in his big chair with a cigar and reeled off the brilliant oratory for which he was famous. 'I have been busy day and night,' Grossman later wrote to Pearl, 'working on the radium case.'

On 28 March 1938, the final brief was filed: after consideration of this, Marvel would deliver his verdict. In it, Grossman denigrated the company's 'shameful, shifting defence' and what he called 'the cesspools in respondent's alibis', and continued: 'Language can coin no fitting words of odium with which to condemn the cool, calculating [Radium Dial Company]. [Workers were] lulled into a false sense of security by dastardly and diabolical false and fraudulent misrepresentations.' The company knew, he wrote, 'the legal duties which it owed to [its employees] and murderously refused them'. Its officials repeatedly lied to Catherine 'to induce her and other employees to remain quiescent and not aroused and not aware of her true condition'. They had, he said, 'betrayed her'.

He did not mince his words. 'I cannot imagine a fiend fresh from the profoundest depths of perdition committing such an unnatural crime as the Radium Dial Company did. My God! Is the radium industry utterly destitute of shame? Is the Radium Dial Company utterly dominated by a beast?

'. . . It is an offense against Morals and Humanity,' he concluded, 'and, just incidentally, against the law.'

He wrote powerfully. The judge had declared that he would not make a final ruling until 10 April – yet on Tuesday 5 April the telephone burred in Grossman's office. He was summoned to the headquarters of the IIC at 205 W Wacker Drive, just around the corner from the Metropolitan Building.

The verdict was in.

54

There was no time to tell the Donohues. Grossman managed to summon those few former dial-painters now living in Chicago – Charlotte Purcell and Helen Munch – and they alone were able to reach the hearing in time; it was held just before noon. Helen was nervously smoking a cigarette as they all crowded into the wood-panelled IIC room to hear the verdict. George Marvel's judgment was read aloud by the chairman of the commission. Both Magid and Grossman stood to hear him speak; the two attorneys sized each other up as the chairman called for silence.

Mrs Donohue, Marvel had written, was suffering from a disease that was 'slow, insidious in its nature, progressive and extending over a long period of years'. He concluded: 'The disablement sustained made Mrs Donohue incapable of following any gainful occupation.' Those in the court shifted restlessly; they knew all this. The question was: would he find the company at fault?

The written judgment proceeded. 'The Industrial Commission finds that a relation of employer and employee existed between the company and the plaintiff ... [Catherine Donohue's] disability *did* arise out of and come in the course of her employment.'

He had found the company guilty.

Charlotte and Helen could not help their reaction: they were jubilant. Helen stretched out a grateful hand to Grossman as he turned to them with an irrepressible grin. 'I am very glad for Mrs Donohue,' Helen breathed, 'it is a *just* decision.'

Marvel awarded Catherine her past medical expenses, back salary for the entire period she had not been able to find employment due to her condition, damages and an annual life pension of $277 ($4,656) for the remainder of her life. It came to a total of some $5,661 ($95,160) and was the maximum possible award the judge could deliver under the provisions of the law.

One suspects that he wished he could have gone further. Marvel was reported to have said after Catherine's collapse in the courthouse: 'It would seem to me [from] what has been disclosed here that there could have been a common law action against these people. There is and has been a gross negligence on the part of [the] Radium Dial Company.'

The company officials were guilty. Guilty of causing Catherine's disablement – and Charlotte's too – but not just that alone. They were guilty of killing Peg Looney, Ella Cruse, Inez Vallat . . . so many more. Those women's lives could not be saved, but their murderers were now unveiled in the cold light of day. 'The whole creation of God,' Grossman had written in his legal brief, 'has neither nook nor corner where the Radium Dial Company can hide the secret of its guilt in this case and escape.' The light of justice now flooded in, leaving the callous killers exposed for what they were. There was no full-page advert to hide behind here; no jolly superintendent smoothing down the girls' furrowed brows; no hidden test results that kept the truth concealed. The truth, after all these years, was finally out.

'Justice has triumphed!' declared Grossman exultantly at the hearing. 'No other decision is possible under the overwhelming weight of evidence. A just award follows the rule of conscience. Thank God for justice to the living dead.'

Charlotte Purcell simply said, thankfully, 'This is the first ray of hope we have had after years of discouragement.'

It had been a long, *long* battle. In many ways, it was a battle that had begun on 5 February 1925, when Marguerite Carlough had first filed suit in New Jersey: the very first dial-painter ever to fight back. Catherine's triumph in court, thirteen *years* later, was one of the first cases in which an employer was made responsible for the health of its employees. What the girls had achieved was astonishing: a ground-breaking, law-changing and life-saving accomplishment. The Attorney General's office, which had followed the case closely, heralded the verdict as 'a great victory'.

It was the *Ottawa Daily Times* which claimed to break the news to Catherine Donohue. As the verdict came down the wire, a reporter raced to 520 East Superior Street to talk to the woman at the centre of it all.

He found her alone; Tom had taken the children for a walk. She was – as she had no choice but to be – lying in bed in the front room, her silver-banded watch still hanging loosely on her wrist. As the reporter excitedly told her that a verdict had been given, five days early, Catherine blinked in surprise. 'I never dreamed of the decision this soon,' she croaked with great effort.

The good news spilled from the journalist's lips: a secret he could not wait to share. Yet Catherine was so ill, she showed little emotion at her victory and did not smile. Tom would later confide that she 'cries but rarely smiles; she has forgotten how to laugh'.

It could have been that she did not quite believe it. 'She half-rose in bed in an effort to look at the compensation decision, which was written out for her', but she did not have the strength to do so fully. She sank back against the pillows; and as the news, too, sank in, her primary thought was for Tom. 'Her first words,' wrote the eager journalist, 'formed a wish that her husband Thomas would quickly hear of the decision.'

'I am glad for the sake of my children and my husband,' Catherine whispered. 'The lump sums will help [Tom], who has been out of work for many months.'

As though remembering, she then said with a weak smile to the reporter: 'This is the second good news we had in a week. My husband just got back on the payroll at the glass factory.' Some workers had been recalled to Libbey-Owens, and Tom had managed to get on the night shift.

As the reporter lingered in the room, hoping for more copy, Catherine spoke on. 'The judge is grand,' she commented. 'He's so wonderful. He's very fair. That means a lot.'

As though the idea of fairness had triggered something in her, an anger blazed within her for a brief moment. 'It should have been done a long time ago,' she said, almost bitterly. 'I've been suffering. I'll have to suffer more ... I wonder if I will live to receive any of the money; I hope so. But I am afraid it will come too late.'

Yet Catherine had not put her life at risk for herself; she had done it for her family and friends. 'Now maybe [Tom] and our two children can really live again,' she said in hope. 'I may not live to enjoy the money myself, but [I hope] it will come in time for the other girls. I hope they get it before their conditions become as bad as mine.'

She added one final comment, a croaked whisper that fell flat in the oddly quiet, stale-aired room, which held none of the jubilation of the courtroom in Chicago.

'I hope the lawyers don't upset it ...' Catherine Donohue said.

Two weeks after the judgment, Radium Dial filed an appeal against the verdict, 'upon contention the award was contrary to the evidence'. Having anticipated such a move, Grossman and the Society of the Living Dead instantly staged a media photo call and launched an appeal for immediate funds for Catherine. 'She has no money, no prospects of getting any through her own efforts [and] mounting doctor's bills,' declared Charlotte Purcell. 'I fear Mrs Donohue will die before her case is adjudicated.'

Catherine was touched by her friends' support, but her over-whelming concern was for Tom. He had taken the news of the appeal hard. 'He doesn't say much,' Catherine confided to Pearl, 'but it has been such a strain on him.'

The women continued to enlist the help of the media in their campaign for justice; the Donohues invited the *Toronto Star* into their home for an interview. 'The eggshell woman in the bed may be dying,' wrote *Star* journalist Frederick Griffin, 'but she is fighting.'

They were all fighting – the women, and their supporters too. As Griffin visited 520 East Superior Street one quiet April evening, he met all the dial-painters who had filed suit, as well as the men standing behind them: Inez's father George; Tom, Al, Clarence and Hobart. This senseless tragedy had affected them as well as their wives and daughters. 'They're scared,' Clarence Witt said of the women, as his wife readied Catherine in the other room. 'Every little ache or pain scares them.'

It was now more than two months since Catherine had battled to give her evidence from her sickbed; the intervening weeks had wrought havoc with her body. 'I looked at the shrunken face, arms, form, the shapeless jaw and mouth,' Griffin remembered as he entered her makeshift bedroom. 'A glance at her skeleton outline beneath the coverlet makes you wonder if she will see the week out.'

Yet as Catherine fluttered open her eyes and fixed them on the reporter, he realised that she had more grit in her than he'd thought. 'Mrs Donohue, this remnant of a woman, took on her role as president of this strange society,' he later wrote. 'She lay motionless, but she was business-like.'

'Please publish this,' she said candidly. 'I want you, when you write about us, to put in a good word for our lawyer, Mr Grossman.'

She was commanding; her voice at this meeting, Griffin said, was 'brisk' and 'strong'. Grossman had paid for the entire legal proceedings himself – including the continuing expenses of the appeal – and Catherine wanted to be sure he would be rewarded with publicity at least.

'You hear the voice of the Society of the Living Dead,' Grossman himself now intoned. 'That is the voice of the ghost women speaking not only here in this room but to the world. This voice is going to strike the shackles off the industrial slaves of America. You girls have rights to better laws. That's what the society is going to work for.'

Griffin interviewed them all; each woman had her own heart-rending story. 'I'd hate to tell you [how I feel now],' Marie sighed. 'My ankles and jaw pain [me] all the time.'

'I don't know what day's going to be my last,' said Olive

anxiously. 'I lie at night staring at the ceiling and thinking maybe it's my last on earth.'

'It is an effort to do things in the ordinary way, to act normally,' confessed Pearl. 'I don't show it, but at present I am nervous and shaking. What I have lost I can never recover.

'I am missing so much,' she almost cried out. 'The chance of being a mother again ... I can never be the mother and wife that my fine husband deserves.'

As for Catherine, she suddenly burst out with just three words: 'All are gone!' Perhaps, like Katherine Schaub, she had a chorus of ghost girls playing in her head: Ella and Peg and Mary and Inez ...

'The words,' remarked Griffin, 'came unexpectedly and strong. There was silence again.'

For Tom Donohue, listening in, it was too much. He spoke up bitterly, his voice quivering. 'We've got humane societies for dogs and cats, but they won't do anything for human beings,' he spat out. 'These women have *souls*.'

Griffin asked one final question before he took his leave. 'How do you keep up your morale?'

It was Catherine who answered, 'unexpectedly with startling effect and strength'. 'By our faith in God!' she said.

But though Catherine's faith was as strong as it had ever been, as the days went by her body weakened. Just a week or so later, she wrote to Pearl: 'Tried to write sooner but somehow I can't write anymore. It is so difficult for me to get up for any length of time and when I do I'm all in for a week afterwards.' The continuing legal trouble was not helping her. 'I only wish my case was through with,' she said wistfully. 'Lord knows I need the medical care, and need it badly.'

Though her friends tried to rally round her – Olive brought fruit and a pail of fresh eggs and Pearl even bought her a new nightgown from the meagre funds she and Hobart had to spare – Catherine's body refused to respond to their comforting gestures. She suffered excruciating, constant pain that required the continuous administration of narcotics. Her jawbone continued to fracture into ever-smaller fragments, each new break more painful than the last, and with the new breaks came a new development.

Catherine started haemorrhaging from her jaw.

She lost approximately one pint of blood each time. Though she wanted to stay at home with Tom, her physician Dr Dunn rushed her to hospital; what Catherine called 'a hurried-up trip'. 'I want to be home,' she wrote forlornly to Pearl from her hospital bed. 'Am so lonesome ... Doctor wants me here; Tom wants a nurse at home. I just don't know what to do. I suffer so much pain.' She begged Pearl to visit her: 'Come over if possible, won't you, as soon as you get this letter? I'm so lonesome and blue.'

Dr Dunn was increasingly concerned for Catherine. Though he kept her in hospital for several weeks, her condition was terminal; she was so weak that he thought the slightest labour could be lethal. He issued a formal statement: 'In my opinion any unusual stress such as a court appearance might prove fatal. I have advised and urged her to forgo any such activity.'

But this was Catherine Donohue he was talking about. No matter what her doctor said, she was determined to fight Radium Dial tooth and nail. The company was not going to get away with it this time. Released from hospital by the start of June 1938, she was home just in time to hold a meeting at her house the day before the appeal hearing. Grossman and the other women were there. 'There's not much hope for me now,' Catherine said to them, acknowledging it. 'I only have to wait a while. It will help [you girls] to win and it will help my children.'

Her children and Tom, she said, 'are worth all the pain and suffering'.

Dr Loffler visited the same day. Her thin body 'barely dented the mattress' as he took her blood, drawing it from 'arms scarcely thicker than fingers'. Catherine was so weak these days she did not wear her glasses anymore, but the watch Tom had given her still encircled her wrist, on the tightest fastening they could find. Whereas once she had dressed smartly in her polka-dot dress for such gatherings, now she wore a starched white cotton nightgown, embroidered with two crucifixes on the pointed collar.

When Dr Loffler weighed her, Catherine knew at once that he would not overrule the veto Dunn had put on her attending tomorrow's hearing. Catherine Donohue now weighed 61 pounds (4 stone); she was not much heavier than her five-year-old son. In truth, even if she had been well enough to attend, it would have been almost impossible for them to transport her. She could not bear the slightest pressure on her body anymore.

Although Catherine was unable to attend the appeal hearing, she trusted Grossman implicitly to represent her interests. 'He is just about the best there is, isn't he?' she said of him. And Grossman was not alone in standing up for her: Pearl, Charlotte, Marie, Olive and the other women were there; and so was Tom Donohue. The hearing was held before a 'capacity crowd' on a Monday afternoon. Having seen Catherine's condition the day before, Grossman now declared the case was a 'race with death'. 'If Mrs Donohue dies before a final ruling,' he said solemnly, 'her estate under the law would receive nothing.'

Perhaps that was why Magid immediately requested a postponement; but it was not granted. Presumably on Catherine's request, Grossman suggested that a bedside hearing be held so she could be present, but this was vigorously contested by the firm. In the end, the judge determined that he would hear the appeal evidence that very afternoon.

The gathered media were speculative about what grounds Radium Dial might have for an appeal. One of the company's arguments was that the IIC was without jurisdiction, but this was immediately dismissed. Another was the statute of limitations (again); and a third argument was something completely different.

For Radium Dial now contested the girls' claims altogether: the firm alleged they were lying. As sworn evidence, Radium Dial submitted to the court a formal statement from one Mr Reed, the girls' former boss.

In it, Reed swore 'he never told anyone, nor ever heard anyone tell, Catherine Donohue or other employees radium wouldn't hurt them'. He also swore 'he was not on the company payroll during the time Catherine was exposed' to radium. His wife, Mercedes Reed, also submitted a signed stipulation. Both she and her husband said they 'would testify that neither of them gave, nor did anyone else in their hearing give, any orders or instructions to Catherine Donohue to insert in [her] mouth the brushes used'.

The girls were stunned. The Reeds were the ones who were lying! Why, you only had to look up their name in the town directory through all the years Catherine was employed there to find Mr Reed's name next to that of Radium Dial; the company and the man were synonymous. How could he claim he was not working there? And as for swearing that nobody told the girls radium wouldn't hurt them – unfortunately for the company, a full-page advert signed by its president and printed in several editions of the local newspaper asserted exactly that.

In response to the Reeds' sworn statements, all the women present that day declared they would testify to the direct opposite. During the hearing, Charlotte and Al Purcell both gave evidence to that effect. Tom Donohue was also on the stand, but this quiet man appears to have been overcome by the occasion; no doubt he was also handicapped by worry for his wife. He 'stumbled in his testimony, his voice being scarcely audible, and so the commissioner ruled out [almost] all of his testimony'.

The so-called evidence of the Reeds was the sole item submitted by the company in its appeal. And so, at 3.30 p.m., the hearing was closed. A five-man committee would judge the final verdict; they promised a decision by 10 July.

Catherine just needed to hang on a little bit longer.

56

In America, religion is king – and in 1938 there was an heir apparent: Father Keane of Chicago. He ran the Sorrowful Mother Novena, a weekly church service attended by more than 200,000 people countrywide, to which worshippers submitted personal requests for aid. Keane prayed publicly for them – in church, on the radio and in a weekly booklet, which was published nationwide so that Catholics across the States could pray for those in need. The novena was a cultural phenomenon.

Catherine didn't have the energy to read anymore, depending instead on Tom, so she probably didn't read those published prayers – but Pearl Payne's sister-in-law did. 'I would suggest all of your girls write to Father Keane,' she encouraged, 'I am sure all of you will benefit greatly and MIRACLES do happen even in this day and age, Pearl, so don't give up hope.'

Catherine had nothing to lose. During every moment she spent with Mary Jane and Tommy, she felt like her heart was breaking. She needed more time ... she needed so much more time with them. And so, at the direction of her dear friend Pearl, on 22 June 1938, Catherine summoned all her courage and her faith, and she wrote from the bottom of her heart.

Dear Father Keane,

The doctors tell me I will die, but I mustn't. I have too much to live for – a husband who loves me and two children I adore. But, the doctors say, radium poisoning is eating away my bones and shrinking my flesh to the point where medical science has given me up as 'one of the living dead'.

They say there is nothing that can save me – nothing but a miracle. And that's what I want – a miracle ... But if that is not God's will, perhaps your prayers will obtain for me the blessing of a happy death.

Please,

Mrs Catherine Wolfe Donohue

That 'Please' said it all. Catherine was begging for help. She had no shame or pride now – she just wanted to survive. Just one month longer. Just one more week. One more day.

Such was her fame as the leader of her Living Dead Society, her letter made front-page news. The reaction to her note was extraordinary, even by the standards of the popular novena. There was 'a sweeping response the length and breadth of the land'. Prayers were said daily for Catherine throughout the nation; hundreds of thousands of people queued in the rain to pray for her. Catherine herself received almost 2,000 letters. 'I would like to answer them all,' she said, quite overwhelmed, 'but of course I can't.'

And even though one has to take the news reports with a pinch of salt, *it worked.* By the following Sunday, Catherine was sitting up and eating her first meal with her family in months.

'Doctors told me today,' Leonard Grossman announced on 3 July, 'they don't know what is keeping her alive. It is fortunate indeed that Catherine finds comfort in prayer. It is fortunate that she is a Christian and may forgive – she can never forget.'

Catherine counted each day as it passed; 10 July was not so very far away. She was living for her children, for Tom – but also for justice. She simply prayed it would be done.

And on 6 July 1938 – four days early – her prayers were answered. On this date, the appeal of the Radium Dial Company was thrown out of court by the IIC. They upheld Catherine's award; and not only that, they added an additional $730 ($12,271) to it, to cover the medical expenses she had incurred since April. It was a unanimous decision from all five members of the adjudicating panel. 'It was,' Catherine wrote with exultant pleasure, 'a wonderful victory.'

'I'm so happy for Catherine,' Pearl wrote excitedly to Grossman after she heard the good news. 'I sincerely hope she benefits at once, so she may enjoy some medical comforts and things she actually wishes for.'

Yet the one thing Catherine truly wished for – the return of her health – seemed, despite all her prayers, to be out of her reach. In the middle of July, she had a 'bad spell' and had to have the doctor, but Catherine Donohue was not done fighting yet. When Olive stopped in to see her a day later, she found Tom asleep from his night shift but Catherine sitting up eating her lunch wearing the pretty nightgown Pearl had given her. 'She did look nice in it,' commented Olive fondly. 'Poor child, my heart goes out to her.'

Catherine was doing so well that on 17 July the women decided to have a reunion to celebrate their success; they had a 'lovely time' talking about their incredible victory. The other girls were full of plans for their own cases. Thanks to Catherine's triumph in court, they too could now bring claims before the IIC; Grossman said he would begin litigating Charlotte's case immediately. The others were having medical examinations in Chicago to support their claims; Pearl began consulting Dr Dalitsch. 'Personally,' she wrote to him, 'I think it was an act of God that sent you to Ottawa in Catherine Donohue's case.'

Pearl felt an unfamiliar sensation these days; she realised with some surprise it was a pleasant anticipation for the future. 'I live,' she said simply, 'in the hope of living.'

Catherine did the same. Yet it was not a smooth life. On Friday 22 July, Tom was so worried about her that he called out Father Griffin to administer last rites. Catherine, lying weakly in bed, 'wistfully' asked her husband, 'Is it that bad?'

Though Tom was unable to answer, in fact, it *wasn't* that bad. Catherine lived on and on, day after day, the verdict in court seeming to buoy her. It gave her another hour, another dawn; one more day in which she could greet Tom in the morning, kiss Mary Jane goodnight, see Tommy draw just one more picture with his watercolour paints. Catherine kept on living.

And then, on 26 July, Radium Dial went above the IIC to file another appeal in the circuit court. They alleged the commission did not take into proper consideration the firm's 'judicial propositions'.

It was a shock: a stab to the happy balloon of hope that Catherine had been carrying. It was a blow from which she found she simply could not recover. 'She had held on,' Grossman said, 'to a slim thread of life as long as she could, but yesterday's move to deprive her of what was legally hers was too much. She had to let go.'

Catherine Wolfe Donohue died at 2.52 a.m. on Wednesday 27 July 1938, the day after Radium Dial filed its latest appeal. She passed away at home on East Superior Street; Tom and the children were by her side. She remained conscious until a short time before her death, and then just slipped away. 'Those who were with her to the end agreed she died a peaceful death.'

She weighed less than 60 pounds.

As was tradition, her family kept her at home with them. They washed and dressed her in a pretty pink gown; looped her precious rosary beads through her still fingers. Her plain grey coffin was an open one, lined with ivory silk and covered with a veil, and as she lay there she did, indeed, look truly peaceful and at rest. Her casket was surrounded by garlands and tall candles, lending light against the darkness as she spent her final few nights in the place she had called home.

Now, the neighbours came. Some of them had shunned her before, but now they came to help. All day long, Eleanor, the housekeeper, took in offers of aid and dishes of food. 'Everyone has been very kind,' she said, perhaps a little tightly. Some of that kindness would have gone further when Catherine was still alive.

Catherine's friends came, too. They brought flowers; they brought their love and grief. Pearl came dressed in the same outfit she had worn when she and Catherine had gone to Chicago on that long-ago summer day when they'd persuaded Grossman to take their case; perhaps it was a symbolic choice, chosen for happier times. Yet it did not work. As Pearl knelt by her friend's coffin to pray for her, she was 'almost hysterical' at her loss.

Tom was strangely stoic, though his head was bowed and his cheeks sunken. Observers said his spirit seemed 'broken', but he had to carry on for the children. He dressed respectfully for Catherine in a black suit and tie, but his shoes were scuffed and unpolished; perhaps the sort of detail to which his wife had once attended. He and Eleanor got the children ready for their day, tying a ribbon in Mary Jane's hair and slicking down Tommy's (it did not work; bits of it kept sticking up). Tom paid his best attention to them, letting Mary Jane fiddle with the unfamiliar suit jacket on her father's shoulders; giving Tommy a hug as his son shyly looped an arm around his father's neck.

The children stood before their mother's casket, but they did not understand. They spoke to her and wondered why she did not reply.

'Why doesn't Mommie talk?' asked Mary Jane innocently.

Tom could not, he just could not answer. He tried, but his words were choked back by tears. He led the children silently away.

That first evening without Catherine, nuns from the St Columba parish school she had attended came to say the rosary by her side. They chanted the prayers, a song of loss and lamentation as they sent her soul on its way. They were still there as the children went through their first nightly routine without their mother and knelt to say their own prayers.

Mary Jane, aged just three, said hers in 'a tiny piping voice' that carried through the quiet house. As her mother lay downstairs – perhaps, to her young mind, merely sleeping – Mary Jane prayed as she had always learned to do.

'God bless Mommie and Daddy.'

The night before Catherine's funeral, as necessitated by Illinois law in cases of poisoning, an inquest was held into her death. Tom and Catherine's friends attended; Grossman was there too. He branded her death 'a cool, calculating, money-making murder'.

As dramatic as Grossman's declaration was, it was Tom's testimony which was most powerful, due to his raw emotion; the inquest was held the day after Catherine died. He was described as 'a weary little man with grey hair, shaken with grief' – but no matter how shaken he was, he had to testify at the inquest. 'He spoke with great difficulty and choked up when he described his wife's death,' said a witness. 'His breathing became greatly laboured and further questioning was cut short. He left the stand in tears.'

The jury of six men stayed silent throughout, as not only Tom but also Dr Dunn and Dr Loffler gave evidence. The jury was instructed by the coroner that they had 'only to find the cause of death and that it was not their province to fix the responsibility for Mrs Donohue's demise'.

But they did anyway. 'We, the jury, find that [Catherine Donohue] died of radium poisoning absorbed while she was employed in an industrial plant in Ottawa.' At Grossman's suggestion, the name of the Radium Dial Company was added to the formal verdict.

'It's the only industrial plant Mrs Donohue ever worked in,' he said sharply.

With the jury's verdict in, Catherine's death certificate was formally signed.

Was death in any way related to occupation of deceased?

Yes.

Catherine Wolfe Donohue was buried on Friday 29 July 1938. Her children were not old enough to attend the funeral, but hundreds of people gathered to pay their respects to this most exceptional woman: a quiet, unassuming person who had only wanted to work hard and love her family, but who made a difference to millions in the way she responded to her own personal tragedy. She was carried from her home by an assorted mix of Wolfe and Donohue relatives; this final journey, at last, causing her no more pain.

Her friends lined the street outside her home to accompany her to church; only Charlotte Purcell was missing, quarantined in Chicago caring for her children, who had caught scarlet fever. The women wore their best clothes; not black attire but floral dresses and coloured gowns. They dipped their heads as Catherine's coffin was carried past, and then they followed her: past Division Street and on to Columbus, where the slow cortège turned left. They followed her all the way to St Columba, which had always been her spiritual home: the place where she had been baptised, where she had married Tom and where, now, she took her final bow.

She had not been back since she fell ill. But on this funeral day Catherine Donohue, once again, made her slow way down the aisle of the church and rested once more in the grace of God beneath that towering, arched ceiling that had been so familiar to her in life, bathed in the coloured light from the stained-glass windows that her husband's family had helped to buy.

Father Griffin led the Mass. He 'spoke of the relief that death made for Mrs Donohue after her long and patient sufferings'. It seemed, to Tom, too short a service – for when it was over, all that was left was the burial. The burial; and then the rest of his life to live without her. He was 'near collapse' as he bade farewell to his wife.

The other mourners joined him in his helpless grief. 'In a silent but impressive few moments,' a witness wrote, 'Catherine's best friends – the girls who worked in the plant with her and contracted the same poisoning – said goodbye. The scene brought to mind the words of the ancient gladiators of glorious Rome: *"Moritamor te salutmamus –* we who are about to die salute thee."'

Their heads and their hearts were filled with Catherine, filled with her even as they left the church and blocked out the sight of the old high school across the road, where she had been poisoned. Their hearts stayed full of her, as Pearl wrote to Grossman later that same day:

When I returned home from Catherine Donohue's funeral with my heart full of her and thoughts of your great work at the inquest and the circuit court, I felt I must drop this little note and let you know my heart is filled with gratitude when I think of the courageous battle you are waging on behalf of us girls.

She closed 'with prayers and best wishes for further successes'.

For even on the day of Catherine's funeral, Grossman was in court, defending her claim. The company had been denied the right of appeal, but they were appealing even that. They appealed over and over and over. In fact, Radium Dial fought the case all the way to the Supreme Court of the United States of America.

Other lawyers might have dropped the case, citing lack of funds – for Grossman was still covering the expenses – but Leonard Grossman had vowed to stand by the women and he did not let them down. 'He just collapsed after over-working on this case,' his wife, Trudel, said. Perhaps Radium Dial were hoping either he or the girls would give up the fight, maybe run out of money, but they were battling in Catherine's memory now and that was a powerful motivator.

Grossman had to get a special licence to be admitted to the Supreme Court. '[That] licence was under glass in our house forever,' said his son. 'He talked about [the case]. He was proud of it and the scrapbooks were always in the middle of the book-shelf. I heard some of the stories over and over again; I grew up with this case.

'When the case went to the Supreme Court,' he went on, 'my parents both went to Washington for it. I looked it up. "Cert denied." The court decided not to hear it. It meant they were upholding the lower court.'

Catherine Wolfe Donohue had won her case. She won it eight times in total. But the final victory came on 23 October 1939.

The papers described her battle for justice as 'one of the most spectacular fights against industrial occupational hazards'. Now, that battle was at an end – *finally* at an end. It was a pure, clean victory, with no clouds or contingencies to sully it.

No settlement. No board of doctors to poke and prod and say that there was no such thing as radium poisoning; no firm reneging on an out-of-court agreement that had been made in good faith. Now there were no more legal machinations; no law-yers' twisting words; no law itself whose unclear wording tied mercy up in knots. It was outright justice, plain and true. The women had been vindicated. The dial-painters had won.

And it was Catherine Wolfe Donohue, in the end, who had led them to victory.

'If there are saints on earth,' one commentator said, 'and you believe in that, I think Catherine Wolfe Donohue was one of them. I really do.'

She was buried in St Columba Cemetery. She has a simple, flat gravestone, as unobtrusive, and as neat and tidy, as she herself had been in life.

EPILOGUE

The radium girls did not die in vain. Although the women could not save themselves from the poison which riddled their bones, in countless ways their sacrifice saved many thousands of others.

Fifty days before the final triumph in Catherine Donohue's case, war was declared in Europe. It meant that there was, once again, an enormous demand for luminous dials to light the dashboards of military machines and the wristwatches of soldiers taking up arms. Yet thanks to Catherine and Grace and their colleagues' bravery in speaking out about what had happened to them, dial-painting was now the most feared occupation among young women. No longer could the government sit idly by: the radium girls' demise demanded a response.

Safety standards were introduced which protected a whole new generation of dial-painters, based entirely on knowledge gained from the bodies of those women who had come before. The standards were set not a moment too soon, for seven months later America formally entered the war. The US radium dial-painting industry exploded, with USRC alone increasing its personnel by 1,600 per cent. Radium dials were even bigger business than the first time round: the United States used more than 190 grams of radium for luminous dials during the Second World War; in contrast, fewer than 30 grams were used world-wide in the earlier conflict.

In addition, a chemist called Glenn Seaborg, who was employed on the most secret mission of them all – the Manhattan Project – wrote in his diary: 'As I was making the rounds of the laboratory rooms this morning, I was suddenly struck by a disturbing vision [of] the workers in the radium dial-painting industry.' Atomic-bomb-making involved wide-spread use of radioactive plutonium and he realised at once that similar hazards faced those working on the project. Seaborg insisted that research be undertaken into plutonium; it was found to be bio-medically very similar to radium, mean-ing it would settle into the bones of anyone exposed to it. The Manhattan Project issued non-negotiable safety guidelines to its workers, based directly on the radium safety standards. Seaborg was determined that the women's ghosts would not be joined by those of his colleagues who were working to win the war.

After the Allies had triumphed – helped by the deployment of those very atomic bombs that the Manhattan Project built – the debt the country owed to the radium girls was acknowledged in full. An official of the US Atomic Energy Commission (AEC) wrote: 'If it hadn't been for those dial-painters, the [Manhattan] project's management could have reasonably rejected the extreme precautions that were urged on it and thousands of workers might well have been, and might still be, in great danger.' The women had been, officials said, 'invaluable'.

Even after the war was over, the dial-painters' legacy contin-ued to save lives, as the world entered the age of atomic energy. 'We were going to live in an era of plutonium,' enthused one man who grew up in 1950s America. 'We'd have plutonium cars, planes ... It was infinite.' The large-scale production of radioactive materials seemed inevitable. 'In the foreseeable future,' wrote the Consumers League, 'millions of workers may be affected by ionising radiations.'

The League was right. Almost immediately, however, it became clear that it was not just employees in the new atomic industries who were at risk: the whole planet was. Less than five years after the Second World War ended, the nuclear arms race began: over the next decade, hundreds of above-ground atomic tests were conducted across the globe.

Each blast, mushroom-clouding bomb debris into the sky, eventually resulted in radioactive fallout drifting back to earth: landing not only on the test site, but raining down upon fields of green grass, wheat and cereal – through which the radio-active isotopes in the fallout entered the human food chain. Just like radium had done in the dial-painters, these isotopes, especially a particularly dangerous, newly created one called strontium-90, began to deposit in human bones. 'Every one of us,' wrote the Consumers League in alarm, 'is a potential victim.'

The AEC dismissed the concerns: the risks, it said, were very small when compared to 'the terrible future we might face if we fell behind in our nuclear defence effort'. Yet their words were not sufficient to calm the troubled public; after all, 'the radium dial-painters' agony [had] alerted the world to the hazards of internal radiation'. '[They] serve as a warning,' railed the Consumers League, 'of the results of carelessness and igno-rance ... a cloud on the horizon, no larger than a man's hand.'

In 1956, growing public unease led the AEC to establish a committee to examine the long-term health risks of atomic tests, specifically the effects of strontium-90. But how, the research-ers thought, could they possibly begin this study for the future health of humanity when they were dealing with such an unknown substance? All they really knew was that strontium-90 was chemically similar to radium ...

'There is only a limited pool of people that have had exposure to internal radiation,' a radioactivity specialist said. 'If anything happens in our up-and-coming nuclear age, these [people] are about the only starting point that anyone has.'

The dial-painters were needed to help once more.

They seemed Cassandra-like in their powers: able to predict for scientists the likely long-term health effects of this new radioactive danger. 'Something that happened far in the past,' an AEC official said, 'is going to give us a look far into the future.' He termed the women of 'incalculable value': their suffering would provide 'vital insight, with implications for hundreds of millions of people all over the world'. In a spookily prophetic letter, Pearl Payne had once written, 'My history is unusual and may be of interest to medical men of the future.' She could never have anticipated just how right she was.

Medical studies began immediately, including in New Jersey and Illinois; later, the research would be amalgamated into the Center for Human Radiobiology (CHR), which was located in a multi-million-dollar clinic called the Argonne National Laboratory, based 75 miles from Ottawa. Here, special lead-lined vaults were constructed, buried under three feet of concrete and ten feet of earth, in which the dial-painters' body burdens (the amount of radium inside them) were measured. The research was designed to help future generations and called 'essential to the security of the nation'. 'If we can determine the long-term effects of radium,' one of the scientists said, 'we're quite sure we can predict the long-term, low-level effects of fallout.' Scientists were seeking to 'give the world an exact guide on safe radioactivity by studying all dial-painters who can be found'.

There were dial-painters still living – albeit with a time bomb ticking in their bones. Dr Martland had already explained why they had survived thus far. Radium was known to settle in the girls' bones and known to cause late-onset sarcomas, but *when* such deadly tumours might begin to grow was the factor that remained mysterious, like a dark trick. Radium had not given away all its secrets just yet.

The hunt to find those living dial-painters now began in

earnest: WANTED: RADIUM WORKERS OF THE ROARING TWEN-
TIES read the headlines. Employment records were procured
and snapshots of those long-ago USRC picnics unearthed; the
company photograph taken on the steps of Radium Dial became
a vital source of clues. The scientists pronounced, 'Each of these
persons is worth [her] weight in gold to science'; the girls were
termed 'a reservoir of scientific information'. In an eerie echo of
the women's treatment when they sued their former firms, private investigators were hired to track them down.

Those they found, they often found willing. 'She said she
would be very happy to do it (anything for science),' a memo
recorded. Those dial-painters who were still working for USRC
participated anonymously, for fear of jeopardising their jobs.

There were some who didn't want to stir things up. 'Miss
Anna Callaghan does not know she has radium poisoning and
her family does not want her to know,' read one note. Another
woman was reluctant to be measured for radium as the scientists 'couldn't do anything about it anyway'.

Even family members of the girls took part. Grace Fryer's
little brother Art was one. They tested him 'because he spent so
much time with her and basically she was radioactive,' said his
son. 'I guess the government was trying to figure out if he was
going to suffer any ill effects.'

Though Art was fine, it was a concern that was not exagger-
ated. Swen Kjaer's notes minuted the death of a dial-painter's
sister: she had 'reportedly died from radiation exposure, but had
never worked at the [USRC] plant. The source of contamination
appears to have been her sister, the dial-painter, with whom she
shared a bed'.

Many of the original girls, of course, were no longer alive to
help with the study. Edna Hussman had died on 30 March 1939;
she was said to have 'maintained her good spirits and courage
until the last'. She died of a sarcoma of the femur, leaving her
husband, Louis, a widower at the age of forty.

Albina Larice, too, had passed away. She died aged fifty-one on 18 November 1946, also of a leg sarcoma. Pictures of her towards the end show her smiling, with no tension in her face. She passed away fourteen days before she and James would have celebrated their twenty-fifth wedding anniversary.

Yet even the deceased dial-painters had something to offer the scientists. Dr Martland had collected tissue and bone samples from the radium girls when he was making his ground-breaking discoveries in the 1920s – and these ended up in the studies' archives. Those contributing to the world's knowledge of radiation included Sarah Maillefer, Ella Eckert, Irene La Porte and many more ... The researchers even went to the Cook County Hospital and brought back Charlotte Purcell's amputated arm; they found it still in its formaldehyde crypt, saved through the decades due to its never-before-seen symptoms.

In 1963, perhaps at least partly in response to the research on the dial-painters, President Kennedy signed the international Limited Test Ban Treaty, which prohibited atomic tests above ground, underwater and in outer space. Strontium-90, it had been determined, was too dangerous for humanity after all. The ban undoubtedly saved lives and, very possibly, the entire human race.

Atomic energy remained part of the world; it is a part of our lives even today, when fifty-six countries operate 240 nuclear reactors, and more still are used to power nuclear ships and submarines. Yet thanks to the radium girls, whose experiences led directly to the regulation of radioactive industries, atomic power is able to be operated, on the whole, in safety.

The study of the dial-painters did not end when the threat of nuclear war subsided. A leading figure in the research, Robley Evans, 'forcefully argued that it was prudent, and indeed a moral obligation to future generations, to learn as much as possible about the effects of radiation'. The AEC agreed and so, through the Center for Human Radiobiology, the dial-painters were studied 'for their full life spans'.

Decade after decade, the radium girls came to CHR to be

tested. They agreed to have bone-marrow biopsies, blood tests, X-rays, physical exams; the women were asked to fast before coming and to wear clothes that they could 'easily slip in and out of'. They were given probing questionnaires about their mental and physical health, they had breath tests and, of course, they had their body burdens measured in the claustrophobic iron rooms beneath the earth. Even after death, some were autopsied; their bodies giving up secrets that the scientists couldn't learn in life. Thousands of women helped with the study, through their forties, fifties, sixties and beyond; their contribution to medical science is incalculable. We all benefit from their sacrifice and courage, every day of our lives.

And among those women who submitted to examination for the good of humanity were some familiar faces. Pearl Payne was one of them. 'I believe I was fortunate,' she once said of her survival, 'in the fact that [my] radium did not become localised in some of the bones of the body which cannot be removed, as is the case with many of the girls that are dead.'

Instead of death, Pearl embraced life. She made curtains and dresses on her sewing machine and 'the best homemade pies', using freshly fallen fruit from the trees in her backyard. Her survival meant she was around when her little sister needed help. 'At the time my dad walked out on my mom,' Pearl's nephew Randy said, 'there was nobody left. Nobody to help us. And so Pearl and Hobart were the best people in our lives. They would take care of us.'

Another dial-painter who came to the Argonne Laboratory was Marie Rossiter. 'The state man came and took me and he said, "I hear you're one of the rare ones,"' she remembered, 'and I said, "Rare or bare?" and he looked at me and he said, "No, I mean it."'

Marie survived to see her son, Bill, marry the girl-next-door, Dolores; and to see her granddaughter Patty grow up to become a dancer. Although for much of her life Marie had 'huge and spotty' legs from the radium, which forced her to walk with a limp, she would dance with Patty regardless. 'She'd always dance with me,' remembered her granddaughter fondly. 'It wasn't that great, but we danced together. She had a wonderful love for life. I used to look upon her like she could do anything.'

Marie simply refused to let the radium rule her life. 'She was in pain,' recalled Dolores. 'Pain to walk. Pain to just stand here, sometimes, it was that bad.' Yet although Marie had her bad times – 'I prayed to die and couldn't die,' she once said. 'Why would I want to live, I had so much pain?' – she added stoically, 'I witnessed the bad times, but you get through them.'

There was a friend of hers who had gotten through the bad times too: Charlotte Purcell. She'd been told in the 1930s that she was the Ottawa dial-painter most likely to die after Catherine Donohue, but thirty years later, she was still living. 'She's a beautiful woman to go on graciously losing an arm, raising a family,' remarked Marie, 'and maybe sometimes God looks at people and when they help somebody else [like she helped Catherine], He's gonna help her too.'

Charlotte had had a sarcoma, back in 1934, but her courage in electing to have an amputation undoubtedly saved her life. She lost all her teeth and had one leg shorter than the other but, like Marie, she refused to let it bring her down. 'I now feel fine, although I'm bothered a bit by arthritis,' she said to a reporter in the 1950s. 'I've gone through all that years ago; I don't like to think about it.' Although it was a time in her life she wanted to forget, when the scientists invited her to Argonne, she answered their call. The doctors had told her that doing so would help others, and Charlotte Purcell was never a woman to turn down an appeal for aid.

The research at Argonne uncovered what happened to the Ottawa women's lawsuits after Catherine Donohue won her test case. Many fought on with Grossman's help after that victory in court – though the small pot of money available meant the payouts were not high; those who did claim were paid only a few hundred dollars each. Charlotte received $300 ($5,000), a negligible sum that made Al Purcell 'very angry'; it paid for the amputation of her arm and that was about it. Others received nothing; Marie was taken to lunch when she went to Argonne and said, 'This is the most we'll probably ever get.' Some dropped their cases: the Glacinski sisters and Helen Munch were among them. Perhaps they'd joined forces for Catherine, and upon her death the fight went out of them. There was very little money anyway; maybe, in the end, it did not seem worth it. It was the judgment they had fought for, and that they had achieved.

As for the companies, eventually the law caught up with them – though by then the damage had been done. In 1979, the US Environmental Protection Agency (EPA) found that the former USRC site in Orange had unacceptable, environmentally hazardous levels of radioactivity: twenty times higher than was safe. There was widespread contamination – and not just of the site, but in those locations where the company had dumped its radioactive waste as landfill. Almost 750 homes had been built on top of that waste; they too needed decontamination. More than 200 acres of land were affected in Orange, some to a depth of more than 15 feet.

The EPA ordered the corporate successor of USRC to perform the clean-up work, but it declined, except for agreeing to erect a new security fence (even this they did not see through; the EPA was forced to complete it). The courts were not forgiving; in 1991 the New Jersey Supreme Court found USRC 'forever' liable for the contamination and declared the firm had had 'constructive knowledge' about the dangers at the time it operated there. Residents sued the firm; after seven years, the cases were eventually settled out of court, costing the company some $14.2 million (almost $24 million) in damages. It reportedly cost the government $144 million ($209 million) to clean up radium-contaminated sites across New Jersey and New York.

As for Radium Dial, despite the wartime boom, it went bust in 1943. The building it left behind in the centre of Ottawa, however, had a legacy that lasted far beyond that. A meat-locker company later operated in its basement: its employees died of cancer, while a family who purchased meat there found that 'every brother got colon cancer within six months of each other'. The building itself was knocked down in 1968. 'They just hauled it down,' remembered Peg Looney's niece Darlene, 'and used it as fill, everywhere.' The waste from the building was dumped around town, including alongside a school field. Later studies showed an above-average cancer rate near the factory as well as across town; people found their pet dogs didn't live to maturity and that the local wildlife developed distressing tumours. 'I noted,' said another niece of Peg, 'that nearly one person in each household of that neighbourhood [where I grew up] had cancer.' Another resident remarked: 'There aren't many families not affected.'

Yet the town officials, in a reprise of their attitude towards Catherine and her friends, did not address the evident problem. When film-maker Carole Langer made a documentary, *Radium City*, which highlighted the town's radioactivity, the mayor declared, 'That lady is trying to destroy us.' He ordered 'everybody not to go to [see] it'.

'Well,' said Marie's daughter-in-law Dolores, 'that was the wrong thing to say. Because it filled up the whole [showing] and they had to have another one.' The film was shown to a standing-room-only crowd of nearly 500 residents.

'People were divided,' recalled Darlene. 'There were people that didn't want to hear about it; they didn't want to believe it. And then there were people who were like, "OK, let's get this cleaned up."'

In the end, they did get it cleaned up. The EPA stepped in and funds were found to begin tackling the dangerous legacy of radioactivity that Radium Dial left Ottawa. As in Orange, the damage plunged many feet deep into the earth. It was an operation that would take decades; in 2015, the clean-up was still going on.

*

The Center for Human Radiobiology (CHR) studied the dial-painters for decades. Its scientists came to learn that radium was a wily, tenacious element. With a half-life of 1,600 years, it had plenty of time to make itself known in those it had infiltrated, inflicting its own, special damage across the decades. As the researchers followed the women through the years, they witnessed what the long-term effects of internal radiation *really* were.

For the dial-painters who had survived did not escape unscathed – far from it. Some women were stricken early but then endured a half-life for decades; one Waterbury girl was bedridden for fifty years. The older the women were when they dial-painted, and the fewer years they worked, the less likely they were to die in the early stages – so they lived on, but the radium lived with them: a marriage from which there was no divorce.

Many suffered significant bone changes and fractures; most lost all their teeth. There were unusually large numbers who developed bone cancer, leukaemia and anaemia; some were given blood transfusions for years. The radium honeycombed the women's bones so that, for example, Charlotte Purcell later developed osteoporosis throughout her spine and suffered a partial collapse of her vertebrae. Like Grace Fryer before her, she ultimately wore a back brace.

Marie Rossiter had at least six leg operations – her swollen legs began turning black – and in the end she had her leg amputated. 'She said,' remembered Dolores, '"Take it off! Right now! I don't want to go home and think about it."'

Marie's remaining leg had a metal bar through it from her knee to her ankle; she became, in her words, 'crippled up' – but it still didn't slow her down. She was the life and soul of the care home she later lived in, whizzing about in her wheelchair.

Having studied these long-term effects of radiation, the CHR scientists – who had, at first, been looking for a magic threshold of radiation exposure, under which no harm was done – ultimately came to agree with Martland, who had warned decades before that 'the normal radioactivity of the human body should not be increased'.

It is impossible to say how many dial-painters were killed by their work: so many were misdiagnosed or never traced that the records simply do not exist. Sometimes the cancer that former workers suffered later in life was never attributed to the job they did in their teens, though it came as a direct result. And the deaths, too, were only one part of it; how many women were crippled or suffered the unique pain of childlessness as a result of their poisoning is also unknown.

The files at Argonne are filled with hundreds and hundreds of dial-painters' names; or rather numbers. Each woman was given a reference number, by which she was always known. The Argonne List of the Doomed makes for chilling reading, charting as it does each woman's suffering with cool detachment. 'Bilateral amputations of both legs; amputation of right knee; died of cancer of ear; brain; hip; cause of death: sarcoma; sarcoma; sarcoma' over and over through the files. Some women survived for forty years or more – but the radium always came calling in the end. The newspapers followed some of the deaths: RADIUM, DORMANT KILLER, AT WORK AGAIN screamed the headlines through the years.

Mercedes Reed is said to have died in 1971; she was eighty-six. 'I'm absolutely unequivocally convinced,' said one researcher, 'that the radium level would be huge in her bones. She supposedly died from colon cancer, but maybe it was misdiagnosed.' The Reeds did not continue their association with Radium Dial, even before it went bust. 'Ultimately, Mr Reed was fired from the plant and it is understood that he [was] bitter about this,' researchers discovered. After his dismissal from the company to whom he had been, some might say, unforgivably loyal, he became a maintenance man for the YMCA.

Reed's former president, Joseph Kelly, died about 1969, after a series of strokes 'reduce[d] [his] mental capacity ... he just got frailer and frailer'. In his final years, he would often say, 'Have you seen so-and-so lately?' of someone who had worked with him in the 1920s. Given his distance from the dial-painters he had sentenced to death, when he signed his name to the advert that told them they were safe, it seems unlikely his damaged mind was being haunted by the ghost girls.

As for those girls he used to employ in Ottawa, against all the odds a few of them had long, good lives. Pearl Payne lived to be ninety-eight years old; she and Hobart embraced the extra time they had unexpectedly been given. 'They travelled all over the world,' revealed their nephew Randy. 'They went to Jerusalem, England ... they were in every state in the union.'

Before she died, Pearl called Randy over to her house one day. 'She asked me to go up in the attic and pick up some boxes and bring 'em down,' he recalled. He picked his way through the items Pearl kept in her loft: a baby stroller, a crib; strange things for an old lady to have in her attic, but perhaps Pearl had found herself unable to let go of these final traces of the many children she had wanted, but could never have. Randy found the boxes she meant: they were full of newspaper cuttings about Catherine Donohue and letters and documents to do with her case.

'This is what happened to us,' Pearl told Randy urgently. And she said emphatically, 'These need to be safeguarded. This stuff is important. Be sure that Pearl [her daughter] gets these if anything happens to me.'

Hobart and Pearl 'were two very fine people,' said Randy. 'I don't [normally] visit graves, but I go to theirs. And I wanna tell you, I say thank you every time I'm there. That's the kind of people they were.'

Charlotte Purcell lived to be eighty-two. She was adored by her grandchildren. 'She was probably one of my most favourite people in the whole world,' raved her granddaughter Jan. 'She was one of the most courageous, loved and influential people in my life. What my grandmother taught me was: it doesn't matter what life throws at you, you can adapt.

'When I asked her to teach me how to jump rope, she said, "Well, I don't think I can teach you because I only have one arm." I suppose that was upsetting to me so she said, "Well, wait." She tied a rope to a chain-link fence and then she jumped rope with one arm and showed me how to jump rope.'

Jan's brother Don added, 'It was nothing out of the ordinary to me [that she didn't have an arm] because she made it that way.'

The kids would chorus, 'Tell us the story of how you got your arm cut off!'

'She would repeat the story,' remembered Jan. 'She would repeat it over and over, any time we asked.'

'When I was a young girl,' Charlotte Purcell would say, 'I got paid a lot of money to paint numbers on watches and clocks.

'We didn't know that the paint was a poison.

'After I had left there, my friend Catherine Donohue got very sick. And a lot of the girls started getting very sick. The poison settled in my arm, but with my friend Catherine it went throughout the body and she died. She died and left her husband and children without a mom.'

She was always 'pretty sad' when she got to that part of the story.

Though Charlotte had been unable to attend Catherine's funeral, her son remembered something from his mother's life that maybe, to a poetic mind, suggested that the friends got to say their own goodbye. 'When the weather was nice,' Donald recalled, 'my mother used to go out to the porch and sit on the glider they had there, and swing back and forth. While she was there, a little yellow-and-black canary used to come and sit on her left shoulder [where her arm was missing] and might stay with her about thirty minutes and then leave. That happened several times. Normally, birds don't have anything to do with people.'

The women didn't talk to their families about the incredible legacy they had given the world. And the radium girls did not simply set safety standards and contribute incalculably to science – they left their mark in legislation too. In the wake of Catherine Donohue's case in 1939, Secretary of Labor Frances Perkins announced that the fight was 'far from won' when it came to workers' compensation. Subsequently, building on what the women had achieved in life, further legal changes were made to protect all employees. The dial-painters' case ultimately led to the establishment of the Occupational Safety and Health Administration, which now works nationally in the United States to ensure safe working conditions. Businesses are required to inform employees if they work with dangerous chemicals; workers certainly won't be told that those corrosive elements will make their cheeks rosy. There are now processes for safe handling, for training, for protection. Workers also now have a legal right to see the results of any medical tests.

To the dial-painters' frustration, however, the results of the exams at Argonne were not shared with them. This secrecy may well have been to do with the highly technical nature of the measurements the researchers were making; perhaps they thought the results would mean nothing to the women, but the women still wanted to know. 'They would never tell [Marie] anything, which made her mad,' recalled Dolores. By 1985, after going there for decades, Charlotte Purcell had had enough of it. When the researchers called her that year, she said she had not been feeling well, 'but why should I discuss it – you people don't help me – I don't get anything out of it – I don't even have any money to go to the doctor.' She refused to go again.

Marie did the same. And it wasn't just the scientists' silence that bothered her; it was her town's continued reaction to what the women had endured. She always thought the whole saga 'will be swept under the carpet ... it will never come to light. You'll never hear about it.' It made her shock all the greater when Carole Langer came to Ottawa to make her movie. Marie told the director, 'God has left me here. I always knew some-one would walk through that door, and I would finally have a chance to tell my story.' Langer dedicated the film to Marie, calling her someone 'who fought against the odds all of her life and never lost her belief in God or her sense of humour'.

When Marie died in 1993, like many dial-painters she donated her body to science. 'She thought maybe she could help other people,' said her granddaughter Patty. 'Maybe they could find exactly what happened and they could find a cure. Maybe she could help other women.' Marie's body would not be the last Ottawa dial-painter's corpse to be studied; nor was it the first. That honour goes to Margaret Looney.

Peg's family had wanted her exhumed for testing as soon as they'd heard about the post-war studies on the dial-painters. At that time, however, research was limited to the living. By the time CHR was established, the remit had broadened. Finally, someone was prepared to investigate what had really killed Peg.

Every one of her nine brothers and sisters signed the necessary forms. 'It was going to help somebody else get better,' said her sister Jean, 'of course we let them do it.'

In 1978, researchers exhumed Peg's body from St Columba Cemetery, where she had been resting alongside her parents. They discovered she had 19,500 microcuries of radium in her bones; one of the highest quantities found. It was more than 1,000 times the amount scientists then considered safe.

They didn't just discover the radium; they discovered that the company doctor had cut her jawbone from her while she lay dead. That was probably how the Looney family found out about it.

'I'm *angry*,' said one of Peg's sisters. 'They knew she was full of radium. And then they lied.'

'Every family has sadness and grief,' Jean said steadily. 'But Margaret's death was unnecessary.'

That was the tragedy. Radium had been known to be harmful since 1901. Every death since was unnecessary.

The researchers exhumed more than a hundred dial-painters;

many tests proving once and for all that radium poisoning, and not syphilis or diphtheria, was the woman's true cause of death. And there was one deceased dial-painter in particular in whom the scientists were very interested: Catherine Donohue. In 1984, CHR wrote to her daughter to request her exhumation.

They wrote to Mary Jane because, by that time, Catherine's devoted husband Tom had died. He passed away on 8 May 1957, aged sixty-two. He had lived the remainder of his life at 520 East Superior Street, never leaving the home he had once shared with Catherine; the home where, when the news of her triumph in court had come through, he and the family had celebrated with pot-luck food. 'We all went down and we celebrated with him,' remembered his niece Mary. 'Because it was such a moral victory. Something that nobody had ever done before.'

Though the money helped a lot, it couldn't bring Catherine back. 'I think it broke him when she died,' said a relative. 'His heart was broken.'

The family rallied round; for a time, Tom's sister Margaret moved in to help with the children. Tom doted on the kids. 'They were all he had left,' Mary said simply.

'As time went on,' she added, 'he healed. He became [a] smiling man; it was really great to see that.' He rarely talked of Catherine but, said Mary, 'It was a painful memory because of the painful death that she had.'

Tom Donohue never remarried. No one could replace Catherine Wolfe Donohue.

As for Mary Jane's brother Tommy, he had gone to fight in the Korean War – and made it home. He married a young woman from Streator and worked in a glass factory, just like his father. But he died shortly after his thirtieth birthday in 1963, of Hodgkin's disease: a type of cancer. Mary Jane had been on her own for a long time now.

She had not had an easy life. The little girl who weighed only 10 pounds on her first birthday had stayed small. 'She was almost childlike,' remembered her cousin Mary. 'She was tiny.'

Yet Mary Jane, showing some of her mother's spirit, rose above the challenges she faced. 'It was really remarkable,' said Mary, 'that she was able to hold down [her] job because she was so small. She was very sweet as an adult; everyone liked her. We tried to invite her to all the functions of the family because of course she had no one.'

When Mary Jane received the request from CHR, she considered it carefully and then wrote back. 'I have really developed a lot of medical problems,' she told the doctors. 'I realise now that most of them are probably a result of my mother's illness. If it is convenient and you wish me to do so, I would like to come up to Argonne Lab. I feel it is important for myself and for research.'

It seems Mary Jane was tested, adding her own contribution to science. On 16 August 1984, she gave permission to the researchers to exhume her mother. 'If this could help one person,' she said, 'it is worth it.'

And so, on 2 October 1984, Catherine Donohue left St Columba Cemetery for an unexpected journey. The scientists ran their tests and she made her unique endowment to medical knowledge. Catherine was reinterred on 16 August 1985 – where she rests, to this day, beside her husband Tom.

When Mary Jane wrote to CHR, she said, in an uncanny echo of her mother's final letter to Father Keane, 'I pray all the time that God will let me live a long life. I certainly try hard enough to fight all the time for a life of fulfilment and happiness.'

But it was not to be. After a life full of physical challenges, Mary Jane Donohue died – from heart failure, according to her relatives – on 17 May 1990. She was fifty-five years old.

For a long time – too long – the legacy of the radium girls was recorded only in the law books and in scientific files. But in 2006, an eighth-grade Illinois student called Madeline Piller read a book on the dial-painters by Dr Ross Mullner. 'No monuments,' he wrote, 'have ever been erected in their memory.'

Madeline determined to change that. 'They deserve to be remembered,' she said. 'Their courage brought forth federal health standards. I want people to know [there] is a memorial to these brave women.'

When she began to champion her cause, she found that Ottawa, at last, was ready to honour its native heroines and their comrades-in-arms. The town held fish-fry fundraisers and staged plays to secure the $80,000 needed. 'The mayor was supportive,' said Len Grossman. 'It was a complete turnaround. That was wonderful to see.'

On 2 September 2011, the bronze statue for the dial-painters was unveiled by the governor in Ottawa, Illinois. It is a statue of a young woman from the 1920s, with a paintbrush in one hand and a tulip in the other, standing on a clock face. Her skirt swishes, as though at any moment she might step down from her time-ticking pedestal and come to life.

'The radium girls,' the governor announced, 'deserve the utmost respect and admiration ... because they battled a dishonest company, an indifferent industry, dismissive courts and the medical community in the face of certain death. I hereby proclaim 2 September 2011 as Radium Girls Day in Illinois, in recognition of the tremendous perseverance, dedication and sense of justice the radium girls exhibited in their fight.'

'If [Marie] saw that memorial down there today,' said Marie Rossiter's daughter-in-law, 'she wouldn't believe it. When I go downtown and I go past, I say, "Well, Marie, they finally did something!" If she was alive today to see the statue, she would have said, "About time."'

The statue is dedicated not only to the Ottawa dial-painters, but also to 'dial-painters who suffered all over the United States'. This bronze radium girl, forever young, forever present, stands for Grace Fryer and Katherine Schaub; for the Maggia and Carlough sisters; for Hazel and Irene and Ella too. She stands for all the dial-painters: whether they lived and died in Orange, in Ottawa, in Waterbury or anywhere else. It is a fitting and most deserving memorial. After all, there is so much to thank the women for.

'The studies of the radium dial workers,' wrote Dr Ross Mullner, 'form the basis of much of the world's present knowledge of the health risks of radioactivity. The suffering and deaths of these workers greatly increased [scientific] knowledge, ultimately saving countless lives of future generations.'

'I always admired their strength,' said Catherine Donohue's great-niece, 'to stand up and unite.'

And, united, they triumphed. Through their friendships, through their refusal to give up and through their sheer spirit, the radium girls left us all an extraordinary legacy. They did not die in vain.

They made every second count.

POSTSCRIPT

'We girls,' said one worker, 'would sit around big tables, laughing and talking and painting. It was fun to work there.'

'I felt lucky to have a job there,' revealed another girl. 'The job paid top dollar for women in this area. All of us got along real good.'

'We slapped the radium around like cake frosting.'

The women wore smocks; washed once a week amidst the family laundry. They drank open cans of soda through their shifts, sourced from the machine in their studio. They worked with bare hands and painted their fingernails with the material 'for kicks'; they were allowed to take radium home to practise painting.

There was radium everywhere in the plant – and outside on the sidewalk. Contaminated rags piled up in the workrooms or were burned outside in the yard; radioactive waste was emptied into the toilet of the men's washroom; ventilation shafts discharged above a nearby children's play area. The women didn't clean their shoes before they left work, so they walked the radium all over town.

'You couldn't work in that plant without getting covered with the stuff,' one dial-painter recalled. 'Sometimes I'd get up in the night and look in the mirror and my hair would be glowing.' The women's hands 'would be bleeding' as they tried to scrub away the supernatural shine.

'The company,' said one girl, 'always led us to believe everything was under control and safe, but I don't think they cared.'

She was right. Before too long, the workers started suffering. 'I had to have a mouth operation,' said one, 'but now my teeth are so loosened that they are probably all going to fall out ... I have a blood disease I can't seem to get rid of.' The women noticed tumours appearing on their feet; their breasts; their legs. 'They kept cutting her leg off,' one woman recalled of her colleague Ruth, 'until they couldn't cut any more. Ruth finally died.'

The women went to their supervisor, worried sick. 'A man from the New York headquarters came out here,' a radium girl remembered, 'and told us [our work] wouldn't hurt us.'

'Breast cancer,' said the executive, 'is thought to be a hormonal problem, not a radioactivity hazard.'

But he was mistaken. 'The link between radiation and breast cancer,' observed a national cancer-institute specialist, 'is one of the best-established relationships there is.'

The executive continued to bluster: 'The plant manager isn't entirely to blame. Employees are responsible for safety too.'

But there were no warning signs in the workrooms. The women had been told that, as long as they didn't lip-point, they would be perfectly safe.

These women worked in a little town called Ottawa, Illinois.

These women worked for Joseph Kelly's firm, Luminous Processes.

The year was 1978.

The original radium girls were indeed Cassandra-like in their powers; and just like Cassandra, their prophecies were not always listened to. Safety standards only keep you safe if the companies you work for use them. Concerns had been raised about the Ottawa plant for decades, but it wasn't until 17 February 1978 that the dangerous studio was finally shut down: inspectors found radiation levels were 1,666 times higher than was safe. The abandoned building became something of a bogeyman for Ottawa residents, who became afraid to walk or even drive past it; it was graffitied with the slogan: DIAL LUMINOUS FOR DEATH.

'A lot of us are dead,' one LP dial-painter stated bluntly. Of a hundred workers she mentioned, sixty-five had died; the cancer rate was twice as high as normal.

Yet Luminous Processes was unapologetic. It wriggled out of paying clean-up costs, contributing approximately $62,000 ($147,500) to the multi-million-dollar bill, while executives used 'doubletalk' to put off the women when they demanded answers. Workers were offered just $100 ($363) in severance pay and had difficulty suing the firm. 'They didn't have any respect for the health of the girls,' one LP worker spat. 'They were just interested in getting the work out.'

'Luminous Processes,' declared the local paper, 'seems to put profits before people.'

How quickly we forget.

ACKNOWLEDGEMENTS

This book was born when I had the privilege of directing the play *These Shining Lives*, which dramatises the experiences of the Ottawa dial-painters. I want to say a huge thank you to play-wright Melanie Marnich, for introducing me to their story, and to my incredible cast – Anna Marx, Cathy Abbott, Darren Evans, David Doyle, James Barton-Steel, Julia Pagett, Lionel Laurent, Mark Ewins, Nick Edwards, Sarah Hudson and William Baltyn – for bringing it to life. Team TSL, our shared passion for telling this story has inspired me endlessly. Thank you for your talent, commitment and enduring support.

I will be eternally grateful to the dial-painters' families for so generously contributing to this book; you have enriched it beyond measure. Thank you for opening up your homes, your hearts and your family albums to me; for guided tours and cemetery visits; for your hospitality and friendship. It has been an honour to meet you all, and I hope I have done your relatives justice. Sincere thanks to Michelle Brasser, Mary Carroll Cassidy, Mary Carroll Walsh, James Donohue, Kathleen Donohue Cofoid, Art Fryer, Patty Gray, Darlene Halm, Felicia Keeton, Randy Pozzi, Donald Purcell, Dolores Rossiter, Jean Schott, Don Torpy

THE RADIUM GIRLS

and Jan Torpy. You all had gems of insight and information to share, and I am extremely thankful for every last one. Special thanks to Darlene and Kathleen for all the additional support you kindly gave.

Len Grossman – what a generous man you are. Thank you for sharing your father's legal briefs and scrapbooks with me, but also for escorting me to Northwestern, providing countless leads for further research and for your own priceless interview.

I am indebted to those authors who came before me on this journey, Claudia Clark and Ross Mullner, whose books were an invaluable resource; thank you also to Ross for sharing research materials and agreeing to an interview. (I was saddened to learn that, like her subjects, Claudia Clark died relatively young.) Librarians and archivists across America were extraordinarily helpful: thanks to Alice, Orange Public Library; Beth Zak-Cohen, Newark Public Library; Doug, Glenn and Sarah, National Archives, Chicago; Ken Snow and Erin Randolph, LaSalle County Historical Society & Museum; and the staff at the Reddick Public Library, Ottawa, the Library of Congress, and the Westclox Museum, Peru, Illinois. The most enormous thank you to Bob Vietrogoski of Rutgers, who incalculably enhanced this book: Bob, you are a legend. Thanks to all those who contributed to my research, including Rainy Dias, Gordon Dutton, Stephanie Jaquins, Stacy Piller, Cindy Pozzi, Amanda Cassidy, D. W. Gregory, Eleanor Flower and Jeralyn Backes. Additional thanks to all those who gave permission for the use of photographs, and to all my American hosts.

Writing this book has proved to be a life-enhancing experience which took over my entire world. Thank you to John and Beth Gribble for their continual cheerleading, to my sisters Penny and Sarah for their support, to Jo Mason for exemplary New York hospitality, to Anna Morris for her wise words and to all my friends for their endless enthusiasm for this project. Thanks also to Natalie Galsworthy, Ed Pickford and Jennifer Rigby for generously sharing their professional advice.

To Duncan Moore, my husband, 'thank you' is simply not sufficient for all you have done for this book. I am indebted to you for your love and support, but most of all for your perceptive guidance and direction, supplied as ever with your innate creative wisdom. Thank you for everything you have done for me and for the women, and for being my first reader.

Finally, I'd like to thank the whole team at Simon & Schuster for their support of this book, in particular Jo Whitford for her editorial feedback, patience and hard work, Jamie Criswell for PR, Sarah Birdsey in rights, and Nicki Crossley for her in-house championing. Very special thanks must go to senior commissioning editor Abigail Bergstrom, who shared my desire to tell this story. Abbie, quite simply, without your publishing vision and belief in this book, it would not exist. Thank you so much not only for what you have done for me, but for what you have done for the dial-painters. Their story has now been told – and it would not have happened without you. Thank you, from the bottom of my heart, for giving them a voice.

Kate Moore, 2016

PICTURE ACKNOWLEDGEMENTS

Page 1: *New York Evening Journal* (top left). CHR, National Archives, Chicago (top right and bottom left). *American Weekly* (bottom right).

Page 2: *New York Evening Journal* (top left). RBP, reel 2 (top right). *American Weekly* (all bottom images).

Page 3: CHR, National Archives, Chicago.

Page 4: Hagley Museum and Library, sourced from *Deadly Glow* by Dr Ross Mullner (American Public Health Association, 1999) (top left). Blackstone Studios, University Archives, Rare Book & Manuscript Library, Columbia University in the City of New York (top right). *Newark Ledger* (middle left). Michael Frunzi, HMP, Rutgers (middle right). RBP, reel 2 (bottom left).

Page 5: Collection of Ross Mullner (top). Lippincott, Williams and Wilkins, sourced from *Deadly Glow* (middle left). *American Weekly* (middle right). CHR, National Archives, Chicago (bottom).

Page 6: Used courtesy of Darlene Halm and the Looney family (top left). Used courtesy of Dolores Rossiter and Patty Gray (top middle). *Chicago Daily Times*, courtesy of Sun-Times Media (top right). CHR, National Archives, Chicago (bottom).

Page 7: Used courtesy of Randy Pozzi, sourced from PPC, LaSalle County Historical Society & Museum (top left). *Chicago Daily Times*, courtesy of Sun-Times Media (all other images on page).

Page 8: *Chicago Daily Times*, courtesy of Sun-Times Media (top images). Sourced from the scrapbooks of Leonard Grossman, lgrossman. com (bottom left). *Chicago Herald-Examiner* (bottom right).

NOTES

Abbreviations

CHR – Health Effects of Exposure to Internally Deposited Radioactivity Projects Case Files. Center for Human Radiobiology, Argonne National Laboratory. General Records of the Department of Energy, Record Group 434. National Archives at Chicago

HMP – Harrison Martland Papers, Special Collections, George F. Smith Library of the Health Sciences, Rutgers Biomedical and Health Sciences, Newark

PPC – Pearl Payne Collection, LaSalle County Historical Society & Museum, Utica, Illinois

RBP – Raymond H. Berry Papers, Library of Congress, Washington, DC

AL – Albina Maggia Larice
CD – Catherine Wolfe Donohue
CP – Charlotte Nevins Purcell
EH – Edna Bolz Hussman
GF – Grace Fryer
KS – Katherine Schaub
MR – Marie Becker Rossiter
PP – Pearl Payne
QM – Quinta Maggia McDonald
VS – Sabin von Sochocky

NCL – National Consumers League
USRC – United States Radium Corporation

Dates of sources are supplied in the Notes where known.

Prologue

'it made an' Eve Curie, *Madame Curie: A Biography*, translated by Vincent Sheean (Read Books, 2007).

'my beautiful radium' Marie Curie, cited in 'The Radium Girls', themedicalbag.com.

'These gleamings' Marie Curie, *Pierre Curie*, translated by C. and V. Kellogg (Macmillan, 1923), p.104.

'it reminds one' US Surgeon General, transcript of the national radium conference, 20 Dec 1928, Raymond H. Berry Papers (RBP), reel 3.

'the unknown god' Cited in Claudia Clark, *Radium Girls* (The University of North Carolina Press, 1997), p.49.

'The gods of' George Bernard Shaw, *The Quintessence of Ibsenism* (Courier Corporation, 1994).

Part One: Knowledge

Chapter One

'A friend of' Katherine Schaub (KS), autobiography (*Survey Graphic*, 1932), p.1.

'a very imaginative' Dr E. B. Krumbhaar, letter to Raymond H. Berry, 21 June 1929, RBP, reel 3.

'a very pretty' *New York Sun*.

'about all the' William D. Sharpe, MD, 'Radium Osteitis with Osteogenic Sarcoma: The Chronology and Natural History of a Fatal Case'.

'All her life' *Popular Science*, July 1929.

'I went to' KS, memo to Berry, RBP, reel 1.

'the greatest find' *Chicago Daily Tribune*, 21 June 1903.

'The sun of' The Bible, Malachi 4:2, cited by Dr Howard Kelly in the *Newark Evening News*, 9 Jan 1914.

'old men young' Mr Smith, court transcript, 26 April 1928.

'Sometimes, I am' JFJ, advertising brochure for the Radiumator, Harrison Martland Papers (HMP), Rutgers Biomedical and Health Sciences.

'like a good' *American* magazine, Jan 1921.

'liquid sunshine' Cited in John Conroy, 'Radium City'.

'The Radium Eclipse' Advertisement, LaSalle County Historical Society & Museum.

'[I was] to see' KS, memo to Berry, RBP, reel 1.

'I had never' Alice Tolan, court transcript, 26 Nov 1934.

'We put the' KS, memo to Berry, RBP, reel 1.

'a little bit' Mae Cubberley Canfield, examination before trial, RBP, reel 2.

'The first thing' Ibid.

'Here in the' KS, autobiography, p.1.

'I was asked' KS, memo to Berry, RBP, reel 1.

Chapter Two

'The war to' President Woodrow Wilson, April 1917.

'The girls' KS, autobiography, p.1.

'When Grace was' Ethelda Bedford, *Newark Ledger*, 24 May 1928.

'a girl enthused' *Newark Ledger*, 24 May 1928.

'She had a' Mary Freedman, court testimony, 26 Nov 1934.

'like a fine' Alice Tolan, court testimony, 26 Nov 1934.

'She told me' KS, court transcript, 25 April 1928.

'lip, dip, paint' Melanie Marnich, *These Shining Lives* (Dramatists Play Service, Inc., 2010).

'I know I' Alice Tolan, court testimony, 26 Nov 1934.

'It didn't taste' Grace Fryer (GF), *Orange Daily Courier*, 30 April 1928.

'Here is a' *Popular Science*, July 1929.

'Dresden Doll' Unknown newspaper, CHR.

'a very nice' Anna Rooney, quoted in USRC memo, 20 July 1927.

'The place was' Florence E. Wall article, Orange Public Library.

'merry giggling' Unknown newspaper, HMP.

'very clumsy' Albina Maggia Larice (AL), court testimony, 25 April 1928.

'I always did' AL, *Sunday Call*.

'pot boiler' Florence E. Wall article, Orange Public Library.

'kindred soul' Ibid.

'remarkable man' Ibid.

'someone that liked' Ibid.

'one of the' *American*, Jan 1921.

'he would not' Cited in Ross Mullner, *Deadly Glow* (American Public Health Association, 1999), p.13.

'an animal had' Florence E. Wall article, Orange Public Library.

'only by taking' Sabin von Sochocky (VS), *American*, Jan 1921.

'There may be' Thomas Edison, *New York Daily News*, 1903, cited in Mullner, *Deadly Glow*, p.17.

Chapter Three

'I don't go' Quinta Maggia McDonald (QM), *New York Sun*.

'inseparable' Unknown newspaper, RBP, reel 2.

'Without so doing' Edna Bolz Hussman (EH) affidavit, 15 July 1927, RBP, reel 1.

'sparkling particles' National Consumers League (NCL) memo, Nov 1959, NCL files, Library of Congress.

'hands, arms, necks' William B. Castle, Katherine R. Drinker and Cecil K. Drinker, 'Necrosis of the Jaw in Workers Employed in Applying a Luminous Paint Containing Radium', *Journal of Industrial Hygiene* (JIH), August 1925.

'When I would' EH affidavit, 15 July 1927, RBP, reel 1.

'you could see' EH, court testimony, 12 Jan 1928.

'like the watches' QM, court testimony, 12 Jan 1928.

'more like matinee' *Orange Daily Courier*, 12 Dec 1929.

'I could do' GF, court testimony, 12 Jan 1928.

'I remember chewing' QM, court testimony, 12 Jan 1928.

'Nasal discharges' GF affidavit, July 1927, RBP, reel 1.

'lively Italian' Cited in Clark, *Radium Girls*, p.16.

'hard-working' Frederick Hoffman, letter to Arthur Roeder, 7 March 1925, RBP, reel 2.

'the type that' USRC memo, 20 July 1927.

'I obtained' KS, autobiography, p.1.

'Do not do', VS, quoted by GF, court testimony, 12 Jan 1928.

'Do not do' Ibid.

'She told me' Ibid.

Chapter Four

'Barker would just' Robley Evans memo on conversation with Wallhausen, cited in 'Historic American Buildings Survey: US Radium Corporation' by the National Park Service, HAER no. NJ-121; HAER NJ 7-ORA, 3-.

'They passed' EH, court testimony, 12 Jan 1928.

'They were taken' Ibid.

'The lips were' EH, court testimony, *Orange Daily Courier*, 13 Jan 1928.

'cracking and' KS medical history, 17 Dec 1927, RBP, reel 2.

'anxious to take' Swen Kjaer, 8 April 1925, cited in Clark, *Radium Girls*, p.15.

'The girls at' KS, autobiography, p.1.

'I wasn't making' Ibid.

'the kind of' *Orange Daily Courier*, 30 April 1928.

'most hygienic' VS, cited in Clark, *Radium Girls*, p.17.

'more beneficial' Ibid.

'I instructed them' KS, court testimony, 25 April 1928.

'Locked up in' VS, *American*, Jan 1921.

'What radium means' Ibid.

Chapter Five

'I treated her' Dr Joseph P. Knef, *Sunday Call*, 14 Oct 1927.

'Instead of' Ibid.

'My sister' QM, *Star-Eagle*.

'extreme methods' Amelia Maggia form, Swen Kjaer study, CHR.

'extraordinary affliction' *American Weekly*, 28 Feb 1926.

'peculiar' Ibid.

'it differed' Ibid.

'Painting numbers' Amelia Maggia, quoted by Knef, unknown newspaper.

'I asked' Knef, *Sunday Call*, 14 Oct 1927.

'I thought' Ibid.

'has unnerved me' QM, *Star-Eagle*.

'not by an' *American Weekly*, 28 Feb 1926.

'Radium may be' *Newark Evening News*, Feb 1922.

'whenever a portion' *American Weekly*, 28 Feb 1926.

'devoted' *Popular Science*, July 1929.

'slowly ate its' Knef, *Star-Eagle*.

'painful and terrible' QM, *Star-Eagle*.

'She died' AL, *Star-Eagle*.

'My elder sister' AL, court testimony, 25 April 1928.

'for Amelia' Bill from Dr Thompson, RBP, reel 2.

Chapter Six

GIRLS WANTED Radium Dial advertisement, *Ottawa Daily Republican-Times*, 16 Sep 1922.

'Several girls' Ibid.

'genuine American community' Ottawa town directory, 1922.

'where friendliness reigns' Slogan for the Merchants and Farmers Trust and Savings Bank, Ibid.

'one block north' Advert for White's Garage, Ibid.

'a small [town]' *Chicago Daily Times*, 17 March 1936.

'The citizens' Ottawa town directory, 1922.

'lung trouble' *Ottawa Free Trader*, 1 Aug 1913.

'It was fascinating' Catherine Wolfe Donohue (CD), *Chicago Daily Times*, 9 July 1937.

'Japanese brushes' CD, court testimony, legal brief.

'Miss Lottie Murray' CD, court testimony, *Chicago Daily Times*, 10 Feb 1938.

'lip, dip' Marnich, *These Shining Lives*.

'18 years or' Radium Dial advertisement, *Ottawa Daily Republican-Times*, 16 Sep 1922.

'she ate the' MIT memo, 6 Dec 1958, CHR.

'When I was' Charlotte Nevins Purcell (CP), *Chicago Daily Times*, 8 July 1937.

'a happy, jolly' *Chicago Daily Times*, 18 March 1936.

'washing was a' Report on dial-painting studios, edited by Edsall and Collis, *JIH*, 1933, CHR.

'The girls were' *Chicago Herald-Examiner*, 27 Feb 1938.

'I used to' Ova Winston, CHR.

'When I went' CD, court testimony, *Chicago Herald-Examiner*, 11 Feb 1938.

'humorously termed' *Chicago Herald-Examiner*, 27 Feb 1938.

'were expected to' Interview with Mary Carroll Cassidy.

'We used to' CD, court testimony, *Denver Colorado Post*, *Chicago Herald-Examiner* and *Chicago Daily Times*.

'We made' CD, court testimony, *Chicago Daily Times*, 11 Feb 1938.

'We were extremely' Unknown dial-painter, *Chicago Daily Times*, 1953.

'We expect you' Westclox Manual for Employees, Westclox Museum.

Chapter Seven

'I immediately' Dr Barry, court testimony, 4 Jan 1928.

'some occupational' Ibid.

'The word radium' KS, court testimony, 25 April 1928.

'garlic odour' Cited in Clark, *Radium Girls*, p.34.

'he has warned' Charles Craster, letter to John Roach, 3 Jan 1923, RBP, reel 3.

'a foreman' Memo on KS visit to the Department of Health, 19 July 1923, RBP, reel 3.

'make a survey' Craster, letter to Roach, 3 Jan 1923, RBP, reel 3.

'no reports' Lillian Erskine, letter to Roach, 25 Jan 1923, RBP, reel 3.

'This case' Ibid.

'It is my' Dr M. Szamatolski, letter to Roach, 30 Jan 1923, RBP, reel 3.

'considerable' USRC 'Memorandum for Scientific Witnesses', 28 Aug 1934.

'one cannot doubt' Cited in Mullner, *Deadly Glow*, p.38.

'I would suggest' Szamatolski, letter to Roach, 30 Jan 1923, RBP, reel 3.

'The reputation for' George Willis, *Journal of the American Medical Association (JAMA)*, cited in Mullner, *Deadly Glow*, p.46.

'I feel quite' Szamatolski, letter to Roach, 6 April 1923, RBP, reel 3.

Chapter Eight

'were what I' Georgia Mann, CHR.

'[One worker]' Ibid.

'Mrs Reed had' Ibid.

'The practice' Bob Bischoff, CHR.

'It was a' Interview with Darlene Halm.

'good money' Darlene Halm, mywebtimes.com, 4 Dec 2010.

'trying to hide' St Xavier's Academy yearbook, 1922.

'delightful sunny' Ibid.

'a rugged' Author unknown, 'The Westclox Story', circa 1930s, handout at the Westclox Museum.

'I was working' Marie Becker Rossiter (MR), *Radium City* (1987; directed by Carole Langer).

'Her attitude' Interview with Patty Gray.

'skinny minny' Ibid.

'closest friend' MR, *Radium City*.

'My first day' Ibid.

'Well, you know' Ibid.

'corsets, gloves' Ottawa town directory, 1922.

'Gee, this week' MR, *Radium City*.

'I bought my' Ibid.

'ruckus' Ibid.

'They were all' Edith Schomas, Ibid.

'Prohibition was huge' Interview with Kathleen Donohue Cofoid.

'Many of the' CD, court testimony, *Chicago Herald-Examiner*, 27 Feb 1938.

'We just figured' MR, *Radium City*.

'turn the lights' CP, quoted in interview with Felicia Keeton.

'You don't see' MR, *Radium City*.

'just for fun' MR to Catherine Quigg, *Learning to Glow: A Deadly Reader*, edited by John Bradley (The University of Arizona Press, 2000).

'We were a' CP, *Ottawa Daily Republican-Times*, 14 March 1936.

'They thought it' Interview with James Donohue.

Chapter Nine

'I thought' GF affidavit, 8 June 1927, RBP, reel 1.

'a progressive' Definition of Vincent's angina, medicinenet.com.

'complete' Irene Rudolph death certificate, RBP, reel 2.

'not decisive' Ibid.

'terrible and' KS, autobiography, p.1.

'Still another' Memo on KS visit to the Department of Health, 19 July 1923, RBP, reel 3.

'They have to' Ibid.

'A foreman there' Ibid.

'Many of the' QM, *Star-Eagle*.

'They were all' QM, *New York Sun*.

'We were all' QM, *World*.

'I hobbled around' QM handwritten notes for Berry, RBP, reel 1.

'I went to' QM, *Star-Eagle*.

'[and] woke up' QM, *New York Sun*.

'I tried to' Lenore Young, court testimony, April 1928.

'I let the' Young, letter to Roach, 2 Feb 1924, RBP, reel 3.

'I began' KS, autobiography, p.1.

'flinty' Barry, court testimony, 4 Jan 1928.

'Patient' Ibid.

'I kept thinking' KS, autobiography, p.1.

'[Irene] had necrosis' KS, memo to Berry, RBP, reel 1.

'seriously shocked' KS legal complaint, 17 Sep 1927, RBP, reel 1.

'My foot was' GF, court testimony, 12 Jan 1928.

'I said nothing' GF affidavit, 8 June 1927, RBP, reel 1.

'Towards the end' Ibid.

'When [the company]' Josephine Smith form, Kjaer study, CHR.

Chapter Ten

'one summer between' Emma Renwick, CHR.

'comely' *Ottawa Daily Republican-Times*, 12 Feb 1938.

'wanted to be' Helen Munch, *Toronto Star*, 23 April 1938.

'fine husband' Pearl Payne (PP), *Toronto Star*, 23 April 1938.

'very knowledgeable' Interview with Randy Pozzi.

'During my' PP, 'Life History of Pearl Payne in brief', Pearl Payne Collection (PPC).

'never an unkind' Interview with Randy Pozzi.

'dearest friend' CD, letter to PP, 9 March 1938, PPC.

'I was [to be]' Unnamed dial-painter, CHR.

'I never heard' Interview with Jean Schott.

'She looked after' Ibid.

'I remember [Peg]' Jane Raub, *Radium City*.

'She was everything' Unnamed Looney sister to Martha Irvine, 'Suffering Endures for Radium Girls', Associated Press (4 Oct 1998).

'She entertained' Interview with Darlene Halm.

'We used to' CD, court testimony, *Chicago Herald-Examiner*, 27 Feb 1938.

Chapter Eleven

'I felt' KS, court testimony, 25 April 1928.

'There recently' USRC, letter to its insurance company, cited in Clark, *Radium Girls*, p.38.

'We do not' Ibid.

'We discussed employment' KS, court testimony, 25 April 1925.

'there was something' Ibid.

'I refused to' Barry, court testimony, 4 Jan 1928.

'After seeing several' Harrison Martland, letter to Andrew McBride, 28 Aug 1925, HMP.

'exceedingly high-grade' USRC memo, July 1927.

'a very peculiar' Dr Humphries, court testimony, 25 April 1928.

'pus bags' Katherine Wiley notes, RBP, reel 3.

'moth-eaten' Frederick Flinn, medical publication (possibly *JAMA*, vol. 96, no. 21), HMP.

'poisoning by a' Hazel Kuser form, Kjaer study, CHR.

'there is little' Young advising what Hazel's family have told her the doctors have said, letter to Roach, 8 Feb 1924, RBP, reel 3.

'negligent' Young, letter to Roach, 2 Feb 1924, RBP, reel 3.

'considerable difficulty' USRC production manager's report, cited in Clark, *Radium Girls*, p.65.

'she was about' Undated USRC internal memo.

Chapter Twelve

'We must determine' Roeder, letter to Cecil K. Drinker, 12 March 1924, RBP, reel 1.

'very interesting' Drinker, letter to Roeder, 15 March 1924, RBP, reel 1.

'very much improved' Roeder, letter to Drinker, 12 March 1924, RBP, reel 1.

'I have been' Ibid.

'We are inclined' Drinker, letter to Roeder, 15 March 1924, RBP, reel 1.

'At the same' Ibid.

'chronic infectious' Drinker notes, RBP, reel 1.

'I have been' GF, *Graphic*.

'psychological and' Roeder, letter to Harold Viedt, 19 March 1924.

'The individuals' Viedt, letter to Roeder, 12 March 1924.

'did not reflect' Cited in judgment, La Porte vs USRC, 17 Dec 1935.

'just as I' Roeder, letter to Viedt, 14 March 1924.

'I do not feel' Viedt, letter to Roeder, 18 March 1924.

'We should create' Roeder to Viedt, 19 March 1924.

'the most important' Ibid.

'You ought to' Dr Davidson, quoted in Wiley notes on her interview with the dentist, RBP, reel 3.

'If I could' Ibid.

'The authorities' Young, cited in Clark, *Radium Girls*, p.66.

'pains in the' Marguerite Carlough form, Kjaer study, CHR.

'serious lesions' Drinker, letter to McBride, 30 June 1925, RBP, reel 1.

'scoffed at' Ibid.

'characteristic of' Ibid.

'There seemed' Ibid.

'no malignant' Ibid.

'examined a number' Katherine Drinker, court testimony, 14 Nov 1927.

'persisted in the' Castle, Drinker and Drinker, 'Necrosis of the Jaw', *JIH*, Aug 1925.

'recovered satisfactorily' Ibid.

'from the [constant]' Ibid.

'At least once' Knef, transcript of conversation with USRC, 19 May 1926, RBP, reel 3.

'I have been' Ibid., reel 2.

'fogged' *American Weekly*, 28 Feb 1926.

Chapter Thirteen

'of late she' Castle, Drinker and Drinker, 'Necrosis of the Jaw', *JIH*, Aug 1925.

'a statement not' Ibid.

'this poor sick' Wiley, cited in Clark, *Radium Girls*, p.66.

'After seeing one' Wiley, letter to Roeder, 17 Jan 1925, RBP, reel 3.

'to stick to' Wiley, cited in Clark, *Radium Girls*, p.66.

'unwilling to discuss' Wiley notes, RBP, reel 3.

'Miss Mead wishes' Ibid.

'When radium poisoning' Ibid.

'They have none' Wiley, letter to Alice Hamilton, 4 March 1925, RBP, reel 3.

'furious' Wiley, cited in Clark, *Radium Girls*, p.85.

'rebuked him' Ibid.

'An investigation' Wiley notes, RBP, reel 3.

'Put it in' Ibid.

'practically normal' Table of results from the Castle, Drinker and Drinker report, enclosed with letter from Viedt to Roach, 18 June 1924, RBP, reel 1.

'I do not' Viedt, letter to Roach, 18 June 1924, RBP, reel 1.

'every girl is' Roeder, cited by Hamilton, letter to Katherine Drinker, 4 April 1925, RBP, reel 1.

'He tells everyone' Ibid.

'Rumours quieted' USRC memo.

'It is not' Theodore Blum, letter to USRC, 14 June 1924, RBP, reel 3.

'a precedent' USRC, letter to Blum, 18 June 1924, RBP, reel 3.

'The results of' Ibid.

'We are sorry' Ibid.

'I was only' Blum, letter to USRC, 20 June 1924, RBP, reel 3.

Chapter Fourteen

'nervous case' Wiley notes, RBP, reel 3.

'I could not' KS, autobiography, p.1.

'The pain' KS, *American Weekly*, 28 Feb 1926.

'advised work' Wiley notes, RBP, reel 3.

'I had stopped' KS, autobiography, p.1.

'Why should I' KS, *Graphic*.

'It seemed to' QM affidavit, 29 Aug 1927, RBP, reel 1.

'could not move' Humphries, court testimony, 25 April 1928.

'white shadow' Ibid.

'a white mottling' Ibid., 27 Nov 1934.

'The whole situation' Roach, cited in 'Occupational Diseases – Radium Necrosis', information secured by Miss E. P. Ward, CHR.

'Such trouble as' Szamatolski, letter to Roach, 6 April 1923, RBP, reel 3.

'radium jaw' Blum, address to the American Dental Association, Sep 1924.

'all necessary' KS, memo to Berry, RBP, reel 1.

'They told me' QM affidavit, 29 Aug 1927, RBP, reel 1.

'I could still' Ibid.

'That cast eased' QM, *Star-Eagle*.

'one leg was' Ibid.

'suffered so frightfully' Wiley notes, RBP, reel 3.

'She suffered' Karl Quimby, letter to Martland, 23 June 1925, HMP.

'vigorously' Hamilton, letter to Wiley, 30 Jan 1925, RBP, reel 3.

'From what I' Ibid.

'special investigator' Ibid.

'a lamentable case' Hoffman, letter to Roeder, 13 Dec 1924, RBP, reel 2.

'If the disease' Ibid., 29 Dec 1924, RBP, reel 2.

'That it will' Ibid.

Chapter Fifteen

'I began to' CD, court testimony, *Ottawa Daily Republican-Times*, 10 Feb 1938.

'No strenuous' Rufus Fordyce, cited in Kjaer study, CHR.

'And the cold' Ottawa High School yearbook, 1925.

'He was big' Interview with Jean Schott.

'a devil' Interview with Dolores Rossiter.

'He used to' Ibid.

'The whole family!' Interview with Jean Schott.

'requested to' Kjaer study, CHR.

Following quotations, Ibid.

'Radium paints' Ethelbert Stewart, *World*, 17 July 1928.

'I abandoned' Stewart, letter to Hamilton, 17 Dec 1927, RBP, reel 3.

Chapter Sixteen

'totally incapacitated' *Newark Evening News*, 9 March 1925.

'seriously injured' Ibid.

'became frightened' Cited in Clark, *Radium Girls*, p.109.

'sure' Ibid.

'did not dare' Ibid.

'women's clubs' Roeder, quoted in meeting with Swen Kjaer, cited in Ibid., p.100.

'unusual interest' Roeder, letter to Wiley, 3 March 1925, RBP, reel 3.

'perfectly proper' Ibid., 9 Jan 1925.

'nothing could be' Hoffman, letter to Roeder, 7 March 1925, RBP, reel 2.

'truly pitiable' Ibid.

'that a presentation' Roeder, letter to Drinker, 25 Feb 1925, RBP, reel 1.

'such an investigation' Ibid.

'I heard the' Hoffman, court testimony, 12 Jan 1928.

'We sincerely believe' Roeder, letter to Hoffman, 17 Dec 1924, RBP, reel 2.

'the disease in' VS, letter to Hoffman, 14 Feb 1925, RBP, reel 3.

'My dear Mr' Wiley, letter to Roeder, 2 March 1925, RBP, reel 1.

'We believe that' Drinker, letter to Roeder, 3 June 1924, RBP, reel 1.

'it would seem' Drinker, letter to Viedt, 29 April 1924, cited in Mullner, *Deadly Glow*, p.58.

'In our opinion' Drinker report, 3 June 1924, RBP, reel 1.

'ample evidence' Ibid.

'The only' Ibid.

'similar chemical' Ibid.

'if absorbed' Ibid.

'Radium, once' Drinker report, 3 June 1924, RBP, reel 1.

'No blood' Ibid.

'practically normal' Table of results from the Castle, Drinker and Drinker
    report, RBP, reel 1.

'It seems to' Drinker report, 3 June 1924, RBP, reel 1.

'to call your' Ibid.

'precautions' Drinker, letter to Roeder, 3 June 1924, RBP, reel 1.

'This is much' Roeder, memo to Viedt, 9 April 1925, USRC files.

'mystified' Roeder, letter to Drinker, 6 June 1924, RBP, reel 1.

'reconcile' Ibid.

'contemplating' Ibid.

'Your preliminary' Ibid., 18 June 1924.

'I am sorry' Drinker, letter to Roeder, 20 June 1924, RBP, reel 1.

'We found blood' Ibid.

'I still feel' Roeder, letter to Drinker, 1 July 1924, RBP, reel 1.

'The unfortunate' Drinker, cited in Mullner, *Deadly Glow*, p.59.

'It does not' Drinker, cited in Clark, *Radium Girls*, p.125.

'Are we in' Cited in Ibid., p.95.

'the almost complete' Roeder, letter to Drinker, 16 July 1924, RBP, reel 1.

'showed a very' Wiley, cited in Clark, *Radium Girls*, p.96.

'dishonest' Ibid.

'Do you suppose' Hamilton, letter to Katherine Drinker, 4 April 1925,
    RBP, reel 1.

'very indignant' Katherine Drinker, letter to Hamilton, 17 April 1925, RBP,
    reel 1.

'he has proved' Ibid.

'it can only' Cecil K. Drinker, letter to Roeder, 17 Feb 1925, RBP, reel 1.

'stupid enough' Hamilton, letter to Wiley, 2 Feb 1925, RBP, reel 3.

Chapter Seventeen

'They had impressed' Hoffman, court testimony, 12 Jan 1928.

'I wish that' Roeder, letter to Hoffman, 2 April 1925, RBP, reel 2.

'opportunity' Ibid.

'I wish to' Hoffman, letter to Roeder, 17 April 1925, RBP, reel 2.

'I find on' Ibid.

'thought that the' Roeder, cited in Kjaer study, RBP, reel 2.

'The matter has' Roeder, letter to Roach, 22 April 1925, RBP, reel 1.

'In view of' Roeder, letter to Drinker, 9 April 1925, RBP, reel 1.

'We have [both]' Drinker, letter to Roach, 29 May 1925, RBP, reel 1.

'felt the conduct' Ibid.

'assured [him]' Ibid.

'I will do' Drinker, letter to Roeder, 29 May 1925, RBP, reel 1.

'I made my' Flinn, 'Outline of F. B. Flinn's Association with Radium Research', USRC files.

'not know directly' EH affidavit, 15 July 1927, RBP, reel 1.

'did not take' Ibid.

'[He] told me' Ibid.

'a very depressing' KS, autobiography, p.2.

'Since [my] first' KS, *Graphic*.

'To be under' KS, autobiography, p.2.

'I have been' GF, *Newark Ledger*.

'became so painful' GF affidavit, 8 June 1927, RBP, reel 1.

'extremely rotten' Martland, court testimony, 26 April 1928.

'I feel absolutely' Berry, notes following interview with VS, RBP, reel 3.

'small, rapid and' Marguerite Carlough form, Kjaer study, CHR.

'The women were' Hoffman, court testimony, 12 Jan 1928.

'The cumulative effect' Ibid.

'We are dealing' Hoffman's report, cited in Mullner, *Deadly Glow*, p.66.

'Miss Carlough may' Hamilton, letter to Katherine Drinker, 4 April 1925, RBP, reel 1.

'There was none' Hoffman, court testimony, 12 Jan 1928.

'The most sinister' Hoffman, speech to the American Medical Association, 25 May 1925, CHR.

'[Von Sochocky] gave' Hoffman, court testimony, 12 Jan 1928.

'It has seemed' Hoffman, *JAMA*, 1925.

'It is a' William Bailey, cited in Mullner, *Deadly Glow*, p.114.

'It seems to' Hamilton, letter to Wiley, 7 March 1925, RBP, reel 3.

Chapter Eighteen

'lost interest' Martland, letter to McBride, 28 Aug 1925, HMP.

'no difference' Biography 1940, HMP.

'heavy but' Ibid.

'*sans* tie' Ibid.

'did his' Samuel Berg MD, *Harrison Stanford Martland: The Story of a Physician, a Hospital, an Era* (Vantage Press, 1978), p.15.

'One of the' Martland, 'The Danger of Increasing the Normal Radioactivity of the Human Body', HMP.

'The first case' Martland, court testimony, 6 Dec 1934.

'scoffed' Cecil K. Drinker, letter to McBride, 30 June 1925, RBP, reel 1.

'her palate had' Clark, *Radium Girls*, p.103.

'went bad quite' Martland, court testimony, 26 April 1928.

'profoundly toxic' Sarah Maillefer form, Kjaer study, CHR.

'removed to receive' *Newark Evening News*, 19 June 1925.
'The patient was' Martland, court testimony, 26 April 1928.
'She couldn't' Ibid.
'delirious' Sarah Maillefer form, Kjaer study, CHR.
'I have nothing' Martland, unknown newspaper, HMP.
'This poisoning' Ibid.
'We have nothing' Martland, *Newark Evening News*, 19 June 1925.
'No,' *Newark Evening News*, 19 June 1925.
'Why?' Ibid.
'She is said' Ibid.

Chapter Nineteen

'small possibility' Viedt, cited in unknown newspaper, HMP.
'We have engaged' Ibid.
'nothing was found' Ibid.
'absurd to think' Ibid.
'filled with old' Sarah Maillefer form, Kjaer study, CHR.
'dark-red marrow' Ibid.
'by its gamma' Elizabeth Hughes, court testimony, 25 April 1928.
'physiologically' Martland, court testimony, 26 April 1928.
'The distance' Ibid.
'whirling, powerful' VS, *American*, Jan 1921.
'a type of' Martland, 'The Danger of', HMP.
'considerable' Sarah Maillefer form, Kjaer study, CHR.
'I then took' Martland, court testimony, 26 April 1928.
'every day' Ibid.
'There is nothing' Martland, letter to McBride, 28 Aug 1925, HMP.
'Radium is indestructible' Knef, *Star-Eagle*.
'If this is' Ibid.
'There is not' Martland, letter to McBride, 28 Aug 1925, HMP.
'Would it be' Flinn, letter to Martland, 20 June 1925, HMP.
'It will be' Stryker, quoted by Roeder, letter to Drinker, 22 June 1925, RBP, reel 1.
'If the Department' Ibid.
'furious' Wiley, cited in Clark, *Radium Girls*, p.85.
'I am arranging' Drinker, letter to Roeder, 18 June 1925, RBP, reel 1.
'tell 'em' Philip Drinker, quoting his lawyer brother, letter to Hamilton, 1952, cited in 'Historic American Buildings Survey'.
'This report' Josephine Goldmark, cited in Mullner, *Deadly Glow*, p.60.
'Doctors, who' Bailey, cited in Clark, *Radium Girls*, p.174.
'take in one' Bailey, unknown newspaper, HMP.

'Radium, because' Clarence B. Lee, *Newark Evening News*, 21 June 1925.

'unfit' Roeder, press release, cited in Kenneth A. DeVille and Mark E. Steiner, 'New Jersey Radium Dial Workers and the Dynamics of Occupational Disease Litigation in the Early Twentieth Century', *Missouri Law Review* (vol. 62, issue 2, spring 1997).

'was not in' USRC company official, *New York Times*, cited in Florence E. Wall article, Orange Public Library.

'did not agree' McBride, letter to Drinker, 25 June 1925, RBP, reel 3.

'human life' Ibid.

'I would issue' Ibid.

'I can scarcely' Quimby, letter to Martland, 23 June 1925, HMP.

'a terrible' Martland, court testimony, 26 April 1928.

'In the midst' KS, autobiography, p.2.

'It must be' Ibid.

'Her general' Martland, cited in Clark, *Radium Girls*, p.103.

'I can stand' QM, *Star-Eagle*.

'He told me' GF affidavit, 18 July 1927, RBP, reel 1.

'He told me that my trouble' QM affidavit, 29 Aug 1927, RBP, reel 1.

'When I first' GF, *Graphic*.

'I was horror-stricken' Ibid.

'When the doctors' KS, autobiography, p.2.

'The medical examiner's' Ibid.

'gave me *hope*' Ibid.

Part Two: Power

Chapter Twenty

'Do not do' VS, quoted by GF, court testimony, 12 Jan 1928.

'*all your trouble*' GF affidavit, 18 July 1927, RBP, reel 1.

'Why didn't you' Question derived from Berry's evidence that GF 'asked [VS] why it was [he] hadn't informed them' of the danger, court transcript, April 1928.

'aware of these' Court transcript, 27 April 1928.

'warned other members' Berry, citing statement of VS to Martland and Hoffman, legal notes, RBP, reel 1.

'endeavoured' Berry, letter to Hoffman, summarising Hoffman's statement of what VS had said, 3 Jan 1928, RBP, reel 3.

'The matter was' Court transcript, 27 April 1928.

'I don't believe' GF, *Graphic*.

'brave and smiling' Ethel Brelitz, unknown newspaper, RBP, reel 2.

'She often' Ibid.

'Nothing could' QM, quoted in AL complaint, RBP, reel 1.

'When I realise' GF, *Graphic*.

'considerable money' Berry, summary of girls' cases, RBP, reel 1.

'[But] I had' GF affidavit, 18 July 1927, RBP, reel 1.

'Each of the' KS, autobiography, p.2.

'palm off' Roeder, cited in Kjaer study, RBP, reel 2.

'1. Helen Quinlan' The List of the Doomed, HMP.

'I know I' KS, *Graphic*.

'Her face' *American Weekly*, 28 Feb 1926.

'in a very' Hoffman, letter to Roeder, 8 Dec 1925, RBP, reel 2.

'mentally deranged' USRC memo, 20 July 1927.

'When you're sick' KS, *Sunday Call*.

'very badly ill' Berry, summary of KS illness, RBP, reel 1.

'She is not' Josephine Schaub, *New York Sun*.

'Night and rainy' KS, *Sunday Call*.

'to a mere' *American Weekly*, 28 Feb 1926.

Chapter Twenty-One

'I got rid' AL, court testimony, 25 April 1928.

'I am' AL, *Star-Eagle*.

'I know' Ibid.

'I first started' EH, court testimony, 12 Jan 1928.

'She had a' Humphries, court testimony, 25 April 1928.

'By that time' Ibid.

'They put me' EH, court testimony, 12 Jan 1928.

'I went to' Wiley, letter to Attorney General, 23 Feb 1928, RBP, reel 3.

'asked Dr Flinn' Ibid.

'My dear Miss' Flinn, letter to KS, 7 Dec 1925, RBP, reel 3.

'a terrible' Flinn, 'Outline'.

'I was ill' KS, court testimony, 25 April 1928.

'I never replied' Flinn, 'Outline'.

'I [have] examined' Flinn, 'Newer Industrial Hazards', possibly *JAMA* (vol. 197, no. 28).

'a hitherto unrecognised' Martland, *JAMA*, Dec 1925.

'We are a' James Ewing, cited in Clark, *Radium Girls*, p.104.

'none of the' Martland, cited in Ibid., p.105.

'The original' Martland, 'The Danger of', HMP.

'automatically reduced' Radium Ore Revigator Company, letter to Martland, 8 April 1926, HMP.

'Our friend' Howard Barker, cited in Clark, *Radium Girls*, p.109.
'I am rather' Ibid.
'Though I am' Flinn, letter to Drinker, 16 Jan 1926, RBP, reel 3.
'half dead' Hoffman, *Orange Daily Courier*, 9 June 1928.
'the most tragic' Ibid.
'beautiful concentrations' Martland, court testimony, 26 April 1928.

Chapter Twenty-Two

'I feel better' GF, unknown newspaper, RBP, reel 2.
'so she won't' QM, *New York Sun*.
'He cheers me' AL, *Star-Eagle*.
'I'm such' Ibid.
'While other girls' KS, *Graphic*.
'You don't know' Ibid.
'It's pretty hard' Josephine Schaub, *New York Sun*.
'unbias opinion' Flinn, letter to KS, 7 Dec 1925, RBP, reel 3.
'radium could' Berry, legal notes, RBP, reel 1.
'[We] all thought' AL complaint, RBP, reel 1.
SUITS ARE SETTLED *Newark Evening News*, 4 May 1926
'Mr Carlough' Ibid.
'Gentlemen' Gottfried, letter to USRC, 6 May 1926, RBP, reel 3.
'I have your' Stryker, letter to Gottfried, 15 June 1926, RBP, reel 1.
'refused to do' Berry, legal notes, RBP, reel 1.
'not unpopular' Cited in Clark, *Radium Girls*, p.116.
'hand to mouth' Cited in 'Historic American Buildings Survey'.
'jumpy and nervous' Cited in *Newark Ledger*.
'I feel we' GF, *Graphic*.
'An industrial' Flinn, 'Radioactive Material: An Industrial Hazard?' *JAMA*, Dec 1926.
'more bias' Hoffman, deposition, 25 Aug 1927, RBP, reel 1.
'I cannot but' Flinn, letter to Drinker, 16 Jan 1926, RBP, reel 3.
'Your blood' GF affidavit, 18 July 1927, RBP, reel 1.
'He told me' GF, court testimony, 12 Jan 1928.
'If you will' Dialogue from USRC transcripts of this meeting, 19 May 1926, RBP, reels 2 and 3.
All remaining quotations in chapter, Ibid.

Chapter Twenty-Three

'They were best' Interview with Don Torpy.
'real quiet' Interview with James Donohue.

'He was the' Interview with Kathleen Donohue Cofoid.
'People didn't travel' Interview with James Donohue.
'not an extrovert' Interview with Mary Carroll Cassidy.
'They were both' Ibid.
'I quit ten' Elizabeth Frenna, CHR.
'I remember when' Mr Callahan, CHR.
'face [that] broke' Edna Mansfield, CHR.
'I quit because' Goldie King, CHR.
'terrible, excruciating' Mrs Clarence Monsen, CHR.
'Though we had' Ibid.
'I had friends' Ibid.
'awkward' CD, court testimony, *Chicago Herald-Examiner*, 12 Feb 1938.
'clumsy to handle' CD, court testimony, *Ottawa Daily Republican-Times*, 11 Feb 1938.
'We were watched' Hazel McClean, CHR.
'Supervisor wasn't' Ida Zusman, CHR.
'went back' Hazel McClean, CHR.
'We had our' CD, court testimony, legal brief.
'Those who were' Dr Brues, CHR.
'The company left' CD, court testimony, *Chicago Herald-Examiner*, 12 Feb 1938.

Chapter Twenty-Four

'in rather a' Martland, cited by Hoffman, letter to GF, 9 Dec 26, RBP, reel 3.
'in a spirit' Hoffman, letter to Roeder, 6 Nov 26, RBP, reel 2.
'Mr Roeder is' Lee, letter to Hoffman, 16 Nov 26, RBP, reel 2.
'You must take' Hoffman, letter to GF, 9 Dec 26, RBP, reel 3.
'X-rays taken' Humphries, court testimony, 25 April 1928.
'destruction of' GF medical notes, RBP, reel 2.
'crushing and' GF medical history, RBP, reel 3.
'Radium eats' GF, *Orange Daily Courier*.
'I can hardly' Ibid.
'We regret' Hood, Lafferty and Campbell, letter to GF, 24 March 27, RBP, reel 2.
'[Dr Martland] agrees' Hoffman, letter to GF, 9 Dec 1926, RBP, reel 3.
'It's awfully hard' GF, *Orange Daily Courier*.
'a very estimable' Hoffman, letter to Roeder, 6 Nov 1926, RBP, reel 2.
'carelessly and' GF legal complaint, 18 May 1927, RBP, reel 1.
'became impregnated' Ibid.
'continually' Ibid.
'Plaintiff demands' Ibid.

HER BODY *Newark Evening News*, 20 May 1927.
'James McDonald' QM and James McDonald legal complaint, RBP, reel 1.
'I do what' QM, *World*.
'Life' AL, *Newark Ledger*.
'I heard' EH, court testimony, 12 Jan 1928.
'peaceful and resigned' *World*.
'I'm religious' EH, *World*.
'[I] feel' Ibid.

Chapter Twenty-Five

'conspiracy' USRC memo.
'Berry's outfit' Ibid.
'guilty of contributory' USRC legal reply, 20 July 1927, RBP, reel 1.
'it gave no' Ibid.
'denie[d] that' Ibid.
'considerable information' USRC memo, 20 July 1927.
'Miss Rooney says' Ibid.
'Miss Rooney seems' Ibid.
'Miss Rooney claimed' Ibid., 25 July 1927.
'You know how' Interview with Art Fryer.
'Kindly let us' USRC lawyers Collins & Collins, letter to Berry, 7 June 1927, RBP, reel 3.
'So far as' Berry, letter to Collins & Collins, 8 June 1927, RBP, reel 3.
'Although its employees' Wiley affidavit, July 1927, RBP, reel 1.
'He does not' Secretary of Cecil K. Drinker, letter to Berry, 7 July 1927, RBP, reel 3.
'While I thoroughly' Martland, letter, 3 Nov 1927, HMP.
'Had anybody' Workmen's Compensation Commission, Connecticut, letter to Berry, 27 Dec 1927, RBP, reel 3.
'willing to play' Clark, *Radium Girls*, p.141.
'In these negotiations' DeVille and Steiner, 'New Jersey Radium Dial Workers'.
'double-dealing' Martland's opinion, cited by Berry, letter to Hamilton, 6 Jan 1928, RBP, reel 3.
'[I] understood' Dr McCaffrey, cited by Berry, letter to Dr St George, 24 Feb 1928, RBP, reel 3.
'Our records' New Jersey Board of Medical Examiners, letter to Berry, 29 Sep 1927, RBP, reel 3.
'a fraud' Consumers League, letter to *World*, 25 March 1929, NCL files, Library of Congress.

Chapter Twenty-Six

'strong and robust' George Weeks, letter to Kjaer, CHR.

'I never wanted' Nellie Cruse, *Chicago Daily Times*, 18 March 1936.

'popular young' Ella Cruse obituary, *Ottawa Daily Republican-Times*, 6 Sep 1927.

'The next day' Nellie Cruse, *Chicago Daily Times*, 18 March 1936.

'That's all bunk' Cruse family doctor, quoted in Ibid.

'The next day' Ibid.

'She suffered' Ibid.

'Streptococcic poisoning' Ella Cruse death certificate.

'Miss Cruse's' Ella Cruse obituary, *Ottawa Daily Republican-Times*, 6 Sep 1927.

'Life has never' Nellie and James Cruse, *Chicago Daily Times*, 18 March 1936.

'She had been' Ella Cruse obituary, *Ottawa Daily Republican-Times*, 6 Sep 1927.

Chapter Twenty-Seven

'a real villain' Wiley, cited in Clark, *Radium Girls*, p.111.

'consider very' Hamilton, cited in Ibid., p.110.

'What you mean' Flinn, cited in Ibid.

'impossible to' Hamilton, cited in Ibid., p.111.

'never lack' *World* mission statement, cited in Mark Neuzil and William Kovarik, *Mass Media and Environmental Conflict* (Sage Publications, 1996, revised 2002).

'intolerable' *World*, 20 July 1927.

'despicable' Ibid.

'It is scarcely' Ibid.

'the Court of' Unknown newspaper, HMP.

'We should also' Collins & Collins, letter to Berry, 7 June 1927, RBP, reel 3.

'The deposit' Martland, unknown newspaper, RBP, reel 2.

'to free it' *Orange Daily Courier*, 17 Oct 1927.

'The outer box' Ibid.

'unmistakeable signs' Florence E. Wall article, Orange Public Library.

'The body' *Orange Daily Courier*, 17 Oct 1927.

'washed [her bones]' Amelia Maggia form, Kjaer study, CHR.

'No evidence' Initial Amelia Maggia autopsy report, 3 Nov 1927, RBP, reel 3.

'Each and every' Amelia Maggia form, Kjaer study, CHR.

'I have spent' Ella Eckert, *New York Evening Journal*.

'badly swollen' Ella Eckert form, Kjaer study, CHR.
'We object' Court transcript, 14 Nov 1927.
Following quotations, Ibid.
'near death' Berry, letter to Hamilton, 7 Dec 1927, RBP, reel 3.
'This case is' Martland, *Newark Evening News*.
'calcareous formation' Ella Eckert form, Kjaer study, CHR.
'considerable size' Berry, letter to Hamilton, 15 Dec 1927, RBP, reel 3.

Chapter Twenty-Eight

'I could hardly' KS, autobiography, p.2.
'The condition of' Berry, letter to Hamilton, 7 Dec 1927, RBP, reel 3.
'I have lost' AL, *Sunday Call*.
'therapeutically induced' AL medical history, 17 Dec 1927, RBP, reel 2.
'I've been so' AL, *Sunday Call*.
'destroys' Humphries, *New York Telegram*.
'abnormal angle' EH form, Kjaer study, CHR.
'I cannot keep' EH, *Newark Ledger*.
'I do what' EH, *Sunday Call*.
'The worst thing' Ibid.
'Recovery?' Court transcript, 4 Jan 1928.
'Isn't the' Ibid.
'Of what importance?' Ibid., 12 Jan 1928.
Following quotations in courtroom, Ibid.
'chameleon-hued' Berry, letter to Hamilton, 21 Feb 1928, RBP, reel 3.
'Grace is so' *Orange Daily Courier*, 30 April 1928.
'We were instructed' Court transcript, 12 Jan 1928.
Following quotations in courtroom, Ibid.
'Everything was going' KS, autobiography, p.2.
'I notice you' Court transcript, 12 Jan 1928.
'Trouble with my' Ibid.
'Tomorrow there would' KS, autobiography, p.2.
'I was awakened' Ibid.
'A Russian doctor' GF, *Sunday Call*.
'I face the' QM, *Star-Eagle*.
'to turn aside' Unknown newspaper, Dec 1929, CHR.
'I haven't had' KS, letter to Berry, Feb 1928, RBP, reel 3.
'safe from intrusion' Ibid.
'really a friend' Berry, letter to Hamilton, 21 Feb 1928, RBP, reel 3.
'clandestine' Berry, letter to Flinn, 1 Feb 1928, RBP, reel 3.
'impudent' Flinn, letter to Berry, 2 Feb 1928, RBP, reel 3.
'The other' Ibid.

'very close' Berry, letter to Krumbhaar, 23 Nov 1928, RBP, reel 3.
'paper-thin skin' Dial-painter's relative, capitolfax.com, 6 Sep 2011.
'so that the' Berry, letter to Krumbhaar, 21 Nov 1928, RBP, reel 3.
'held the instrument' Ibid.

Chapter Twenty-Nine

'I ascended' KS, autobiography, p.3.
'The young lady' Court transcript, 25 April 1928.
Following quotations in courtroom, Ibid.
'sadly smiling' *Star-Eagle*.
'maintained an attitude' Unknown newspaper, RBP, reel 2.
'The [women] listened' *Star-Eagle*.
'Were you ever' Court transcript, 25 April 1928.
Following quotations in courtroom, Ibid.
'She exhibited a' *Orange Daily Courier*, 25 April 1928.
'What is your' Court transcript, 25 April 1928.
Following quotations in courtroom, Ibid.
'I thought it' KS, autobiography, p.3.
'It had to' Ibid.
'I do not' Court transcript, 25 April 1928.
Following quotations in courtroom, Ibid., 25–26 April 1928.
'His forthright' *Star-Eagle*.
'star witness' Ibid.
'excruciating' Ibid.
'As she listened' Ibid.
'out of two' Court transcript, 26 April 1928.
Following quotations in courtroom, Ibid., 26–27 April 1928.
'We should' USRC memo, 20 July 1927.
'Well, Mr Berry' Court transcript, 27 April 1928.
Following quotations in courtroom, Ibid.
'heartless and' KS, autobiography, p.3.

Chapter Thirty

'down on the' *Newark Ledger*.
'Grace' Grace Fryer Snr., quoted in Ibid.
'it would hardly' *Newark Evening News*.
'Girl Alone' *Orange Daily Courier*.
'impossible' Edward Markley, letter to Berry, 5 May 1928, RBP, reel 3.
'going abroad' Ibid.
'I am sure' Berry, letter to Markley, 10 May 1928, RBP, reel 3.

'far from' Berry, letter to Charles Norris, 5 May 1928, RBP, reel 3.
'These girls are' Dr Gettler, sworn statement, 9 May 1928, RBP, reel 1.
'kept under a' *World*.
'We confidently' Ibid.
'Open the courts' Allan C. Dalzell, *News*, HMP.
'the conscience' Cited in Mullner, *Deadly Glow*, p.83.
'vivid example' Norman Thomas, unknown newspaper, RBP, reel 2.
'Everywhere' KS, autobiography, p.3.
'Letters came' Ibid.
'Radium could not' T. F. V. Curran, letter to QM, 24 May 1928, RBP, reel 1.
'For $1,000' Missouri woman, cited in *Star-Eagle*.
'scientific baths' Ibid.
'If not' Ibid.
'It is not' Electric blanket firm, letter to Berry, 31 May 1928, RBP, reel 3.
'Don't write all' QM, *New York Sun*.
'still living' GF, *Graphic*.
'I am facing' Ibid.
'Don't think' KS, *Sunday Call*.
'seemed to destroy' Clark, *Radium Girls*, p.130.
'Personally' Markley, letter to Berry, 12 May 1928, RBP, reel 3.
'I am surprised' Berry, letter to Markley, 14 May 1928, RBP, reel 3.
'there is no' Flinn, *World*.
'all the appearance' *World*.
'It is not' Ibid.
'exploited by a' Markley, Ibid.
'When I die' KS, unknown newspaper, HMP.
'So many of' KS, *Graphic*.
'I couldn't say' GF, Ibid.
'My body means' Ibid.
'Can't you' Ibid.
'It is not' *Newark Ledger*.
'therefore, the' Clark, *Radium Girls*, p.134.
'I will set' Judge Mountain, letter to Berry, 28 May 1928, RBP, reel 3.

Chapter Thirty-One

'To Judge' Berry's diary, 23 May 1928, CHR.
'I know nothing' Berry, unknown newspaper, RBP, reel 2.
'more determined' Ibid.
'not physically' Humphries, *Newark Ledger*.
'I don't dare' GF, unknown newspaper, CHR.
'Pains like streaks' KS, *Graphic*.

'Just because I' Judge Clark, *Newark Evening News*.

'the directors' Barker, Ibid.

'We absolutely' Barker, *Newark Ledger*.

'cleverly designed' USRC, cited in Clark, *Radium Girls*, p.136.

'the human aspect' Ibid.

'There is no' Judge Clark, unknown newspaper, RBP, reel 2.

RADIUM VICTIMS *Orange Daily Courier*, 2 June 1928.

'I will not' GF, *Star-Eagle*.

'I have two' QM, Ibid.

'absolutely refuse' GF, Ibid.

'smiling sorority' Ibid.

'all aisles and' Ibid.

'You can say' Judge Clark, *Newark Evening News*.

'[The firm] was' Markley, letter to Judge Clark, unknown newspaper, RBP, reel 2.

'humanitarian' Ibid.

'[USRC] hopes' Ibid.

'If any two' Berry, letter to Hamilton, 6 June 1928, RBP, reel 3.

'I fully believe' Ibid.

'a very honourable' Berry, letter to Drinker, 6 June 1928, RBP, reel 3.

'was friendly' Berry, letter to Goldmark, 12 Dec 1947, RBP, reel 3.

'possibly had some' Ibid.

'is, or was' Berry, letter to Drinker, 6 June 1928, RBP, reel 3.

'I have' Berry, letter to Norris, 6 June 1928, RBP, reel 3.

'I want to' Judge Clark, letter to five women, unknown newspaper, RBP, reel 2.

'I am glad' AL, *Newark Ledger*.

'The settlement' QM, *Newark Ledger* and *World*.

'dissatisfied' QM, *Star-Eagle*.

'I am glad' QM, Ibid.

'I think Mr' EH, Ibid. and *Newark Ledger*.

'God has heard' KS, *Newark Ledger*.

'quite pleased' GF, *World*.

'I'd like to' GF, *Newark Ledger*.

'It is not' GF, unknown newspaper, RBP, reel 2.

'You see' GF, *Graphic*.

Chapter Thirty-Two

MORE DEATHS *Ottawa Daily Republican-Times*, 4 June 1928.

'The girls became' CD, *Ottawa Daily Republican-Times*, 14 March 1936.

'they separated' MR, quoted in interview with Dolores Rossiter.

'When I asked' CD, court testimony, unknown newspaper, lgrossman.com.

'Why, my dear' Mr Reed, cited in CD testimony, *Chicago Daily Tribune*, 11 Feb 1938.

'Neither of us' CD, court testimony, *Chicago Herald-Examiner*, 11 Feb 1938.

'Don't worry' Mr Reed, cited in Ibid. and legal brief.

'Are the workers' MR, court testimony, *Chicago Daily Tribune*, 12 Feb 1938.

'You don't' Mr Reed, cited in Ibid.

'We have at' Statement by the Radium Dial Company, *Ottawa Daily Republican-Times*, 7 June 1928.

'If their reports' Ibid.

In view of . . . Ibid.

'expert' Ibid.

'The tests would' Berry, letter to Hoffman, 2 June 1928, RBP, reel 3.

'He said that' CD, court testimony, *Chicago Herald-Examiner*, 12 Feb 1938.

'Radium will put' Mr Reed, cited in MR testimony, Ibid.

'Radium will make' Mr Reed, cited in Marguerite Glacinski testimony, Ibid.

'ever-watchful' *Ottawa Daily Republican-Times*, 11 June 1928.

'They went to' Interview with Dolores Rossiter.

'The girls' Mr Etheridge, cited in Stevie Croisant, Abby Morris, Isaac Piller, Madeline Piller, Haley Sack, 'Radium Girls: The Society of the Living Dead', LaSalle County Historical Society & Museum.

Chapter Thirty-Three

'I could find' KS, autobiography, p.3.

'like Cinderella' Ibid.

'I bought the' Ibid.

'Not a cent' GF, *Popular Science*, July 1929.

'What for?' Journalist, Ibid.

'For the future!' GF, Ibid.

'In my opinion' VS, *Orange Daily Courier*, 9 June 1928.

'two kinds of' Clark, *Radium Girls*, p.118.

'survived the early' Ibid.

'I am of' Martland, letter to Robley Evans, 13 June 1928, HMP.

'Someone may find' GF, *Graphic*.

'the dream of' *Popular Science*, July 1929.

'a long, leisurely' Ibid.

'We have a' EH, postcard to Berry, 20 June 1928, RBP, reel 1.

'There is no' Stewart, *World*, 17 July 1928.

'The new methods' Ibid.

'real country life' KS, autobiography, p.3.

'splendid' Berry, letter to KS, 11 Sep 1928, RBP, reel 1.

'a vacation like' KS, letter to Berry, 15 July 1928, RBP, reel 3.

'I loved to' KS, autobiography, p.3.

'I myself know' KS, letter to Berry, 15 July 1928, RBP, reel 3.

'I am writing' KS, letter to Martland, 28 June 1928, HMP.

'I think that' Berry, letter to Hughes, 14 June 1928, RBP, reel 3.

'so-called radium' Lee, letter to Department of Health, 18 June 1928.

'[He] must be' Norris, letter to Berry, 2 July 1928, RBP, reel 3.

'closely allied' Martland, letter to Norris, 3 July 1928, HMP.

'impossible' Ibid.

'The damage is' Ibid.

'obviously very' Krumbhaar, court testimony, 27 June 1929.

'distinct limitation' Ibid.

'exposed' Ibid.

'Mrs Larice' Ibid.

'proved positive' Ewing, letter to Krumbhaar, 19 Oct 1928, RBP, reel 2.

'The question' Ibid.

'intimate friend' Berry, letter to USRC, 23 Nov 1928, RBP, reel 3.

'assisted' Court transcript, 27 June 1929.

'great suspicion' Berry, letter to USRC, 23 Nov 1928, RBP, reel 3.

'constitutes a breach' Ibid.

'All five patients' Dr Failla, test results, 20 Nov 1928, RBP, reel 2.

'felt funny' Mae Cubberley Canfield, examination before trial, 17 Jan 1929, RBP, reel 2.

'improper diet' Flinn, 'Elimination of Radium', *JAMA* (vol. 96, no. 21, 23 May 1931).

'tendency' Ibid.

'Without his' Martland, *Star-Eagle*.

'veritable Frankenstein' Unknown newspaper, RBP, reel 2.

'He died' Martland, *Star-Eagle*.

'anything we draw' US Surgeon General, transcript of the national radium conference, 20 Dec 1928, RBP, reel 3.

'a whitewash' Florence Kelley, letter to Wiley, 2 Jan 1929, NCL files, Library of Congress.

'The luminous watch' Stewart, radium-conference transcript, 20 Dec 1928, RBP, reel 3.

'cold-blooded' Kelley, letter to Wiley, 2 Jan 1929, NCL files, Library of Congress.

'The Radium Corp.' Berry, letter to Krumbhaar, 5 Dec 1928, RBP, reel 3.

'My advice to' Delegate, radium-conference transcript, 20 Dec 1928, RBP, reel 3.

Chapter Thirty-Four

'Only once' CD, court testimony, *Ottawa Daily Republican-Times*, 10 Feb 1938.

'Further steps' Kjaer report, CHR.

'intention' Joseph A. Kelly, letter to Kjaer, 22 March 1929, CHR.

'to assist you' Ibid.

'I feel that' Kjaer, letter to Weeks, 15 March 1929, CHR.

One dial-painter ... Kjaer, 'Radium Poisoning: Industrial Poisoning from Radioactive Substances', *Monthly Labor Review* (date unknown), CHR.

'a high standard' Statement by the Radium Dial Company, *Ottawa Daily Republican-Times*, 7 June 1928.

'Chuck used to' Edith Schomas, *Radium City*.

'Chuck felt awful' Jane Raub, Ibid.

'We'd all be' Interview with Jean Schott.

'My parents took' Jean Schott, *Radium City*.

'She knew she' Ethel Looney, *Chicago Herald-Examiner*, 18 March 1936.

'Well, Mother' Margaret Looney, cited in Ibid.

'We had no' Edith Schomas, *Radium City*.

'Radium Dial probably' Interview with Darlene Halm.

'I went one' Jane Raub, *Radium City*.

'No way is' Jack White, quoted by Edith Schomas, Ibid.

'They wanted the' Darlene Halm, *Buffalo News*, 11 Oct 1998.

'the flat bones' Margaret Looney exhumation autopsy report, 1978, CHR.

'very strongly' Ibid.

'removed by' Ibid.

'The teeth are' Margaret Looney original autopsy report, 27 Aug 1929.

The young woman's ... Margaret Looney obituary, *Ottawa Daily Republican-Times*, 16 Aug 1929.

'Miss Looney's' Ibid.

'It just killed' Interview with Jean Schott.

'She knew' Interview with Darlene Halm.

Chapter Thirty-Five

'unsatisfactory' Ewing, letter to Krumbhaar, 30 April 1929, RBP, reel 2.

'these women are' Ibid.

'the potent argument' Berry, letter to Krumbhaar, 18 June 1929, RBP, reel 3.

'purely a' Ibid.

'cautious about' Ewing, letter to Krumbhaar, 17 April 1929, RBP, reel 2.

'hostile attitude' Berry, letter to Krumbhaar, 5 Dec 1928, RBP, reel 3.

'We are quite' Ewing, draft letter to Berry, 7 May 1929, RBP, reel 2.

'it is very' Louis Hussman, cited by Berry, letter to Krumbhaar, 18 June
    1929, RBP, reel 3.
'People are now' GF, *Star-Eagle*.
'The doctors don't' KS, letter to Berry, 8 Nov 1928, RBP, reel 3.
'the jewel' KS, letter to Berry, 7 March 1929, RBP, reel 3.
'hollyhocks' KS, autobiography, p.3.
'the picture of' *Popular Science*, July 1929.
'in no way' Markley, court transcript, 27 June 1929.
'furious' Server of legal papers, Miller & Chevalier, letter to Berry, 22 June
    1929, RBP, reel 3.
'paper fortunes' *Chicago Daily Tribune*, Oct 1929.
'Wall Street' *New York Times*, 30 Oct 1929.
'in a dying' *Newark Ledger*.
'She was a' Ethel Brelitz, unknown newspaper, RBP, reel 2.
'Her one thought' James McDonald, Ibid.
'My husband' QM, *World*.
'For the past' Ethel Brelitz, unknown newspaper, RBP, reel 2.
'Doctors told me' GF, *Newark Ledger*.
'Each time' GF, *Orange Daily Courier*.
'I'm tired' QM, cited in *Newark Ledger*.
'Would you mind' QM, cited in unknown newspaper, RBP, reel 2.
'tears streaming' Unknown newspaper, CHR.
'I am heartbroken' James McDonald, unknown newspaper, RBP, reel 2.
'to find forgetfulness' KS, autobiography, p.4.
'For a time' Ibid.
'walked briskly' *Newark Evening News*, 10 Dec 1929.
'seemed to' Ibid.
'apparently' Ibid.
'kept close' *Newark Ledger*.
'she could thus' Ethel Brelitz, unknown newspaper, RBP, reel 2.
'The bones of' Martland, cited in *Newark Ledger*.
'I agree not' Terms of Mae Cubberley Canfield settlement, 8 March 1930,
    RBP, reel 1.

Chapter Thirty-Six

'luminous bright' CD, court testimony, *Ottawa Daily Republican-Times*, 10
    Feb 1938.
'I believe' Bob Bischoff, CHR.
'I was excluded' CD, court testimony, legal brief.
'I knew I' Ibid.
'The family' Catherine White, conversation with Brues, 29 Jan 1973, CHR.

'He couldn't do' Jane Raub, *Radium City*.

'My dad said' Interview with Jean Schott.

'Just forget it' Michael Looney, quoted by Jane Raub, *Radium City*.

'The doctor said' Alphonse Vicini, *Chicago Daily Times*, 18 March 1936.

'She suffered' Ibid.

'We thought' Ibid.

Chapter Thirty-Seven

'didn't knit' KS, letter to Martland, 5 Oct 1930, HMP.

'A lump came' KS, autobiography, p.4.

'having difficulty' KS, letter to Martland, 5 Oct 1930, HMP.

'My head' Ibid.

'this craving' Ibid.

'They say' Ibid.

'Relations are far' Krumbhaar, letter to Berry, 2 Oct 1930, RBP, reel 3.

'It might be' Ibid.

'kicking' Ewing, letter to Krumbhaar, 17 April 1929, RBP, reel 2.

'My impression' Art Fryer Snr., interviewed by his grandson, quoted by Art Fryer Jnr. in interview.

'I work' GF, *Graphic*.

'Her whole leg' Mary Freedman, court testimony, 26 Nov 1934.

'a vaginal examination' Martland, court testimony, 6 Dec 1934.

'terrific' Dr Shack, court testimony, 26 Nov 1934.

'I found a' Martland, court testimony, 6 Dec 1934.

'He told me' Vincent La Porte, court testimony, 26 Nov 1934.

'She was always' Shack, court testimony, 26 Nov 1934.

'When I first' Martland, 'The Danger of', HMP.

'I am now' Martland, cited in Mullner, *Deadly Glow*, p.72.

'a huge growth' Martland, court testimony, 6 Dec 1934.

'you couldn't take' Ibid.

'I believe' Martland, letter to Sir Humphrey Rolleston, 2 April 1931, HMP.

Chapter Thirty-Eight

'causing talk' CD, court testimony, multiple newspaper sources.

'I'm sorry' Cited in CD, court testimony, legal brief.

All remaining quotations in chapter, Ibid.

Chapter Thirty-Nine

'I have been' KS, autobiography, p.4.

'We wish to' Ewing, letter to KS, GF, AL and EH, 10 Feb 1932, RBP, reel 3.
'we do not' Ibid.
'to prevent this' Krumbhaar, letter to Berry, 12 Feb 1932, RBP, reel 3.
'I have suffered' KS, letter to Humphries, 28 Oct 1931, HMP.
'[She is] one' Dr May, letter to Dr Craver, 7 Nov 1931, cited in Sharpe, 'Radium Osteitis'.
'I have made' Humphries, letter to Craver, 6 Dec 1932, cited in Ibid.
'the very depressed' Ewing, letter to KS, GF, AL and EH, 10 Feb 1932, RBP, reel 3.
'magnificent home' Unknown newspaper, HMP.
THE RADIUM WATER *Wall Street Journal.*
'It is my' KS, autobiography, p.4.
'There is no' Humphries, letter to Craver, 17 Feb 1933, cited in Sharpe, 'Radium Osteitis'.
'adventure' KS, autobiography, p.4.

Chapter Forty

'I feel better' GF, *Newark Evening News.*
'Home always' EH, Ibid.
'quite a few' EH, *New York Sunday News*, 13 Feb 1938.
'He helps me' EH, unknown newspaper, CHR.
'What good would' EH, *Newark Evening News.*
'I know they' AL, *Sunday Call.*
'I am not' GF, *Newark Ledger.*
'It was not' Grace Fryer Snr., unknown newspaper, CHR.
'radium sarcoma' GF death certificate.
'The family' Interview with Art Fryer Jnr.
'I will never' Art Fryer Snr., interviewed by his grandson, quoted by Art Fryer Jnr. in interview.

Part Three: Justice

Chapter Forty-One

'pure radium only' Statement by the Radium Dial Company, *Ottawa Daily Republican-Times*, 7 June 1928.
'no visible indication' Margaret Looney obituary, *Ottawa Daily Republican-Times*, 16 Aug 1929.
'God has sure' CD, letter to PP, 29 April 1938, PPC.
'None of Tom's' Interview with Mary Carroll Cassidy.
'one of the' *Ottawa Daily Republican-Times*, 22 Jan 1932.

'They were just' Interview with Don Torpy.
'It has never' Dr Pettit, letter to Hamilton, 22 June 1931, HMP.
'It wasn't a' Interview with James Donohue.
'We were so' Tom Donohue, *Chicago Daily Times*, 11 Feb 1938.
'repeatedly advised her' Leonard Grossman, legal brief.
'scoffed' Susie Duffy, *Chicago Daily Times*, 18 March 1936.
'none of the' *Chicago Herald-Examiner*, 18 March 1936.
'They didn't want' Interview with Don Torpy.
'They were all' Interview with Jean Schott.
'It was confusing' Interview with Mary Carroll Cassidy.

Chapter Forty-Two

'The local doctors' Al Purcell, *Chicago Daily Times*, 19 March 1936.
'sharp, knife-like' CP account of illness, medical notes, CHR.
'in a frenzy' Grossman, legal brief.
'reputable medical' Ibid.
'a toxic quality' Ibid.
'terrible pain' Al Purcell, *Chicago Daily Times*, 19 March 1936.
'fifteen Chicago' Ibid.
'She got them' Interview with Felicia Keeton.
'hollow' Helen Munch, *Toronto Star*, 23 April 1938.
'wanted to be' Ibid.
'Now I have' Ibid.
'I'll tell you' Olive Witt, Ibid.
'she could move' *Chicago Daily Times*, 17 March 1936.
'would love to' MR, *Toronto Star*, 23 April 1938.
'Charlotte never felt' Interview with Jan Torpy.
'We finally took' Al Purcell, *Chicago Daily Times*, 19 March 1936.
'There was no' Interview with Jan Torpy.
'Dr Davison says' Al Purcell, *Chicago Daily Times*, 19 March 1936.
'helpless' *Toronto Star*, 23 April 1938.
'My husband' CP, Ibid.
'She still feels' Al Purcell, *Chicago Daily Times*, 19 March 1936.
'There is some' Ibid.

Chapter Forty-Three

'Tom was' Interview with Mary Carroll Cassidy.
'After that' Tom Donohue, speaking at CD inquest, 28 July 1938.
'I firmly believe' PP, letter to Catherine O'Donnell, 23 June 1938, PPC.

'the first time' Tom Donohue, court testimony, cited in Clark, *Radium Girls*, p.184.

'I saw him' Ibid.

'very angry' Interview with Jan Torpy.

'I told him' Dr Charles Loffler, court testimony, cited in Mullner, *Deadly Glow*, p.105.

'Nothing even approaching' Statement by the Radium Dial Company, *Ottawa Daily Republican-Times*, 7 June 1928.

'Very suspicious' Radium Dial 1928 test results, CHR.

'the knowledge of' Bureau of Labor Statistics report, 1929, cited in Conroy and Mullner.

'virtually on charity' *Chicago Daily Times*, 7 July 1937.

'The Dial people' Susie Duffy, Ibid., 18 March 1936.

'spokesmen for' CD, court testimony, *Chicago Herald-Examiner*, 12 Feb 1938.

'She never felt' Interview with Jan Torpy.

'I can't do' CP, *Toronto Star*, 23 April 1938.

'I have received' CD, court testimony, legal brief.

'cultured voice' *Chicago Daily Times*, 13 Feb 1938.

'He has come' CD, court testimony, legal brief.

'We have' Ibid.

'Having consulted' Ibid.

'I don't think' Mr Reed, quoted in CD testimony, *Chicago Herald-Examiner*, 12 Feb 1938.

'There is nothing' Mr Reed, quoted in CD testimony, cited in Mullner, *Deadly Glow*, p.101.

'He refused' CD, court testimony, *Chicago Herald-Examiner*, 12 Feb 1938.

'Mary's was the' Susie Duffy, *Chicago Daily Times*, 18 March 1936.

'steadfastly refused' Mullner, *Deadly Glow*, p.100.

'no' Mary Robinson death certificate.

'Was disease' Ibid.

'bitterly resented' *Chicago Daily Times*, 7 July 1937.

'They weren't' Interview with Dolores Rossiter.

'Margaret Looney was' Bob Bischoff, CHR.

'Some of them' Olive Witt, *Chicago Daily Times*, 28 July 1938.

'business interests' Hobart Payne, letter to Clarence Darrow, 17 May 1937, PPC.

Chapter Forty-Four

'It came to' PP, 'History of Record of Illness' and 'Life History', PPC.

'There were big' Hobart Payne, *Toronto Star*, 23 April 1938.

'for drainage' PP, 'History of Record of Illness', PPC.
'one side of my' Ibid.
'curettement' Ibid.
'During this time' Ibid.
'I knew this' Ibid.
'five years of' PP, letter to Catherine O'Donnell, 23 June 1938, PPC.
'Dearest Sweetheart' PP, letter to Hobart Payne, 15 June 1932, PPC.
'I am unable' PP, 'History of Record of Illness', PPC.
'belonged to a' Ibid.
'I notified my' Ibid.
'[In] July 1933' Ibid.
'I was attacked' Ibid.
'I believed' PP, 'Life History', PPC.
'There were very' Warren Holm, CHR.
'informed them that' Quigg, *Learning to Glow*.
'a fiercely' Holm, CHR.
'We always called' Interview with Mary Carroll Cassidy.

Chapter Forty-Five

'She was a' CD, *Chicago Herald-Examiner*, 18 March 1936.
'vague, indefinite' Supreme Court judgment, Inez Vallat vs Radium Dial
    Company, 1935.
'When Attorney Cook' *Chicago Daily Times*, 7 July 1937.
'The court ruled' *Ottawa Daily Republican-Times*, 18 April 1935.
'an almost' *Chicago Daily Times*, cited in Mullner, *Deadly Glow*, p.102.
'There was never' *Chicago Daily Times*, 8 July 1937.
'I hated to' Jay Cook, Ibid., 9 July 1937.
'There are medicines' CD, *American*, 6 April 1938.
'We never talk' Tom Donohue, *Chicago Daily Times*, 19 March 1936.
'We're so happy' CD, Ibid., 13 Feb 1938.
'fainted during the' Dr Walter Dalitsch, court testimony, legal brief.
'she was looking' Interview with Ross Mullner.
'[Once Irene's]' Mr Smith, court transcript, 3 Dec 1934.
Naturally ... Judgment, Irene La Porte vs USRC, 17 Dec 1935.
'The [case] must' Ibid.

Chapter Forty-Six

'haemorrhage' Inez Vallat death certificate.
'Mr Vallat' Frances O'Connell, CHR.
'rapidly becoming' *Chicago Daily Times*, 22 May 1936.

'The Suicide Club' Used by several newspapers nationwide, including *Ottawa Daily Republican-Times* and *Chicago Daily Times*.

'Unfortunately, any' Senator Mason, *Ottawa Daily Republican-Times*, 14 March 1936.

'We'll always be' PP, *Chicago Daily Times*, 11 Feb 1938.

'Chicago's Picture' *Chicago Daily Times* banner heading, 1930s.

'They shoot to' Mary Doty, *Chicago Daily Times*, 17 March 1936.

'dying off' Ibid.

'Some [girls]' Ibid.

'wizened little' Ibid., 19 March 1936.

'match-thin' Ibid.

'Her parents' Ibid.

'I am in' CD, *Chicago Herald-Examiner*, 18 March 1936.

'it brought tears' Ibid.

'I'm frightened' MR, Ibid.

'the [Chicago]' MR, *Toronto Star*.

'Having three babies' Interview with Jan Torpy.

'She waits hopefully' Doty, *Chicago Daily Times*, 19 March 1936.

'[It] will never' Tom Donohue, Ibid.

'Mom used to' Interview with Donald Purcell.

'telephone conversations' *Chicago Daily Times*, 22 May 1936.

'This day' Tom Donohue, court testimony, cited in Clark, *Radium Girls*, p.184.

Following quotations, Ibid.

'Irish temper' Interview with Kathleen Donohue Cofoid.

'I don't think' Ibid.

'I swang at' Tom Donohue, court testimony, cited in Clark, *Radium Girls*, p.184.

'got excited' Ibid.

'fisticuff encounter' *Chicago Daily Times*, 22 May 1936.

Chapter Forty-Seven

'controlling interests' Hobart Payne, letter to Clarence Darrow, 17 May 1937, PPC.

'vigorously opposed' Ibid.

'persecuted' Ibid.

'typical of' Interview with Mary Carroll Cassidy.

'They know that' Ibid.

'run out of' Andrew Stehney, CHR.

'They struggled' Interview with Jan Torpy.

'You took whatever' Interview with Mary Carroll Cassidy.

'We used to have' Interview with Donald Purcell.
'steady work' PP, letter to Catherine O'Donnell, July 1938, PPC.
'Tom was nearly' Brother-in-law, cited in Mullner, *Deadly Glow*, p.106.
'They were on' Interview with Mary Carroll Cassidy.
'timid-looking' *Ottawa Daily Republican-Times*, 5 April 1938.
'Part of her' Hobart Payne, letter to Darrow, 17 May 1937, PPC.
'It was so' Interview with Mary Carroll Cassidy.
'[Catherine's] husband' MR, *Radium City*.
'would always' Interview with Patty Gray.
'If she [thought]' Ibid.
'I said' MR, *Radium City*.
'All through it' Ibid.
'And they said' CP, Ibid.
'They pull back' Olive Witt, *Chicago Daily Times*, 28 July 1938.
'She used to' Interview with Dolores Rossiter.
'We got a' MR, *Radium City*.
'always took the' Interview with Len Grossman.
'Dear Sir' Hobart Payne, letter to Darrow, 17 May 1937, PPC.
RADIUM DEATH *Chicago Daily Times*, 7 July 1937.
'lives in daily' Ibid.
'their last stand' Ibid.
'Without a lawyer' Ibid., 9 July 1937.
'That's what' CD, Ibid.
'The Radium Dial' Ibid., 7 July 1937.
'This is a' Cook, Ibid.
'All they've' Ibid.
OTTAWA RADIUM COMPANY Ibid., 8 July 1937.
'These women's claims' William Ganley, Ibid.
'I can't recall' Ibid.

Chapter Forty-Eight
'formed an organisation' *Ottawa Daily Republican-Times*, 23 July 1937.
200 WOMEN Cited in interview with Len Grossman.
'He had a' Interview with Len Grossman.
Following quotations, Ibid.
'We were at' CD, *Toronto Star*, 23 April 1938.
'My heart is' Grossman, letter to PP, 15 Oct 1938, PPC.
'presided over' *Ottawa Daily Republican-Times*, 23 July 1937.
'familiarise himself' *Chicago Daily Times*, 23 July 1937.
'contend the paint' *Daily Pantagraph*, 24 July 1937.
'a silver-tongued' Interview with Len Grossman.

'weeping' *Daily Pantagraph*, 24 July 1937.
'We should have' Grossman, courtroom proclamation, quoted in Ibid.
'We do not' Ibid.
'It is a' Ibid.

Chapter Forty-Nine

'For God's sake' Grossman, quoted by CD, letter to PP, 7 Dec 1937, PPC.
'I have written' CD, Ibid.
'*Please* help' PP, letter to Dr Elliston, 1 Feb 1938, PPC.
'produce [the results of]' Grossman, legal brief.
'At a great' PP, letter to Grossman, 13 Feb 1938, PPC.
'my next best' Grossman, *Chicago Daily Tribune*, 13 Feb 1938.
'She hasn't long' PP, *Chicago Herald-Examiner*, 27 Feb 1938.
'The strength' Interview with Kathleen Donohue Cofoid.
'My hip is' CD, letter to PP, 7 Dec 1937, PPC.
'Well, I took' Ibid.
'well-worn' *Ottawa Daily Republican-Times*, 5 April 1938.
'timid-looking' Ibid.
'It had good' Interview with Mary Carroll Cassidy.
'Even now' CD, court testimony, *Newark Evening News*, 11 Feb 1938.
'You could see' Interview with James Donohue.
'You don't see' MR, *Radium City*.
'People are afraid' CD, *Chicago Daily Times*, 13 Feb 1938.
'nearly all of' *Ottawa Daily Republican-Times*, 11 Feb 1938.
'He used to' Interview with James Donohue.
'I think it' Interview with Mary Carroll Cassidy.
'They were almost' Ibid.
'I suffer so' CD, letter to PP, 7 Dec 1937, PPC.
'I have so' Ibid.
'It has indeed' Ibid.
'As to my' Ibid.
'We have not' Ibid.
'This is the' Grossman, *Ottawa Daily Republican-Times*, 12 Feb 1938.
'with every' Grossman, Season's Greetings card to PP, Dec 1937, PPC.
'It makes it' CD, letter to PP, 7 Dec 1937, PPC.
'They went back' Interview with Len Grossman.
'a woman appearing' Dr Weiner, court testimony, legal brief.
'destruction' Dalitsch, court testimony, legal brief.
'right through' Ibid.
'displacement of' Ibid.
'considerable discharge' Ibid.

'an alarming' Loffler, court testimony, *Chicago Herald-Examiner*, 11 Feb 1938.

'near death' Loffler, court testimony, *Ottawa Daily Republican-Times*, 11 Feb 1938.

'about the size' Weiner, court testimony, legal brief.

Chapter Fifty

'Mrs Donohue' *Chicago Herald-Examiner*, 11 Feb 1938.

'toothpick woman' *Chicago Daily Times*, 10 Feb 1938.

'We do not' Grossman, legal brief.

'weak and' Unknown newspaper, collection of Ross Mullner.

'faltering' *Chicago Daily Times*, 10 Feb 1938.

'barely audible' *Chicago Herald-Examiner*, 11 Feb 1938.

'That's the way' CD, court testimony, *Chicago Daily Times*, 10 Feb 1938.

'After those' CD, court testimony, *Chicago Herald-Examiner*, 12 Feb 1938.

'After Miss Marie' CD, court testimony, Ibid., 11 Feb 1938.

'paled at' Ibid.

'Oh!' MR, Ibid.

'That is the' CD, court testimony, *Chicago Daily Tribune*, 11 Feb 1938.

'assumed responsibility' Frances Salawa, CHR.

We have not ... Kelly, letter to IIC, 2 Nov 1928, cited in legal brief.

'You can readily' Ibid.

'The only thing' IIC, letter to Kelly, 5 Nov 1928, Ibid.

'was forced to' Ibid.

'cold-blooded' Florence Kelley, letter to Wiley, 2 Jan 1929, NCL files, Library of Congress.

'Mr Reed said' CD, court testimony, *Chicago Herald-Examiner*, 12 Feb 1938.

'He refused' Ibid.

'Her emaciated' *Chicago Daily Times*, 10 Feb 1938.

'After two years' CD, court testimony, *Ottawa Daily Republican-Times*, 10 Feb 1938.

'These are my' CD, court testimony, various newspapers.

Chapter Fifty-One

'shuddered' *Chicago Daily Times*, 10 Feb 1938.

'became sick' Dalitsch, court testimony, legal brief.

'The doctor is' Marvel, court transcript, legal brief.

'The condition' Dalitsch, Ibid.

Following quotations, Ibid.

'glanced meaningfully' *Chicago Daily Tribune*, 11 Feb 1938.

'permanent, incurable' Legal brief.

'In her' Dalitsch, court testimony, legal brief.

'sobbed, slipped' *Chicago Herald-Examiner*, 11 Feb 1938.

'screamed in' *Chicago Daily Times*, 10 Feb 1938.

'broke down' *Chicago Daily Tribune*, 11 Feb 1938.

'would have' Ibid.

'The woman's sobs' *Chicago Herald-Examiner*, 11 Feb 1938.

'feebly wavering' *Chicago Daily Times*, 11 Feb 1938.

'Don't leave me' CD, Ibid.

'She is in' Attending physician, Ibid., 10 Feb 1938.

DEATH IS THE *Detroit Michigan Times*, 14 Feb 1938.

'Is her condition' Court transcript, legal brief.

Following quotations, Ibid.

'She is beyond' Weiner, court testimony, *Ottawa Daily Republican-Times*, 11 Feb 1938.

'She has but' Loffler, court testimony, *Chicago Daily Tribune*, 11 Feb 1938.

'There is a' Loffler, court testimony, legal brief.

'Radioactive substances' Arthur Magid, court transcript, *Chicago Daily Tribune*, 11 Feb 1938.

'The company's' Ibid.

'not liable' *Washington Herald*, 12 Feb 1938.

'phrase' Magid, cited in legal brief.

'merely a' Ibid.

'The radioactive' Loffler, court testimony, Ibid.

'brilliant sophistry' Grossman, Ibid.

'past master' Ibid.

'For evidence' Ibid.

'in a state' Ibid.

'That is' Grossman, *Chicago Herald-Examiner*, 10 Feb 1938.

Chapter Fifty-Two

'unsettled' Weather report, *Ottawa Daily Republican-Times*, 11 Feb 1938.

'It's too' CD, *Chicago Daily Times* and *American*.

'All this is' Tom Donohue, *Chicago Daily Times*, 11 Feb 1938.

'We've had' Ibid.

'pathetic spectacle' Unknown newspaper, collection of Ross Mullner.

'Weak but determined' *Chicago Daily Times*, 11 Feb 1938.

'through closed' Legal brief.

'Show us' Grossman, court transcript, *Chicago Daily Tribune*, 12 Feb 1938.

'Objection' Magid, Ibid.

'Is there one' Marvel, Ibid.

'Yes' Grossman, Ibid.

'It was decided' Ibid.

'Here's how' CD, court testimony, *Dubuque Iowa Herald*, 13 Feb 1938.

'We dipped' CD, court testimony, *Chicago Daily Tribune*, 12 Feb 1938.

'Then shaped' Ibid.

'a shudder ran' *Dubuque Iowa Herald*, 13 Feb 1938.

'drawn with' *St Louis Times*, 12 Feb 1938.

'I did this' CD, court testimony, *Chicago Herald-Examiner* and *Denver Colorado Post*.

'Did any' Grossman, court transcript, *Ottawa Daily Republican-Times*, 11 Feb 1938.

Following quotations, Ibid.

'We even ate' CD, court testimony, *Ottawa Daily Republican-Times*, *Chicago Daily Times*, *Chicago Herald-Examiner*.

'This is a' Grossman, court transcript, *Chicago Daily Tribune*, 12 Feb 1938.

'were not permitted' *Chicago Daily Tribune*, 12 Feb 1938.

'hung limply' Ibid.

'Were you employed' Court transcript, Ibid.

Following quotations, Ibid.

'the loss of' Unknown newspaper, lgrossman.com.

'Mr Reed said' MR, court testimony, *Chicago Daily Tribune*, 12 Feb 1938.

'so sympathetic' CD, *American*, 6 April 1938.

'Suddenly' Helen McKenna, *Chicago Daily Times*, 13 Feb 1938.

Chapter Fifty-Three

SPRING IS *Chicago Daily Times*, 13 Feb 1938.

'what slim thread' Ibid.

'It's the fighting' CD, Ibid.

'never leave her' *Washington Herald*, 14 Feb 1938.

'there was no' *Chicago Herald-Examiner*, 14 Feb 1938.

'Each declared' Ibid.

Sensing your . . . PP, letter to Grossman, 13 Feb 1938, PPC.

'The purpose of' Grossman, *Chicago Herald-Examiner*, 26 Feb 1938.

'He loved the' Interview with Len Grossman.

'The circumstances call' Grossman, legal brief.

'criminally careless' *Springfield Illinois State Register*, 22 Feb 1938.

'sent medical' *Ottawa Daily Republican-Times*, 28 Feb 1938.

'floundering' Maury Maverick, speech to House, cited in Robert Higgs, 'America's Depression Within a Depression, 1937–39', fee.org.

'We have pulled' Ibid.

'I am hoping' CD, *Toronto Star.*
'Once that' Interview with Mary Carroll Cassidy.
'She was so' Ibid.
'a smell of' Ibid.
'I just remember' Ibid.
'She was falling' Agnes Donohue Miller, mywebtimes.com, 17 Nov 2010.
'the boss of' Interview with James Donohue.
'She was the' Ibid.
'She'd go take' Interview with Kathleen Donohue Cofoid.
'It is like' CD, letter to PP, 9 March 1938, PPC.
'lovely letters' CD, letter to PP, 29 April 1938, PPC.
'my letter will' Wisconsin 'farm matron', cited in *Ottawa Daily Republican-Times*, March 1938.
'You have my' Letter to CD, Ibid.
'brought me a' CD, letter to PP, 9 March 1938, PPC.
'I'm sitting' Ibid.
'He worked' Interview with Len Grossman.
'I have been' Grossman, letter to PP, 15 Oct 1938, PPC.
'shameful' Grossman, legal brief.
All remaining quotations in chapter, Ibid.

Chapter Fifty-Four

'slow, insidious' Marvel's judgment, *Chicago Herald-Examiner*, 6 April 1938.
'The disablement' Ibid.
'The Industrial' Marvel's judgment, *American* and *Chicago Daily Times.*
'I am very' Helen Munch, unknown newspaper, collection of Ross Mullner.
'It would seem' Marvel, court transcript, 10 Feb 1938, legal brief.
'The whole creation' Grossman, legal brief.
'Justice has' Grossman, unknown newspaper, lgrossman.com.
'This is the' CP, *Flint Michigan Journal*, 6 April 1938.
'a great victory' Attorney-General's office, *American.*
'I never dreamed' CD, *Ottawa Daily Republican-Times*, 5 April 1938.
'cries but' Tom Donohue, Ibid.
'She half-rose' Ibid.
'Her first words' Ibid.
'I am glad' CD, *American*, 6 April 1938.
'This is the' Ibid.
'The judge is' CD, *Ottawa Daily Republican-Times*, 5 April 1938.
'It should have' Ibid.

'I wonder if' CD, *American*, 6 April 1938.
'Now maybe' CD, unknown newspaper, collection of Ross Mullner.
'I hope the' CD, Ibid.

Chapter Fifty-Five

'upon contention' *LaSalle Daily Post Tribune*.
'She has no' CP, *Chicago Daily Times*, 19 April 1938.
'He doesn't' CD, letter to PP, 29 April 1938, PPC.
'The eggshell' Frederick Griffin, *Toronto Star*, 23 April 1938.
'They're scared' Clarence Witt, Ibid.
'I looked' Griffin, Ibid.
Following quotations, Ibid.
'Tried to write' CD, letter to PP, 29 April 1938, PPC.
'I only wish' Ibid.
'a hurried-up' CD, letter to PP, 18 May 1938, PPC.
'I want to be' Ibid.
'Come over' Ibid.
'In my opinion' Dr Dunn, *Daily News-Herald*, 6 June 1938.
'There's not' CD, *Chicago Daily Times*, 5 June 1938.
'are worth all' CD, letter to PP, 29 April 1938, PPC.
'barely dented' *Chicago Daily Times*, 5 June 1938.
'arms scarcely' Ibid.
'He is just' CD, letter to PP, 29 April 1938, PPC.
'capacity crowd' *Chicago Illinois News*, 6 June 1938.
'race with death' Grossman, *Daily News-Herald*, 6 June 1938.
'If Mrs' Ibid.
'he never told' Reed's sworn statement, *Chicago Daily Times*, 6 June 1938.
'he was not' Ibid.
'would testify' Sworn statements of Mr and Mrs Reed, *Ottawa Daily Republican-Times*, 7 June 1938.
'stumbled in' *Ottawa Daily Republican-Times*, 7 June 1938.

Chapter Fifty-Six

'I would suggest' Catherine O'Donnell, letter to PP, 11 June 1938, PPC.
Dear Father Keane ... CD, letter to Father Keane, *Chicago Daily Times*, 24 June 1938.
'a sweeping response' Ibid., 26 June 1938.
'I would like' CD, Ibid., July 1938.
'Doctors told' Grossman, Ibid.
'It was' CD, letter to PP, PPC.

'I'm so happy' PP, letter to Grossman, 19 July 1938, PPC.

'bad spell' Olive Witt, letter to PP, 18 July 1938, PPC.

'She did look' Ibid.

'lovely time' Ibid.

'Personally' PP, letter to Dalitsch, 9 March 1938, PPC.

'I live' PP, *Toronto Star*, 23 April 1938.

'wistfully' *Chicago Daily Times*, 28 July 1938.

'Is it that' CD, Ibid.

'judicial propositions' *Ottawa Daily Republican-Times*, 27 July 1938.

'She had held' Grossman, *Chicago Daily Times*, 27 July 1938.

'Those who were' Ibid., 28 July 1938.

'Everyone has been' Eleanor Taylor, Ibid.

'almost hysterical' Ibid.

'broken' Ibid.

'Why doesn't' Mary Jane Donohue, Ibid.

'in a tiny' Ibid.

'God bless' Mary Jane Donohue, Ibid.

'a cool, calculating' Grossman, Ibid.

'a weary' Unknown newspaper, possibly *Ottawa Daily Republican-Times*, Catherine Wolfe Donohue Collection, Northwestern University.

'He spoke with' *Chicago Daily Times*, 29 July 1938.

'only to find' Coroner Lester's instructions, unknown newspaper, PPC.

'We, the jury' Unknown newspaper, possibly *Ottawa Daily Republican-Times*, Catherine Wolfe Donohue Collection, Northwestern University.

'It's the only' Grossman, Ibid.

'spoke of the' Ibid.

'near collapse' *Chicago Daily Times*, 29 July 1938.

'In a silent' Ibid.

When I returned ... PP, letter to Grossman, 29 July 1938, PPC.

'with prayers and' Ibid.

'He just collapsed' Trudel Grossman, *Chicago Daily Times*, 29 Sep 1938.

'[That] licence was' Interview with Len Grossman.

'one of the' *Chicago Daily Times*, 29 July 1938.

'If there are' Interview with Ross Mullner.

Epilogue

'As I was' Glenn Seaborg diary, cited in Mullner, *Deadly Glow*, p.125.

'If it hadn't' AEC official, cited in Ibid., p.127.

'invaluable' Ibid.

'We were going' Interview with Ross Mullner.

'In the foreseeable' NCL memo, Nov 1959, NCL files, Library of Congress.

'Every one of' Ibid.

'the terrible future' AEC official, cited in Mullner, *Deadly Glow*, p.134.

'the radium' Tony Bale, cited in Ibid., p.1.

'[They] serve' NCL memo, Nov 1959, NCL files, Library of Congress.

'There is only' Holm, CHR.

'Something that happened' AEC official, cited in Mullner, *Deadly Glow*, p.134.

'incalculable value' Ibid.

'vital insight' Ibid.

'My history is' PP, letter to Dalitsch, 9 March 1938, PPC.

'essential to the' Argonne National Laboratory press release, cited in *Radium City*.

'If we can' Lester Barrer, *Newark Evening News*, 15 July 1962.

'give the world' *Plainfield Courier*, 21 March 1959.

WANTED *Today's Health*, Nov 1959.

'Each of these' John Rose, *Chicago Sunday Tribune*, 18 March 1959.

'a reservoir' Roscoe Kandle, CHR.

'She said she' CHR memo.

'Miss Anna Callaghan' Martland, letter to MIT, 26 July 1950, CHR.

'couldn't do anything' CHR memo.

'because he' Interview with Art Fryer.

'reportedly died' Kjaer notes, 1925, cited in 'Historic American Buildings Survey'.

'maintained her' *Peoria Illinois Star*, 13 April 1939.

'forcefully argued' Mullner, *Deadly Glow*, p.136.

'for their' Ibid.

'easily slip' CHR medical assistant, letter to CP, 2 Oct 1975, CHR.

'I believe I' PP, 'Life History', PPC.

'the best homemade' Interview with Randy Pozzi.

'At the time' Ibid.

'The state man' MR, *Radium City*.

'huge and spotty' Interview with Jan Torpy.

'She'd always' Interview with Patty Gray.

'She was in' Interview with Dolores Rossiter.

'I prayed to' MR, mywebtimes.com, 13 Nov 2010.

'I witnessed' Ibid.

'She's a beautiful' MR, *Radium City*.

'I now feel' CP, *Chicago Daily News*, 13 June 1953.

'I've gone through' CP, *Chicago Sun-Times*, 29 Sep 1957.

'very angry' Interview with Jan Torpy.

'This is the' MR, cited in interview with Dolores Rossiter.

'forever' New Jersey Supreme Court judgment, cited in Clark, *Radium Girls*, p.201.

'constructive' Ibid., p.202.

'every brother' Interview with Darlene Halm.

'They just hauled' Ibid.

'I noted' Joan Weigers, letter to CHR, 14 Sep 1988.

'There aren't many' Greta Lieske, *Ottawa Delivered*.

'That lady' Mayor George Small, *Ottawa Daily Times*, 12 July 1988.

'everybody not to' Interview with Dolores Rossiter.

'Well' Ibid.

'People were' Interview with Darlene Halm.

'She said' Interview with Dolores Rossiter.

'crippled up' MR, *Radium City*.

'the normal' Martland, cited in Mullner, *Deadly Glow*, p.72.

'Bilateral amputations' CHR files.

RADIUM, DORMANT *Sunday Star-Ledger*, 2 Nov 1958.

'I'm absolutely' Interview with Ross Mullner.

'Ultimately' MIT memo, 6 Dec 1958, CHR.

'reduce[d] [his]' Holm, CHR.

'Have you seen' Joseph Kelly, cited in Ibid.

'They travelled' Interview with Randy Pozzi.

'She asked me' Ibid.

'This is what' PP, Ibid.

'were two' Ibid.

'She was probably' Interview with Jan Torpy.

'It was nothing' Interview with Don Torpy.

'Tell us the' Interview with Jan Torpy.

'She would repeat' Ibid.

'When I was' CP, cited in Ibid.

'pretty sad' Ibid.

'When the weather' Interview with Donald Purcell.

'far from won' Frances Perkins, *Ottawa Daily Republican-Times*, 17 October 1939.

'They would' Interview with Dolores Rossiter.

'but why should' CP, cited in CHR memo, 30 Aug 1985, CHR.

'will be swept' MR, cited in interview with Dolores Rossiter.

'God has left' MR, quoted by Carole Langer, unknown newspaper, Rossiter family scrapbook.

'who fought against' Film dedication to MR, *Radium City*.

'She thought' Interview with Patty Gray.

'It was going' Interview with Jean Schott.

'I'm *angry*' Unnamed Looney sister, Irvine, 'Suffering Endures'.
'Every family' Jean Schott, *New York Times*, 6 Oct 1998.
'We all went' Interview with Mary Carroll Cassidy.
'I think it' Interview with Kathleen Donohue Cofoid.
'They were all' Interview with Mary Carroll Cassidy.
'As time went' Ibid.
'It was a' Ibid.
'She was almost' Ibid.
'It was really' Ibid.
'I have really' Mary Jane Donohue, letter to CHR, 18 July 1979, CHR.
'If this could' Mary Jane Donohue, cited in CHR memo, 16 Aug 1984, CHR.
'I pray all' Mary Jane Donohue, letter to CHR, 18 July 1979, CHR.
'No monuments' Mullner, *Deadly Glow*, p.143.
'They deserve to' Madeline Piller, *Journal Star*, 2007.
'The mayor was' Interview with Len Grossman.
'The Radium Girls' Official proclamation from the State of Illinois Executive Department, Sep 2011, now on display at the Ottawa Historical and Scouting Heritage Museum.
'If [Marie]' Interview with Dolores Rossiter.
'dial-painters who' Radium Girls statue sign, Ottawa, Illinois.
'The studies of' Mullner, *Deadly Glow*, pp.6, 143.
'I always admired' Interview with Kathleen Donohue Cofoid.

Postscript

'We girls' Eleanor Eichelkraut, *Wall Street Journal*, 19 Sep 1983.
'I felt lucky' Beverley Murphy, *Ottawa Daily Times*.
'We slapped' Lee Hougas, quoted in Conroy, 'Radium City'.
'for kicks' Pearl Schott, *Wall Street Journal*, 19 Sep 1983.
'You couldn't' Ibid.
'would be bleeding' Martha, *Radium City*.
'The company' Martha Hartshorn, *Wall Street Journal*, 19 Sep 1983.
'I had to' Pearl, *Village Voice*, 25 Dec 1978.
'They kept cutting' Martha, *Radium City*.
'A man from' Carol Thomas, *Ottawa Daily Times*, 23 Sep 1983.
'Breast cancer' Holm, *Wall Street Journal*, 19 Sep 1983.
'The link between' Charles E. Land, Ibid.
'The plant manager' Holm, *Ottawa Daily Times*, May 1978.
'A lot of' Unnamed dial-painter, *Village Voice*, 25 Dec 1978.
'doubletalk' Jim Ridings, *Ottawa Daily Times*, 1 May 1978.
'They didn't have' Carol Thomas, Ibid., 23 Sep 1983.
'Luminous Processes' Ibid., 1 May 1978.

SELECT BIBLIOGRAPHY

Books

The Age of Radiance: The Epic Rise and Dramatic Fall of the Atomic Era by Craig Nelson (Simon & Schuster, 2014)

CRC Handbook of Management of Radiation Protection Programs, Second Edition by Kenneth L. Miller (CRC Press, 1992)

Deadly Glow: The Radium Dial Worker Tragedy by Dr Ross Mullner (American Public Health Association, 1999)

Harrison Stanford Martland: The Story of a Physician, a Hospital, an Era by Samuel Berg MD (Vantage Press, 1978)

Learning to Glow: A Deadly Reader edited by John Bradley (The University of Arizona Press, 2000)

Madame Curie: A Biography by Eve Curie, translated by Vincent Sheean (Read Books, 2007)

Mass Media and Environmental Conflict by Mark Neuzil and William Kovarik (Sage Publications, 1996, revised 2002)

Mining and Selling Radium and Uranium by Roger F. Robison (Springer, 2014)

A New Jersey Anthology by Maxine N. Lurie (Rutgers University Press, 2010)

Pierre Curie by Marie Curie, translated by C. and V. Kellogg (Macmillan, 1923)

The Quintessence of Ibsenism by George Bernard Shaw (Courier Corporation, 1994)

Radiation Protection and Dosimetry: An Introduction to Health Physics by Michael G. Stabin (Springer, 2007)

Radium Girls: Women and Industrial Health Reform, 1910–1935 by Claudia Clark (University of North Carolina Press, 1997)

Swing City: Newark Nightlife 1925–50 by Barbara J. Kukla (Rutgers University Press, 2002)

Film

Radium City (1987, dir. Carole Langer)

Interviews

Original interviews conducted with the following with my sincere thanks:

Michelle Brasser, Mary Carroll Cassidy, Mary Carroll Walsh, James Donohue, Kathleen Donohue Cofoid, Eleanor Flower, Art Fryer, Patty Gray, Len Grossman, Darlene Halm, Felicia Keeton, Ross Mullner, Randy Pozzi, Donald Purcell, Dolores Rossiter, Jean Schott, Don Torpy and Jan Torpy.

Miscellaneous articles and publications

'Historic American Buildings Survey: US Radium Corporation' by the National Park Service, HAER no. NJ-121; HAER NJ 7-ORA, 3-

'The New Jersey Radium Dial Workers and the Dynamics of Occupational Disease Litigation in the Early Twentieth Century' by Kenneth A. DeVille and Mark E. Steiner (*Missouri Law Review*, vol. 62, issue 2, spring 1997)

'Radium City' by John Conroy

'Radium in Humans: A Review of US Studies' by R. E. Rowland, Argonne National Laboratory

'Radium Osteitis with Osteogenic Sarcoma: The Chronology and Natural History of a Fatal Case' by William D. Sharpe, MD

'Suffering Endures for Radium Girls' by Martha Irvine (Associated Press, 1998)

Newspapers, magazines and periodicals

American
American History
American Weekly
Asbury Park Press

Buffalo News
Chemistry
Chicago Daily Times
Chicago Daily Tribune
Chicago Herald-Examiner
Chicago Illinois News
Chicago Sunday Tribune
Chicago Sun-Times
Daily News-Herald
Daily Pantagraph
Denver Colorado Post
Detroit Michigan Times
Dubuque Iowa Herald
Flint Michigan Journal
Graphic
Journal of the American Medical Association
Journal of Industrial Hygiene
Journal Star
LaSalle Daily Post Tribune
Newark Evening News
Newark Ledger
New York Evening Journal
New York Herald
New York Sun
New York Sunday News
New York Telegram
New York Times
Orange Daily Courier
Ottawa Daily Republican-Times
Ottawa Daily Times
Ottawa Delivered
Ottawa Free Trader
Peoria Illinois Star
Plainfield Courier
Popular Science
Radium
Springfield Illinois State Register
Star-Eagle
St Louis Times
Sunday Call
Sunday Star-Ledger

Survey Graphic
Today's Health
Toronto Star
Village Voice
Wall Street Journal
Washington Herald
Waterbury Observer
World

Special Collections

Catherine Wolfe Donohue Collection, Northwestern University, Chicago, Illinois

Files on the Orange clean-up operation, Orange Public Library, Orange, New Jersey

Harrison Martland Papers, Special Collections, George F. Smith Library of the Health Sciences, Rutgers Biomedical and Health Sciences, Newark, New Jersey

Health Effects of Exposure to Internally Deposited Radioactivity Projects Case Files, Center for Human Radiobiology, Argonne National Laboratory, General Records of the Department of Energy, Record Group 434, National Archives at Chicago, Illinois

National Consumers League files, Library of Congress, Washington, DC

Ottawa High School Yearbook Collection and Ottawa town directories, Reddick Public Library, Ottawa, Illinois

Ottawa Historical and Scouting Heritage Museum, Ottawa, Illinois

Pearl Payne Collection, LaSalle County Historical Society & Museum, Utica, Illinois

Raymond H. Berry Papers, Library of Congress, Washington, DC

Westclox Museum, Peru, Illinois

Websites

ancestry.com (with access to the records of the US Census, town directories, social security records, First World War and Second World War draft registration cards and the Index of Births, Marriages and Deaths)

capitolfax.com

dailykos.com

encyclopedia.com

examiner.com

fee.org

findagrave.com
history.com
lgrossman.com
medicinenet.com
mywebtimes.com
thehistoryvault.co.uk
themedicalbag.com
usinflationcalculator.com
voanews.com

*

For further reading, D. W. Gregory's *Radium Girls* (Dramatic Publishing, 2003) is a play about the Orange women, while *These Shining Lives* by Melanie Marnich (Dramatists Play Service, Inc., 2010) depicts the Ottawa dial-painters.

INDEX